T0325163

Compound Semiconductors: Thin-Film Photovoltaics, LEDs, and Smart Energy Controls

MATERIALS RESEARCH SOCIETY
SYMPOSIUM PROCEEDINGS VOLUME 1538

Compound Semiconductors: Thin-Film Photovoltaics, LEDs, and Smart Energy Controls

Symposia held April 1–5, 2013, San Francisco, California U.S.A.

EDITORS

Mowafak Al-Jassim
National Renewable
Energy Laboratory
Golden, Colorado,
U.S.A.

Clemens Heske
University of Nevada,
Las Vegas
Las Vegas, Nevada,
U.S.A.

Tingkai Li
Gongchuang Photovoltaic Co., Ltd.
Hengyang, P. R. China

Michael Mastro
U. S. Naval Research Laboratory
Washington, D.C., U.S.A.

Cewen Nan
Tsinghua University
Beijing, China

Shigeru Niki
National Institute of Advanced Industrial
Science and Technology
Tsukuba, Ibaraki, Japan

William Shafarman
University of Delaware
Newark, Delaware, U.S.A.

Susanne Siebentritt
University of Luxembourg
Belvaux, Luxembourg

Qi Wang
National Renewable Energy Laboratory
Golden, Colorado, U.S.A.

Materials Research Society
Warrendale, Pennsylvania

CAMBRIDGE
UNIVERSITY PRESS

CAMBRIDGE
UNIVERSITY PRESS

Shaftesbury Road, Cambridge CB2 8EA, United Kingdom

One Liberty Plaza, 20th Floor, New York, NY 10006, USA

477 Williamstown Road, Port Melbourne, VIC 3207, Australia

314–321, 3rd Floor, Plot 3, Splendor Forum, Jasola District Centre, New Delhi – 110025, India

103 Penang Road, #05–06/07, Visioncrest Commercial, Singapore 238467

Cambridge University Press is part of Cambridge University Press & Assessment, a department of the University of Cambridge.

We share the University's mission to contribute to society through the pursuit of education, learning and research at the highest international levels of excellence.

www.cambridge.org
Information on this title: www.cambridge.org/9781605115153

Materials Research Society
506 Keystone Drive, Warrendale, PA 15086
http://www.mrs.org

First published 2013

CODEN: MRSPDH

A catalogue record for this publication is available from the British Library

ISBN   978-1-605-11515-3   Hardback

CONTENTS

*Invited Paper

THIN FILM SOLAR CELLS

*Invited Paper

*Invited Paper

x

*Invited Paper

WIDE BANDGAP MATERIALS

PREFACE

Symposium C, "Thin-Film Compound Semiconductor Photovoltaics" and Symposium FF, "Compound Semiconductors for Generating, Emitting, and Manipulating Energy—II" were held on April 1–5 at the 2013 MRS Spring Meeting in the San Francisco, California. This combined symposia Proceedings represents the latest technical advancements and information on compound semiconductors for generating, emitting, and manipulating energy from universities, national laboratories and industries. It provides insight into emerging trends in these exciting technologies.

The scientific and technological exploration of compound semiconductors was presented in Symposium FF for applications of light emitters, record high efficiency solar cells, and high power devices, and for low cost wide bandgap materials manufacturing. These papers present the current status of compound semiconductor solar cells—from dilute nitrides for the record solar cell efficiencies, photon-recycling for understanding and designing single and multi-junction solar cells, concept and experimental progress of multiple subcells for a full spectrum module, to nano-structure solar cells for high efficiency solar cells at low cost. High power electronics with voltage range from 1 to 100 kV represent a large percentage of the current total power electronics market. High voltage electronics will be essential for next generation high voltage grids for renewable energy such as large-scale wind and solar farms. Several important contributions are given concerning the wide-bandgap materials such as GaN, SiC, and ZnO and the improvements in device processing to yield significant performance improvements. Finally, low cost processes will play a key role for the competitiveness of compound semiconductor devices relative to the Si-based devices. An upsurge of research is occurring and is presented into direct growth of GaN-based devices on large-area Si substrates as well as the incorporation of compound semiconductor-based devices within Si microelectronics.

Symposium C focuses on advances in the materials science, processing, and device issues of thin-film compound semiconductor materials in photovoltaic solar cells and related applications. Relevant materials include chalcogenide semiconductors, such as $Cu(In,Ga)Se2$ and related chalcopyrite alloys, CdTe, CdS, $Cu_2ZnSn(S,Se)_4$, n-type and p-type transparent conducting oxides, and novel materials with importance for thin-film photovoltaics. Among the recent developments that are being highlighted in the proceedings are advances in the characterization of bulk and interface properties in both materials and devices; film deposition and device processing; new earth-abundant materials; fundamentals of defects, grain boundaries, and surfaces; and innovative diagnostic and control tools critical for scale-up and manufacturing.

The organizers of Symposium C would like to thank the National Science Foundation, DuPont Central Research and Development, and GE Global Research for their generous support of the symposium.

Mowafak Al-Jassim
Clemens Heske
Tingkai Li
Michael Mastro
Cewen Nan
Shigeru Niki
William Shafarman
Susanne Siebentritt
Qi Wang

September 2013

MATERIALS RESEARCH SOCIETY SYMPOSIUM PROCEEDINGS

MATERIALS RESEARCH SOCIETY SYMPOSIUM PROCEEDINGS

Volume 1562E — Emerging Materials and Devices for Future Nonvolatile Memories, 2013, Y. Fujisaki, P. Dimitrakis, D. Chu, D. Worledge, ISBN 978-1-60511-539-9

Volume 1563E — Phase-Change Materials for Memory, Reconfigurable Electronics, and Cognitive Applications, 2013, R. Calarco, P. Fons, B.J. Kooi, M. Salinga, ISBN 978-1-60511-540-5

Volume 1564E — Single-Dopant Semiconductor Optoelectronics, 2014, M.E. Flatté, D.D. Awschalom, P.M. Koenraad, ISBN 978-1-60511-541-2

Volume 1565E — Materials for High-Performance Photonics II, 2013, T.M. Cooper, S.R. Flom, M. Bockstaller, C. Lopes, ISBN 978-1-60511-542-9

Volume 1566E — Resonant Optics in Metallic and Dielectric Structures—Fundamentals and Applications, 2013, L. Cao, N. Engheta, J. Munday, S. Zhang, ISBN 978-1-60511-543-6

Volume 1567E — Fundamental Processes in Organic Electronics, 2013, A.J. Moule, ISBN 978-1-60511-544-3

Volume 1568E — Charge and Spin Transport in Organic Semiconductor Materials, 2013, H. Sirringhaus, J. Takeya, A. Facchetti, M. Wohlgenannt, ISBN 978-1-60511-545-0

Volume 1569 — Advanced Materials for Biological and Biomedical Applications, 2013, M. Oyen, A. Lendlein, W.T. Pennington, L. Stanciu, S. Svenson, ISBN 978-1-60511-546-7

Volume 1570E — Adaptive Soft Matter through Molecular Networks, 2013, R. Ulijn, N. Gianneschi, R. Naik, J. van Esch, ISBN 978-1-60511-547-4

Volume 1571E — Lanthanide Nanomaterials for Imaging, Sensing and Optoelectronics, 2013, H. He, Z-N. Chen, N. Robertson, ISBN 978-1-60511-548-1

Volume 1572E — Bioelectronics—Materials, Interfaces and Applications, 2013, A. Noy, N. Ashkenasy, C.F. Blanford, A. Takshi, ISBN 978-1-60511-549-8

Volume 1574E — Plasma and Low-Energy Ion-Beam-Assisted Processing and Synthesis of Energy-Related Materials, 2013, G. Abrasonis, ISBN 978-1-60511-551-1

Volume 1575E — Materials Applications of Ionic Liquids, 2013, D. Jiang, ISBN 978-1-60511-552-8

Volume 1576E — Nuclear Radiation Detection Materials, 2014, A. Burger, M. Fiederle, L. Franks, D.L. Perry, ISBN 978-1-60511-553-5

Volume 1577E — Oxide Thin Films and Heterostructures for Advanced Information and Energy Technologies, 2013, G. Herranz, H-N. Lee, J. Kreisel, H. Ohta, ISBN 978-1-60511-554-2

Volume 1578E — Titanium Dioxide—Fundamentals and Applications, 2013, A. Selloni, ISBN 978-1-60511-555-9

Volume 1579E — Superconducting Materials—From Basic Science to Deployment, 2013, Q. Li, K. Sato, L. Cooley, B. Holzapfel, ISBN 978-1-60511-556-6

Volume 1580E — Size-Dependent and Coupled Properties of Materials, 2013, B.G. Clark, D. Kiener, G.M. Pharr, A.S. Schneider, ISBN 978-1-60511-557-3

Volume 1581E — Novel Functionality by Reversible Phase Transformation, 2013, R.D. James, S. Fähler, A. Planes, I. Takeuchi, ISBN 978-1-60511-558-0

Volume 1582E — Extreme Environments—A Route to Novel Materials, 2013, A. Goncharov, ISBN 978-1-60511-559-7

Volume 1583E — Materials Education—Toward a Lab-to-Classroom Initiative, 2013, E.M. Campo, C.C. Broadbridge, K. Hollar, C. Constantin, ISBN 978-1-60511-560-3

Prior Materials Research Symposium Proceedings available by contacting Materials Research Society

CIGS Growth

Mater. Res. Soc. Symp. Proc. Vol. 1538 © 2013 Materials Research Society
DOI: 10.1557/opl.2013.997

The effect of a high temperature reaction of Cu-In-Ga metallic precursors on the formation of Cu(In,Ga)(Se,S)$_2$

Dominik M. Berg, Christopher P. Thompson, William N. Shafarman
Institute of Energy Conversion, University of Delaware, Newark, DE 19716, U.S.A.

ABSTRACT

The influence of higher processing temperatures on the formation reaction of Cu(In,Ga)(Se,S)$_2$ thin films using a three step reactive annealing process and on the device performance has been investigated. High process temperatures generally lead to the formation of larger grains, decrease the amount of void formation and their distribution at the back Mo/Cu(In,Ga)(Se,S)$_2$ interface, and lead to a much faster formation reaction that shortens the overall reaction process. However, high temperature processing also leads to a decrease in device performance. A loss in open circuit voltage and fill factor could be attributed to enhanced interface recombination processes for the samples fabricated at higher process temperatures, which itself may be caused by a lack of Na and subsequent poor passivation of interface defect states. The lack of Na resulted in a decrease in free charge carrier concentration by two orders of magnitude.

INTRODUCTION

Reactive annealing of Cu-In-Ga precursors in selenium and/or sulfur-containing atmospheres is of high interest for the commercial manufacture of Cu(In,Ga)(Se,S)$_2$ (CIGSS) absorber layers for photovoltaic devices, which have already achieved power conversion efficiencies of 17.5 % for sub-modules [1]. To date, this process has been predominantly conducted using soda-lime glass (SLG) as the substrate, which limits the reaction temperature for industrial purposes to around 520°C due to its low strain point [2]. However, photovoltaic specialty glasses are under development, which permit processing temperatures up to 650°C. Recent results for Cu(In,Ga)Se$_2$ (CIGS) co-evaporation, at temperatures between 600°C and 650°C, have demonstrated more uniform Ga-profiles, enhanced grain size, and improved open circuit voltage (Voc) and fill factor (FF) leading to increased device efficiencies [3, 4]. The reactive annealing of CIGSS has a number of commonly observed issues at the CIGSS/Mo interface, including poor adhesion, Ga accumulation, and void formation [5]. While these issues can be partly controlled at lower temperatures on SLG, increasing processing temperature offers an additional pathway to address these issues. In the present work, we examine whether higher processing temperatures enabled by alkali-containing high temperature specialty glass substrates (HTG), are also beneficial for the reactive annealing process.

EXPERIMENT

For the conducted experiments, Cu-In-Ga metallic precursors (thickness ≈ 650 nm) were prepared by sputtering multiple alternating layers of Cu$_{0.77}$Ga$_{0.23}$ alloy and elemental In onto Mo-coated (thickness ≈ 700 nm) HTG substrates (4" x 4"). To form CIGSS, the precursors were reacted using a three-step process in $H_2Se/Ar/H_2S$ atmospheres (see Figure 1) as described by Kim et al. [5]. This process is similar to that used by Showa Shell Sekiyu K. K., now Solar Frontier [1]. Within the scope of this work, T_1 for the H_2Se reaction step was kept at 400°C for $t_1 = 40 - 50$ min, while temperatures T_2 and T_3 for the Ar anneal and H_2S reaction steps, respectively, were varied between 540, 580, 600, and 650°C with $t_2 = 0 - 20$ min, and $t_3 = 0 - 10$ min. The x's in Figure 1 indicate the point where the process was stopped, following which the samples (A to I) were quickly cooled (10 – 15 °/min).

Solar cells were fabricated using a conventional glass/Mo/CIGSS/CdS/i-ZnO/ITO/Ni-Al structure [5]. The 50 nm thick CdS layer was deposited by chemical bath deposition and i-ZnO (50 nm) and ITO (150 nm) layers were deposited by RF magnetron sputtering. The Ni(50 nm)-Al(3000 nm) front contact was deposited by e-beam evaporation. The solar cells described in this study were mechanically scribed with an area of 0.4 or 1 cm^2.

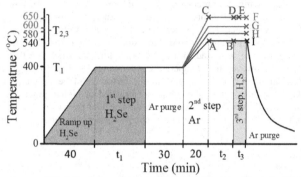

Figure 1 Temperature profile of the 3-step selenization process utilized in this study [5]. The x's indicate temperatures and times where the process was stopped and samples (A to I) were cooled.

Scanning electron microscope (SEM) analyses were performed on a JEOL JSM-7400 F and on an Amray model 1810 microscope. The latter was equipped with an Oxford Instruments PentaFET® 6900 EDX detector for compositional analysis performed with 20 kV acceleration potential. Further compositional analyses were performed using an X-ray fluorescence (XRF) system from Oxford Instruments (model X-Strata). Secondary Ion Mass Spectrometry (SIMS) depth profiles were performed by Evans Analytical Group LLC. Symmetric X-ray diffraction (XRD) scans were performed on a Philips/Norelco unit (1948 XRG, RM 190) employing a Cu X-ray source, with an acceleration potential of 35 kV and a filament compensating current of 20 mA, at a step size of 0.04° and 2 s/step integration time. Current-voltage (JV) characteristics were measured under AM 1.5 illumination at 25°C using a class A Oriel simulator. For the temperature dependent JV (JVT) measurements, a liquid nitrogen cooled Linkam cryostat (L75 920E-P) was used together with an ELH quartz halide lamp. Quantum efficiency (QE) investigations were performed using a 200 W quartz tungsten halogen projector lamp, an Oriel monochromator, a light chopper (operating at 75 Hz), and a metal halide bias lamp. To study the charge density and the width of the space charge region as a function of applied DC voltage, drive level capacitance profiling (DLCP) [6] was performed using an Agilent 4284A LCR meter and an external DC bias source.

RESULTS AND DISCUSSION

Influence of high temperature on the formation reaction

In order to study the influence of high processing temperatures and times on the formation reaction of CIGSS, samples A to E were processed as shown in Figure 1 and were subsequently cooled down at the maximum rate (10 – 15 °/min). In Figure 2 (a), the results of *ex-situ* XRD analyses of samples A to E are shown. In samples A and B, processed for different times ($t_{2,A} = 0$ min, $t_{2,B} = 20$ min) at lower temperatures $T_{2,3} = 540°C$, the presence of an InSe phase is observed as reported previously, at the back interface of CIGS and Mo [5]. Comparing samples A and B shows that the amount of InSe relative to CIGS does not vary over the 20 min anneal at 540°C under Ar atmosphere. At 650°C, samples C and D ($t_{2,C} = 0$ min, $t_{2,D} = 20$ min) still exhibit the InSe phase, but the relative amounts of this second phase are significantly smaller at $t_2 = 0$ min, and decreases after 20 min. In addition, the longer the film is reacted at 650°C in Ar atmosphere, the more InSe is dissolved into the CIGS film, in contrast to processing at 540°C. These results suggest that the reaction of low Ga-CIGS with Cu_9Ga_4 intermetallic, as formed in the first step at 400°C, to produce uniformly Ga-graded CIGS and the InSe secondary phase during the second anneal step [5], proceeds quicker at 650°C than 540°C. This allows a shortening of the process time for step 2. Similarly, elevated process temperatures affect the reaction rate of process step 3. Figure 2(a) shows the XRD scan of sample E ($t_{2,E} = 20$ min, $t_{3,E} = 2$ min). No secondary InSe phase is observed at 2 min into the third step (sample E), indicating that the InSe phase was dissolved into the CIGSS film. At the same time, the targeted average sulfur content of $S/VI \approx 0.1$ could already be reached at $t_3 = 2$ min, as observed from EDX investigations (not shown here). This result shows the quick reaction speed at elevated temperatures (650°C) which allows a short process time for step 3 of $t_{3,650C} = 2$ min.

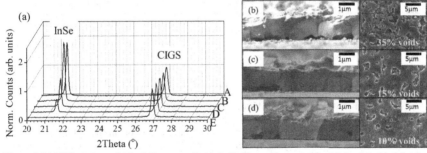

Figure 2 (a) XRD pattern obtained from samples A to E. The peaks were identified using the following JCPDS cards of the ICDD database: InSe (00-42-0919), CIGS (00-35-1102). Figures (b), (c), and (d) show SEM cross section images (left) and top view images of the back CIGSS/Mo interface after peeling (right) of samples that were fully reacted at $T_{2,3} =$ (b) 540°C, (c) 600°C, and (d) 650°C (with $t_2 = 20$ min, $t_3 = 10$ min).

Elevated reaction temperatures also have an effect on the film morphology. Figures 2(b), (c), and (d) show SEM cross-section images (left) and plan-view images of the back of the CIGSS films after peeling from the Mo (right) of samples that were fully reacted at $T_{2,3} =$ (b) 540°C, (c) 600°C, and (d) 650°C (with $t_2 = 20$ min, $t_3 = 10$ min). The cross section in Figure 2(b) looks very typical for films processed at 540°C [5]. The film consists of large columnar grains with a width of 1 - 2 μm with voids at the back interface. The back surface SEM image (Figure

2(b) on the right) shows that 1 - 3 µm sized voids are distributed over ~ 35 % of the total area of the back interface. Samples processed at higher temperatures of (c) 600°C and (d) 650°C show a decrease in the total void fraction to ~ 10 % for the 650°C sample. At the same time, the lateral grain size increases by a factor of 2 to 3, leading to columnar grains with a width of 3 - 5 µm.

Influence of high reaction temperature on device performance

To study the influence of the high reaction temperature on device performance, metallic precursors were deposited on HTG and subsequently reacted in a shortened process with $t_1 = 50$ min and $t_2 = t_3 = 2$ min at four different reaction temperatures: $T_2 = T_3 = 650°C$ (sample F), 600°C (sample G), 580°C (sample H), and 540°C (sample I). Table 1 summarizes the device performance results, composition of the absorber layers, and the activation energy, E_A, obtained from temperature dependent V_{OC} measurements (not shown here) [7]. Even though the compositions of different absorber layers are identical, the efficiency decreases systematically from 13.1 % to 3.9 % for the samples reacted at higher temperatures. The 13.1 % device obtained at 540°C is close to the results obtained on SLG, as shown in a previous work of our group, where a 14.2 % efficient device was reported [5]. The decrease in efficiency can be attributed to a drop in V_{OC} and FF, while the short circuit current (J_{SC}) is independent of the reaction temperature. At the same time, E_A, which is indicative of the dominant recombination process, decreases from $E_A = 1.16$ eV (the value of the CIGSS bandgap as obtained from QE measurements, see Figure 4(c)) for sample I to $E_A = 0.63$ eV for sample F. This result suggests that the devices reacted at higher temperatures are limited by interface recombination [8].

Table 1. Summary of device performance (after 2 min of heat treatment at 200°C in air), film composition obtained by XRF (Cu/III = Cu/(In+Ga), Ga/III = Ga/(In+Ga)) and EDX (S/VI = S/(S+Se)), and activation energies E_A of recombination processes as obtained from temperature dependent JV measurements of samples F to I. The samples were processed at different temperatures $T_{2,3}$ with $t_2 = 2$ min and $t_3 = 2$ min.

Sample	$T_{2,3}$ (°C)	Eff. (%)	V_{OC} (mV)	J_{SC} (mA/cm²)	FF (%)	Cu/III	Ga/III	S/VI	E_A (eV)
F	650	3.9	240	32.1	50.6	0.94	0.26	0.11	0.63 ± 0.05
G	600	11.0	504	31.5	69.3	0.94	0.26	0.11	0.94 ± 0.05
H	580	11.9	528	32.0	70.5	0.95	0.26	0.10	0.98 ± 0.05
I	540	13.1	575	31.8	71.6	0.95	0.26	0.10	1.16 ± 0.05

To explain the JV results, SIMS depth profiles were obtained after CdS deposition. Figure 3(a) shows the composition ratios of samples F (reacted at 650°C) and I (reacted at 540°C). These show no significant differences at the surface or in the bulk. Also, an improvement in Ga homogeneity at higher temperature was not observed unlike reported elsewhere [3, 4]. This is mainly due to the already homogeneous Ga distribution that occurs at the low reaction temperature, 540°C. Figure 3(b) depicts the depth dependence of the Na concentration for samples F to I. While the glass substrates used contain Na concentrations comparable to SLG, the absorber layers reacted at higher temperatures contain progressively less Na than the film reacted at 540°C. The difference of nearly one order of magnitude explains the drop in V_{OC} by enhanced interface recombination with a lack of passivation of interface defect states [8, 9]. The cause for the difference in Na concentration for samples F to I is unresolved and a link to the enhanced grain size and, thus, lower density of grain boundaries, can only be hypothesized. Unlike Na, the amount of K throughout the depth of the films was unchanged (not shown).

6

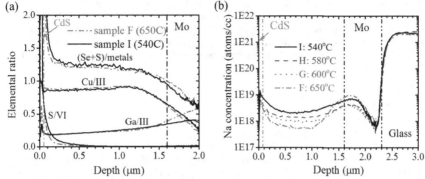

Figure 3 SIMS depth profiling of samples F to I. Graph (a) shows the elemental ratios obtained from samples F ($T_{2,3}$ = 650°C) and I ($T_{2,3}$ = 540°C). Graph (b) shows the depth profile of the Na concentration in the film, the Mo back contact, and the HTG substrate for samples F to I.

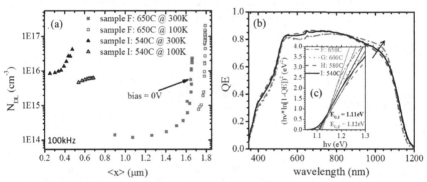

Figure 4 (a) Plot of N_{DL} over the first moment of charge response <x> as obtained from DLCP measurements performed on samples F (red/square symbols) and I (black/triangle symbols), as measured at room temperature (filled symbols) and 100 K (open symbols). (b) QE results as obtained from samples F to I at 0 V bias and in the dark. The black arrow indicates the trend for higher reaction temperatures. (c) Extrapolation of the band gap (x-axis intercept) of films F to I based on the QE results.

A difference in Na concentration of one order of magnitude in bulk CIGSS also has an influence on the free charge carrier concentration [9, 10]. In Figure 4(a), the results of DLCP measurements at room temperature (filled symbols) and 100 K (open symbols) from samples F (650°C, red symbols) and I (540°C, black symbols) are displayed. The drive level charge density (N_{DL}) at room temperature describes the sum of the defect density and the free charge carrier density N_A [6]. For sample F, N_{DL} is two orders of magnitude lower than for sample I. At T = 100 K, where $N_{DL} \approx N_A$, a similar difference of two orders of magnitude between samples F and I is suggested, although measurements on sample F in further forward bias were not possible. A decrease in N_A by two orders of magnitude, here likely related with the drop in Na concentration, can lead to a decrease in V_{OC} in CIGSS devices by itself by narrowing the quasi Fermi level splitting. Another interesting observation from Figure 4(a) is that at 0 V bias, the whole width of the absorber layer is depleted. This result correlates well with the enhanced charge carrier collection at long wavelengths in the samples reacted at higher temperatures, as shown in the QE results in Figure 4(b). Bandgap values of $E_G \approx 1.12$ eV have been observed for all samples I to F.

CONCLUSIONS

High processing temperatures were shown to result in CIGSS thin films with much larger grain size and reduced void formation at the back Mo/CIGSS interface, while the vertical compositional distribution, and especially the Ga-grading, was unchanged compared to a sample made under standard conditions. While the high temperatures lead to a faster formation reaction and, thus, shortening the overall processing time, the device performances of the samples progressively decreased as T increased from 540 to 650°C. The loss in V_{OC} and FF for the high temperature samples was concluded to be caused by enhanced interface recombination, likely originating from a lack of Na in the CIGSS material and, hence, the lack of passivation of interface defect states. Furthermore, the low Na was shown to result in a decrease in N_A.

ACKNOWLEDGMENTS

The authors would like to acknowledge John Elliot, Kevin Hart, Evan Kimberly, and Dan Ryan for their technical support, as well as Brian McCandless, Kihwan Kim, Gregory Hanket, Kevin Dobson, and Jes Larsen for fruitful discussions.

REFERENCES

[1] M. Nakamura et al., "Achievement of 17.5% Efficiency with 30x30cm^2-Sized Cu(In,Ga)(Se,S)$_2$ Submodules", in 38th IEEE Photovoltaic Specialists Conference, 2012
[2] W. Mannstadt et al., "New glass substrate enabling high performance CIGS solar cells", in 25th EU-PVSEC, 2010, p. 3516-3518
[3] J. Haarstrich et al., Sol. Energ. Mat. Sol. C. 95, 1028-1030 (2011)
[4] M.A. Contreras et al., "Improved energy conversion efficiency in wide-bandgap Cu(In,Ga)Se$_2$ solar cells", in 37th IEEE Photovoltaic Specialists Conference, 2011
[5] K. Kim et al., J. Appl. Phys. 111, 083710 (2012)
[6] C.E. Michelson et al., Appl. Phys. Lett. 47, 412 (1985)
[7] A.L. Fahrenbruch and R.H. Bude, "Fundamentals of Solar Cells", Academic press, New York, 1983, p. 236-239
[8] C.P. Thompson et al., "Temperature dependence of V_{OC} in CdTe and Cu(InGa)(SeS)$_2$-based solar cells", in 33rd IEEE Photovoltaic Specialists Conference, 2008
[9] P.T. Erslev et al., Thin Solid Films 517, 2277-2281 (2009)
[10] M. A. Contreras et al., "On the role of Na and modifications to Cu(In,Ga)Se$_2$ absorber materials using thin-MF (M = Na, K, Cs) precursor layers", in 26th IEEE Photovoltaic Specialists Conference, 1997

Mater. Res. Soc. Symp. Proc. Vol. 1538 © 2013 Materials Research Society
DOI: 10.1557/opl.2013.998

Analysis of NaF precursor layers during the different stages of the Cu(In,Ga)Se$_2$ co-evaporation process

M. Edoff[1], P.M.P. Salomé[1], A. Hultqvist[1], V. Fjällström[1]
[1]Ångström Solar Center, Department of Engineering Sciences, Uppsala University
P.O. Box 534, SE-751 21 Uppsala, Sweden

ABSTRACT

NaF precursor layers used for providing Na to Cu(In,Ga)Se$_2$ (CIGS) grown on Na-free substrates have been studied. The NaF layers were deposited on top of the Mo back contact prior to the CIGS co-evaporation process. The co-evaporation process was interrupted after the preheating steps, and after part of the CIGS layer was grown. Completed samples were also studied. After the preheating, the NaF layers were analyzed with X-ray Photoelectron Spectroscopy and after growing part and all of the CIGS film, the Mo/NaF/CIGS stack was characterized using transmission electron microscopy (TEM) and secondary ion mass spectrometry (SIMS). The NaF layers were found to be stable in thickness and composition during the pre-heating in selenium containing atmosphere before the CIGS process. The TEM analyses on the partly grown samples show a layer at the CIGS/Mo interface, which we interpret as a partly consumed NaF layer. This is corroborated by the SIMS analysis. In finalized samples the results are less clear, but TEM images show an increased porosity at the position of the NaF layer.

INTRODUCTION

Na is an important component in Cu(In,Ga)Se$_2$ (CIGS)-based solar cells. In addition to using NaF precursor layers [1] Na in-diffusion from the glass substrate can be used, but there are also other ways to include Na, as e.g. having a Na containing back contact material [2], or adding Na as a post-deposition treatment [3]. In a recent publication, modeling in combination with varying of the thickness of NaF precursor layers and thereby the Na concentration indicated that an increase of Na led to both an increased net carrier concentration as well as reduced defect density [4]. We also observed an increase of solar cell efficiency up to a certain level of Na concentration and then a saturation of the effect. In this work NaF precursor layers deposited on top of the Mo layer are studied by subsequent interruptions of the CIGS deposition process during its various stages. The Na-free substrate/Mo/NaF and Na-free substrate/Mo/NaF/CIGS stacks are analyzed with X-ray Photoelectron spectroscopy (XPS), Secondary Electron Microscopy (SEM) and Transmission Electron Microscopy (TEM) and compared to samples on soda lime glass without NaF. The objective with this study is to find out at which stages the decomposition of NaF occurs and if the NaF is stable upon heating and if it can withstand exposure to Se vapor all important input to process optimization. Samples were fabricated on soda-lime glass and on polished sintered alumina substrates. The choice of sintered alumina substrates was made in order to completely avoid the risk of Na contamination from the substrate and still get a thermal expansion coefficient which is close to that of CIGS. Soda-lime glass is the reference substrate used in our laboratory baseline process.

EXPERIMENTAL DETAILS

The substrates were coated with a bi-layer Mo coating consisting of a thin layer (25 nm) of Mo, DC sputtered at 12 mT, followed by a 380 nm thick Mo layer sputtered at 6 mT, previously described in [5].

NaF precursor layers were deposited onto the Mo-coated substrates by thermal evaporation from a resistively heated temperature controlled source. A liner made from pyrolytic boron nitride was used to hold the NaF source material. The thicknesses of the NaF precursor layers were controlled by a quartz crystal monitor. In addition, thickness measurements were obtained from SEM cross section images. No intentional heating of the substrate was used during the NaF evaporation. After the NaF evaporation, care was taken to avoid air exposure. Samples were either stored in vacuum, in nitrogen or packed in vacuum sealed bags for transport to the analysis.

The CIGS deposition was performed by co-evaporation using two dedicated CIGS co-evaporation systems. One is a true in-line process with stable evaporation rates and moving substrates, described in [6], the other is a system with a stationary substrate arrangement, but where the evaporation rates can be controlled using a mass-spectrometer feed-back loop to mimic almost any dynamic CIGS deposition process. In the stationary co-evaporation we used a process that we believe is close to the in-line process regarding evaporation time, substrate temperature and evaporation rates. The evaporation details are illustrated in figure 1, which shows a process log from one of the runs. The left image shows the substrate and selenium source temperatures and the right image shows the evaporation rates, calculated from the mass-spectrometer signals.

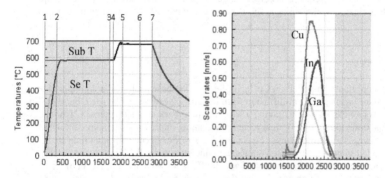

Figure 1. Process logs. Left: The heater (marked Sub T) and selenium (marked Se T) temperature profiles of the CIGS deposition process used in the stationary tool. The positions where the process was interrupted are indicated with numbers. In the gray areas the shutter is closed, while in the white area the shutter is open and the deposition of CIGS takes place. Note that the corresponding substrate temperature is lower than the heater temperature (450°C for the lower temperature region and 540 °C for the higher). Right: The evaporation rates for Cu, In and Ga.

In order to closely follow the development of the NaF precursor layers during the evaporation we made several runs, where the deposition process was interrupted. The full CIGS process including the positions during the run, where the process was interrupted is shown in figure 1. As can be seen, the process was interrupted also before the shutter to the sources opened, in order to check the stability of the NaF/Mo layer stack when exposed to heat and selenium (point 1-3). For the samples interrupted during the CIGS process, we measured the compositions and thicknesses using XRF and profilometry as shown in table 1. Point 1-3, 5 and 7 are discussed more in detail in the following paragraphs. Material characterization by XPS, SIMS, SEM and TEM was made at Evans Analytical. TEM was also performed at the Angstrom laboratory.

RESULTS AND DISCUSSION

Analysis of preheated NaF precursor layers

SEM cross sections of as deposited (point 1) and pre-heated NaF (points 2 and 3) on top of Mo coated alumina substrates are shown in Figure 2. After the first heat up the thickness is unchanged, but there is a tendency of forming a more compact film with defined grains, no change is observed by SEM after heating longer at 450°C in selenium background atmosphere.

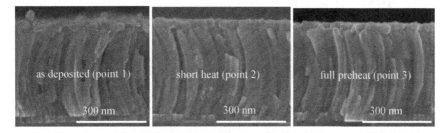

Figure 2. SEM cross section images of NaF films on top of Mo coated alumina substrates. The point notation refers to the different process interruption points shown in figure 1. The Mo grains are curved since the Mo layers are deposited in an in-line tool.

An XPS analysis was performed at point 2 and at point 3. In figure 3, the surface composition of the samples which were just heated up (point 2) of Mo/SLG and NaF/Mo/alumina are shown together with the same surfaces after the full preheating step (point 3). The NaF sample with the full heat treatment is the same as is the one shown to the right in figure 2. Although the samples were transported in evacuated air-tight bags and analyzed within one week from deposition, air exposure could not completely be avoided.

As evidenced from figure 3, there are considerable amounts of carbon and oxygen on the SLG samples, indicating that the Mo is a very reactive surface. Small, but significant amounts of Na have diffused from the SLG through the Mo and ended up on the surface. Some selenium is seen on the surface, even if the shutter that is blocking the evaporation vapors from reaching the sample is closed. This is not surprising. Selenium has a high vapor pressure and the selenium and metal sources were on during the pre-heating leading to heating of the chamber walls. The Mo is mostly in metallic state and only slightly oxidized, judging from the shift in the Mo3d peak in the XPS analysis (not shown).

The NaF/Mo/alumina samples give a completely different impression. We conclude that the NaF layer is almost completely covering the Mo surface, since the Mo signal is visible, but very small. The ratio between Na and F atomic concentrations calculated from integrated Na1s and F1s XPS peaks corrected with sensitivity factors is 1.3 both before and after the heat treatment, so we conclude that there is no decomposition of NaF. It is difficult to say if the slightly increased Mo signal after the preheating is caused by loss of NaF or not, but we believe that it is coupled to a recrystallization and grain growth of the NaF, which leaves some of the Mo between the grains exposed, since there is no evidence of NaF thickness reduction in the SEM images. Furthermore, the exposed Mo is proposed to be highly oxidized.

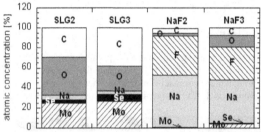

Figure 3. Results from XPS compositional surface analysis of Mo/SLG (SLG) and NaF/Mo/alumina (NaF) samples. The numbers denote the interruption points in figure 1, i.e. SLG1 means the Mo/SLG before the heat treatment (point 1) and SLG3 is analyzed after preheating (point 3) and similarly for the NaF/Mo/alumina samples.

The Mo/NaF/CIGS interface after part of the CIGS evaporation

At point 5 of the evaporation, an approximately 350 nm thick CIGS layer has grown and the substrate temperature has been increased from 450°C to the final temperature of 540°C. The samples from point 5 with NaF were analyzed using TEM and compared to reference samples with no NaF layer. In figure 4a, a layer consisting of 20-40 nm sized grains between the CIGS and the Mo can clearly be seen. In figure 4b it can be seen that the film is not fully covering the CIGS/Mo interface. The sample shown in figure 4c is deposited on glass without any NaF precursor layer and is shown for comparison. In Figure 5, the corresponding SIMS analyses are shown. The results indicate an increased concentration of Na and F at the CIGS/Mo interface, which supports the hypothesis of a remaining NaF layer at this interface after part of the CIGS has been grown.

The effect of NaF precursor layers thickness

In reference [4], the NaF precursor layer thickness was varied between 0 nm and 30 nm and the influences on the Na content, grain size and solar cell parameters were studied. The

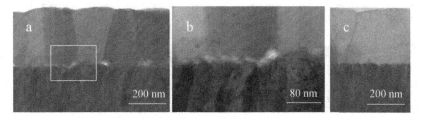

Figure 4. Samples from interrupting the CIGS process at point 5. Images 4a and 4b: CIGS/NaF/Mo/alumina, shown at different magnification. The rectangle in 4a is the same area as is shown in 4b. 4c is a CIGS/Mo/glass structure without NaF and shown for comparison.

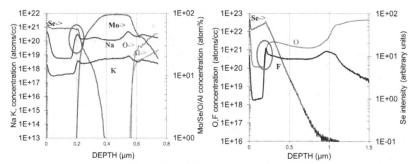

Figure 5. SIMS analyses of the CIGS/NaF/Mo/alumina sample stack after the interrupt at point 5. To the left, the Na and K levels together with Se, O and Al for locating the different layers. To the right, O and F are analyzed. Also here Se is shown. The Na and F profiles at the CIGS/Mo interface are marked with ovals.

sample with a NaF thickness of 30 nm would correspond to point 7 in figure 1, i.e. the full process. The conversion efficiency increased with NaF precursor thickness from 12 % for no NaF to 16 % for 7.5 nm of NaF and then saturated at an efficiency of 17 % for 15-23 nm of NaF, which was the same efficiency as for the reference on soda-lime glass.

With a 30 nm NaF layer we observed a slight efficiency decrease and using even thicker NaF precursor layers, the CIGS delaminated from the Mo layer in the wet chemical CdS buffer process. One reason for this delamination may be remaining NaF at the CIGS/Mo interface. The NaF is water-soluble and will dissolve in a wet chemical process. We were, however, not able to detect any NaF by TEM-EELS analysis at Mo/CIGS interface and the SIMS analysis did not show any clear evidence of a segregated layer of NaF. The TEM micrographs in figure 6 show an increased porosity at the Mo/CIGS interface for the sample with the thicker NaF precursor layer as compared to the thinner and as compared to the reference SLG. An increased porosity at the interface in combination with the stress implied on the sample stack in the wet chemical process would also be a probable cause for delamination.

Figure 6. TEM cross sections of solar cells with varying thickness of the NaF precursor layer. 6a: 15 nm NaF precursor thickness and 6b: 20 nm NaF precursor thickness. 6c: a SLG reference without NaF precursor,

CONCLUSIONS

In this work NaF precursors layers have been fabricated using evaporation from a resistively heated source. The samples with NaF layers have been studied during various stages in a CIGS co-evaporation process. The evaporated NaF layer was found to be close to fully covering the Mo layer and stable upon heating up to 450° C in vacuum with a background pressure of selenium. Increasing the thickness of the NaF precursor leads to beneficial effects on the solar cell parameters, but excessive thickness leads to delamination either caused by remainders of the NaF precursor being dissolved by the wet chemical CdS process or by a very high porosity at the CIGS/Mo interface. We interpret the results from this study as a gradual consumption of the NaF precursor layer, which takes place during the CIGS co-evaporation process. When the process is interrupted, SIMS analysis indicates presence of NaF at the CIGS/Mo interface

ACKNOWLEDGMENTS

This work was funded by the Swedish Energy Agency project Ångström Thin Film Solar Center and by STandUP for Energy. Timo Wätjen is acknowledged for TEM analysis.

REFERENCES

1. M. Bodegård, K. Granath and L. Stolt, Thin Solid films, vol 361-362, pp. 1-16, 2000.
2. J.H. Yun, K.H. Kim, M.S. Kim, B.T. Ahn, J.C. Lee and K.H. Yoon, Thin Solid Films, vol 515, pp. 5876-5879
3. D. Rudmann, A.F. da Cunha, M. Kaelin, F. Kurdesau, H. Zogg, A.N. Tiwari and G. Bilger, Appl. Phys. Lett. Vol 84, no 7, pp. 1129-1131, 2004
4. P.M.P. Salomé, A. Hultqvist, V. Fjällström, M. Edoff, B. Aitken, K. Vaidyanathan, K. Zhang, K. Fuller and C. Kosik Williams, accepted for publication in IEEE journal of photovoltaics
5. M. Edoff, N. Viard, T.Wätjen, S. Schleussner, P.-O. Westin and K. Leifer Proc. 24th EUPVSEC, 2009,
6. M. Edoff, S. Woldegiorgis, P. Neretnieks, M. Ruth, J. Kessler and L. Stolt, Proc. 19th EUPVSEC 2004, p 1690-1693.

Mater. Res. Soc. Symp. Proc. Vol. 1538 © 2013 Materials Research Society
DOI: 10.1557/opl.2013.999

Incorporation of Sb, Bi, and Te Interlayers at the Mo/Cu-In-Ga Interface for the Reaction of Cu(In,Ga)(Se,S)$_2$

Kihwan Kim[1], Jaesung Han[1,2] and William N. Shafarman[1]
[1] Institute of Energy Conversion, University of Delaware, Newark, DE 19716, USA
[2] Yeungnam University, Gyeongsan, Gyeongbuk 712-749, Republic of Korea

ABSTRACT

In this work, we investigate the effects of Sb, Bi, or Te interlayers at the Mo/Cu-In-Ga interface on the reaction to form Cu(In,Ga)(Se,S)$_2$ in order to control void formation and improve adhesion. Interlayers with 10 nm thickness were evaporated onto the Mo back contact prior to sputtering the metal precursors. CIGSS absorber layers were formed by a three-step H$_2$Se/Ar/H$_2$S reaction and solar cells were fabricated. The influences of each interlayer were characterized in the precursor and reacted films in terms of the density of the void formation, film structure and morphology, adhesion, and device performance.

INTRODUCTION

Selenization/sulfization of Cu-In-Ga metal precursors is being developed as a commercial-scale method of fabricating CuInSe$_2$-based solar cells. Reaction in H$_2$Se and H$_2$S produced 30 x 30 cm^2 sub-modules with efficiency > 17% [1]. Critical issues in the reaction of metal precursors in Se- and/or S- containing atmospheres to form Cu(In,Ga)Se$_2$ (CIGS, or when S is included, CIGSS) include void formation and poor adhesion at the CIGSS/Mo interface. These may result in reduced device performance yield and a narrower process window compared to co-evaporated CIGS cells. It has been proposed that the voids result from the agglomeration of slow reacting Cu-Ga intermetallic phases during selenization [2].

The group Va elements, Sb and Bi, have been reported to enhance recrystallization of CIGSS via a surfactant effect [3,4] and Te has been shown to induce wetting of metallic species on Mo [5]. In this work, the incorporation of Sb, Bi or Te interlayers before the metal precursor was investigated to promote greater intermixing/dispersing of the Cu-Ga intermetallic, induce more uniform selenization and minimize void formation by reducing the slow-reacting Cu$_x$Ga intermetallic. With each interlayer the density of the void formation, film structure and morphology, adhesion, and device performance were characterized.

EXPERIMENT

Three types of interlayers were deposited onto Mo-coated soda-lime glass prior to metal precursor deposition. Using electron-beam evaporation, 10 nm-thick Sb, Bi, and Te thin layers were deposited onto the Mo back contact. Then, CuInGa metal precursors were deposited onto SLG/Mo substrate by sputtering of Cu$_{0.77}$Ga$_{0.23}$ and In targets. With a rotating substrate platen, ~700 alternating Cu$_{0.77}$Ga$_{0.23}$ and In layers yield 650-nm thick precursors with Ga/(Ga+In) $\approx$ 0.26 and Cu/(Ga+In) $\approx$ 0.91, as measured by X-ray fluorescence (XRF).

The metal precursors with and without interlayers were reacted to form CIGSS absorbers by a three-step $H_2Se/Ar/H_2S$ reaction which consists of selenization at 400 °C for 50 min (1st step), Ar annealing at 550 °C for 20 min (2nd step) and sulfization at 550 °C for 10 min (3rd step). Details of this sputtering and reaction processes have been discussed previously [2].

Materials characterizations were performed with secondary electron microscope (SEM), X-ray fluorescence (XRF), x-ray diffraction (XRD), and secondary ion mass spectroscopy (SIMS) analyses. To evaluate CIGSS film adhesion to the Mo contact, a pull (delamination) test was carried out. Solar cells were fabricated with a structure of SLG/Mo/CIGSS/CdS/i-ZnO/ITO/Ni-Al and characterized by J-V and quantum efficiency (QE) measurements.

RESULTS and DISCUSSION

Material characterization

Figure 1 shows SEM cross-sectional images of the metal precursors: (a) control (no interlayer), (b) 10 nm Sb interlayer, (c) 10 nm Bi interlayer and (d) 10 nm Te interlayer. As discussed in our previous studies [2,6], In-rich nodules are clearly distinguishable from the smooth Cu-rich background in the control sample. However the Sb-, Bi- and Te-incorporated precursors seem to have less distinction between the In-rich nodules and the Cu-rich background.

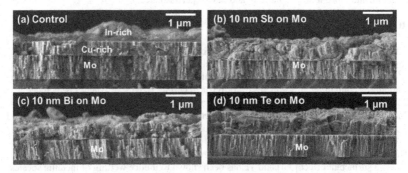

Figure 1. Cross-sectional SEM images of metal precursors with or without interlayer: (a) control (no interlayer), (b) Sb interlayer, (c) Bi interlayer and (d) Te interlayer.

Grazing-incident angle XRD (GIXRD) measurements revealed that In and $Cu_9(Ga,In)_4$ phases were found in the all the samples. But, as shown in Figure 2, the GIXRD patterns from only the Te-incorporated precursor are relatively independent of the incident angle from 0.5° to 4°. The other samples, the films with Sb and Bi interlayers, have slightly weaker peaks of In from the surface but they are still comparable to the control film (not shown). This indicates that the Te interlayer helps to enhance intermixing which should be advantageous since it may help In and Ga to be reacted with H_2Se more homogeneously. This may reduce the agglomeration of the slow reacting $Cu_9(Ga,In)_4$ phase which is considered to be the primary cause of the void formation.

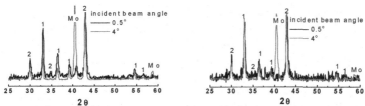

Figure 2. GI-XRD patterns of metal precursors without interlayer (a) and with Te interlayer (b). Phases: 1 - In; 2 - $Cu_9(In,Ga)_4$.

Figure 3 shows cross-sectional SEM images of the reacted CIGSS films. The Sb and Te incorporations seem to enhance recrystallization, whereas the Bi incorporation does not have any noticeable difference in terms of grain size. All the samples had voids at the $Cu(InGa)(SeS)_2$/Mo interface but the Te and Bi interlayer reduced void density by about 30 % compared to the control. The void densities of each sample were defined by the ratio of void area/Mo-CIGS contact area after observing CIGS back side [2]. The Bi-interlayer resulted in a very porous interface as seen in Figure 3(c) and the films had poor adhesion due to delamination of the CIGSS absorber from the Mo back contact.

The adhesion of each CIGSS absorber was evaluated by a delamination test, measuring the force to peel a CIGSS absorber off from the Mo back contact using a stub with 1/8"-diameter glued to the film. Table I gives the applied forces to delaminate the CIGSS absorbers. The Sb-incorporation turned out to significantly improve the adhesion of the CIGSS film, while the Bi-incorporation gave the smallest force for delamination. Even though the Te-incorporated CIGSS film had smaller void density and greater grain size than the control, it did not improve the adhesion of the CIGSS film.

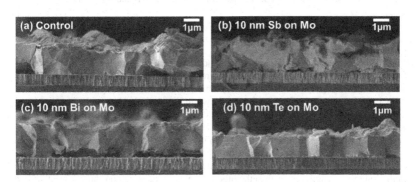

Figure 3. Cross-sectional SEM images of reacted CIGSS absorbers with or without interlayer: (a) control (no interlayer), (b) Sb interlayer, (c) Bi interlayer and (d) Te interlayer.

Table I. Adhesion measurements for each absorber. Uncertainty is determined by the standard deviation from at least 4 tests. The detection limit is ~ 1500 psi.

Interlayer	Control	Sb	Bi	Te
Applied force/ unit area, (psi)	1160 ± 180	> 1500	760 ± 200	1030 ± 200

Figure 4 shows SIMS profiles of the CIGSS samples deposited on different interlayers. The Sb- and Te-incorporated CIGSS films exhibited a steeper Ga grading than the control film, suggesting the 1st-step selenization was enhanced by inserting the Sb- or Te-interlayer. The Sb-incorporated CIGSS absorber also exhibited a S accumulation near the Mo back contact. In contrast with the main constituents, the Na profiles were found to be more affected by the interlayers. The Te-incorporated CIGSS absorber exhibited similar Na concentration [Na] to the control sample, while the Sb- and Bi-incorporated CIGSS absorbers had lower [Na]. The Bi-incorporated CIGSS film had the lowest [Na] which in this case, as will be shown below, apparently affected the device performance.

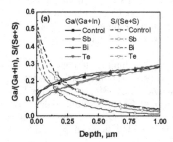 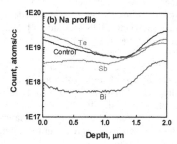

Figure 4. SIMS profiles of reacted CIGSS absorbers: (a) Ga/(Ga+In) and S/(Se+S) ratios and (b) Na contents.

Device characterization

JV results of each cell are shown in Figure 5 and their related parameters are summarized in Table II. Sb lowered all the device parameters, whereas Te and Bi mainly degraded V_{OC}. The V_{OC} drop in the Te-incorporated cell is relatively small, 40 mV, and this might be ascribed to the smaller Ga and S content at the front surface as shown in Figure 4(a). This can be improved with modification of the time/temperature profile to reduce the extent of the selenization reaction in the 1st step. In contrast, the Bi-incorporated cell exhibited significantly lower V_{OC} than the Te-incorporated cell, which is presumably attributed to the lower Na content in the CIGSS absorber as shown in Figure 4(b). The poor device performance of the Sb-incorporated cell is not yet fully understood. However, the external quantum efficiency of the Sb-incorporated cell is significantly lower in the long wavelength region than the other samples, suggesting poor bulk properties (e.g. higher defect density).

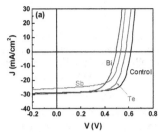 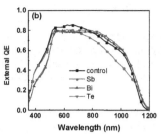

Figure 5. J-V characteristics (a) and external quantum efficiencies (b) of CIGSS cells with or without interlayer.

Table II. J-V values of the cells shown in Figure 5.

Interlayer	V_{OC} (mV)	J_{SC} (mA/cm^2)	FF (%)	η (%)
Control	612	30.3	71.9	13.3
Sb	534	25.7	66.1	9.1
Bi	494	29.4	64.0	9.3
Te	574	29.5	68.5	11.6

CONCLUSIONS

Sb, Bi, or Te interlayers were introduced between the CuGaIn precursor and Mo back contact as a means to control void formation during reaction to form CIGSS. With the interlayers, precursor morphologies were greatly affected and in particular the Te incorporation appeared to mitigate compositional/phase non-homogeneity. The reacted CIGSS absorbers were also greatly affected by the interlayers. Especially, the Sb and Te layers were found to yield greater grain size. None of interlayer did not eliminate void formation at the CIGSS/Mo back contact; however, the Te incorporation appeared to reduce void formation and the Sb incorporation significantly improved the adhesion of CIGSS to the Mo surface. Compared to a CIGSS cell with no interlayer, CIGSS cells with the Sb or Bi interlayer exhibited poor device performances due to reduced V_{OC}. The Te incorporation also yielded slightly reduced V_{OC} which may be attributed to un-optimized reaction conditions.

ACKNOWLEDGMENTS

The authors would like to gratefully acknowledge the contributions of Kevin Hart, Dan Ryan and John Elliot for device fabrication, and Christopher Thompson for electrical measurements and valuable discussions. This work was partially supported by the Foundational Program to Advance Cell Efficiency (F-PACE) funded by the Department of Energy of the USA (No. DE-EE0005407).

Disclaimer: "This report was prepared as an account of work sponsored by an agency of the United States Government. Neither the United States Government nor any agency thereof, nor any of their employees, makes any warranty, express or implied, or assumes any legal liability or responsibility for the accuracy, completeness, or usefulness of any information, apparatus, product, or process disclosed, or represents that its use would not infringe privately owned rights. Reference herein to any specific commercial product, process, or service by trade name, trademark, manufacturer, or otherwise does not necessarily constitute or imply its endorsement, recommendation, or favoring by the United States Government or any agency thereof. The view and opinions of authors expressed herein do not necessarily state or reflect those of the United States Government or any agency thereof."

REFERENCES

[1]   H. Sugimoto, T. Yagioka, M. Nagahashi, Y. Yasaki, Y. Kawaguchi, T. Morimoto, Y. Chiba, T. Aramoto, Y. Tanaka, H. Hakuma, S. Kuriyagawa, and K. Kushiya, "Achievement of over 17% efficiency with 30x30cm^2-sized Cu(InGa)(SeS)$_2$ submodules," *Conference Record of the 37th IEEE Photovoltaic Specialists Conference (PVSC)*, 2011, pp. 003420-003423.

[2]   K. Kim, G. M. Hanket, T. Huynh, and W. N. Shafarman, "Three-step H$_2$Se/Ar/H$_2$S reaction of Cu-In-Ga precursors for controlled composition and adhesion of Cu(In,Ga)(Se,S)$_2$ thin films," *Journal of Applied Physics*, vol. 111, p. 083710, 2012.

[3]   M. Yuan, D. B. Mitzi, O. Gunawan, A. J. Kellock, S. J. Chey, and V. R. Deline, "Antimony assisted low-temperature processing of CuIn$_{1-x}$Ga$_x$Se$_{2-y}$S$_y$ solar cells," *Thin Solid Films*, vol. 519, pp. 852-856, 2010.

[4]   T. Nakada, Y. Honishi, Y. Yatsushiro, and H. Nakakoba, "Impacts of Sb and Bi incorporations on CIGS thin films and solar cells," *Conference Record of the 37th IEEE Photovoltaic Specialists Conference (PVSC)*, 2011, pp. 003527-003531.

[5]   B. M. Basol, V. K. Kapur, and R. J. Matson, "Control of CuInSe$_2$ film quality by substrate surface modifications in a two-stage process," *Conference Record of the 22nd IEEE Photovoltaic Specialists Conference (PVSC)*, 1991, pp. 1179-1184.

[6]   K. Kim, H. Park, W. K. Kim, G. M. Hanket, and W. N. Shafarman, "Effect of Reduced Cu(InGa)(SeS)$_2$ Thickness Using Three-Step H$_2$Se/Ar/H$_2$S Reaction of Cu-In-Ga Metal Precursor," *IEEE Journal of Photovoltaics*, vol. 3, pp. 446-450, 2013.

Mater. Res. Soc. Symp. Proc. Vol. 1538 © 2013 Materials Research Society
DOI: 10.1557/opl.2013.1026

First-Principles Study on Diffusion of Cd in CuInSe$_2$

Tsuyoshi Maeda and Takahiro Wada
Department of Materials Chemistry, Ryukoku University, Seta, Otsu 520-2194, Japan

ABSTRACT

We have investigated the migration energy of Cd atom in CuInSe$_2$ (CIS) with a Cu vacancy by first-principles calculations. The activation energy of Cd migration in CIS and migration pathways are obtained by means of the combination of linear and quadratic synchronous transit (LST/QST) methods and nudged elastic band (NEB) method. The theoretical migration energy of Cd atom in CIS is 0.99 eV. The migration energy of Cd atom (Cd→V$_{Cu}$) in CIS is comparable to that of Cu migration (Cu→V$_{Cu}$) in CIS (1.06 eV). This result indicates that Cd diffusion in CIS easily occurs like Cu diffusion.

INTRODUCTION

CuInSe$_2$ (CIS) and Cu(In,Ga)Se$_2$ (CIGS) are successfully applied for thin-film photovoltaic devices. CIGS solar cells with high conversion efficiency have a typical device structure of TCO/ZnO/CdS/CIGS/Mo/soda-lime glass. The CdS layer is usually deposited by chemical bath deposition (CBD). At the CdS/CIGS interface, some of the Cd would be doped into the CIGS layer. In previous studies, it was demonstrated that a Cd-doped CIGS layer was formed at the CdS/CIGS interface during the CBD process [1-3]. The Cd-doped CIGS layer would show n-type conduction and form a buried p-n junction in the CIGS film.

Recently, we calculated the substitution energies of a Cd atom for a Cu or In atom in CIS [4]. We found that the substitution energy of a Cd atom for a Cu atom (Cd$_{Cu}$) in CIS is smaller than that for an In atom (Cd$_{In}$). The formation energy of the charge-neutral (Cd$_{Cu}$ + V$_{Cu}$) pair is lowest; therefore, the (Cd$_{Cu}$ + V$_{Cu}$) pair is easily formed during CBD of the CdS layer on the CIS layer, and a small amount of n-type Cd$_{Cu}$ is also formed. The SIMS depth profile after the CdS layer was deposited on the CIGS film by CBD showed the decrease of Cu concentration near the surface of CIGS film [1]. Cd diffusion in the CIGS layer was observed.

Most recently, we reported the migration mechanism of Cu and In in Cu-poor CIS by first-principles calculations [5]. Activation energy of Cu migration (Cu → V$_{Cu}$) in CIS (1.06 eV) is considerably lower than that of In migration (In → V$_{Cu}$) in CIS (1.70 eV). Cu migration easily occurs in the CIS crystal. However, there are no theoretical reports on Cd diffusion in CIS. It is important to clarify the diffusion mechanism of Cd atom in CIS. In this study, we calculated the migration energies of Cd atom in CIS by first-principles calculation.

COMPUTATIONAL PROCEDURES

Structural optimization

We performed first-principles calculations within a density functional theory as implemented in the program package DMol^3 in Materials Studio (version 5.5, Accelrys Inc.) [6-8]. All-electron scalar relativistic calculations [9] were performed using generalized gradient approximation (GGA) of the Perdew-Burke-Ernzerhof (PBE) functional as an electron exchange and correlation functional [10, 11]. The wave functions were expanded in terms of a double-numerical quality localized basis set with a real-space cutoff of 5 Å. For primitive cells of CIS, a 5×5×6 k point mesh generated by the Monkhorst-Pack scheme [12] was employed for numerical integrations over the Brillouin Zone. The initial structure of CIS was adopted as the experimental one, and it was obtained from the Inorganic Crystal Structure Database (ICSD). The chalcopyrite-type unit cell (space group: $I\bar{4}2d$) with lattice parameters of a = 5.782 Å and c = 11.620 Å (ICSD #70051) was employed for the initial structure. The lattice parameters of CIS were fixed for the experimental one. Atomic arrangements were fully relaxed. The tolerances of energy, gradient and displacement convergence were 1×10^{-5} Ha (1 Ha = 27.2114 eV), 2×10^{-3} Ha/Å, and 5×10^{-3} Å, respectively. The diffusion models of Cd migrations in CIS were constructed with a supercell with 64 atoms, which was 4 times greater than that of the chalcopyrite-type unit cell.

Calculation of migration energy of Cd atom in CIS

To calculate the migration energies of Cd atom in CIS, we performed the linear (LST) and quadratic synchronous transit methods (QST) combined with conjugate gradient (CG) method refinements to obtain the transition state [13, 14]. The LST gave the maximum pathway along the linear synchronous transit path between the initial and final structures. The CG method provided the saddle point of potential surface from the energy maximum of the LST pathway. The QST allowed the maximum pathway along the quadratic synchronous transit path through the saddle point. The structural refinements were repeated until the transition state of Cd migration in CIS was found at the saddle point. The root mean square convergence of gradients was less than 0.002 Ha/Å. The QST pathway optimized every image without separating by the nudged elastic band (NEB) method [15]. Finally, the NEB pathway indicated a minimum energy path (MEP) in the system. The NEB calculation was performed to find the minimum energy path between the initial and final structures by configured intermediate images, and those images were connected by springs to each other. The schematic images of (a) LST/QST and (b) NEB calculations are illustrated in Fig. 1(a) and 1(b).

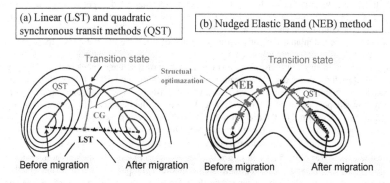

Figure 1 Schematic images of (a) Linear (LST) and Quadratic (QST) Synchronous Transit methods and (b) Nudged Elastic Band (NEB) method

Migration models of Cd atom in CIS
    Figure 2 shows the structures before/during/after Cd migration in CIS crystal. The structures were constructed with a 64-atom supercell, which was 4 times greater than that of the chalcopyrite-type unit cell. The concentration of substituted Cd atom for Cu atom is 1/16 (~6.3%). During the Cd migration, all of the atoms in the supercell were fully relaxed. We assumed that a doped Cd atom migrates via the intrinsic Cu vacancy in CIS. The initial and final structures include the doped Cd atom and Cu vacancy. The doped Cd atom at (0.5, 0.5, 0.5) moves to a position of Cu vacancy at (0.5, 0.75, 0.75). After Cd migration, the Cd atom occupies the Cu vacancy at (0.5, 0.75, 0.75) and leaves the Cu vacancy at (0.5, 0.5, 0.5). This Cd migration process is shown by $Cd \rightarrow V_{Cu}$ for simplicity.

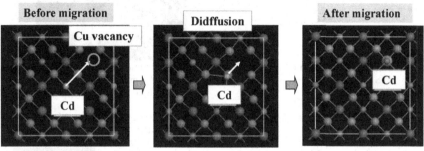

$Cd \rightarrow V_{Cu}$ Cd atom migrates via the intrinsic Cu vacancy in CIS.

Figure 2 Structures before/during/after Cd migration in CIS crystal.

RESULTS AND DISCUSSION

Migration energy of Cd atom in CIS
    Figure 3 shows the calculated migration energy of Cd atoms in CIS. The total energy of the initial structure before Cd migration is set to 0 eV. In the LST path, the maximum energy is 3.4 eV at the transition state. Then, the transition state in the LST path was relaxed by the CG method. The QST path has a transition state determined by the CG method. In the QST path, the maximum energy is 0.99 eV, which is the theoretically determined migration energy of Cd atom in CIS obtained from LST/QST methods. Figure 4 shows the structure at the transition state of Cd migration determined by the NEB method. In the transition state, the migrating Cd atom is located at (0.441, 0.627, 0.626). The transition state of Cd migration in CIS is similar to that of Cu migration (0.428, 0.628, 0.627) [5].
    The theoretical and experimental migration energies of Cd atom in CIS are summarized in Table I. The migration energy of Cu atom in CIS is shown for reference. The migration energy of Cd atom in CIS (0.99 eV) is comparable to that of Cu migration (1.06 eV). Theoretical activation energy of Cd migration in CIS agrees with the experimentally reported value of 1.04 eV [16]. This result indicates that Cd diffusion in CIS easily occurs like Cu diffusion.

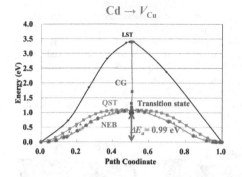

Transition state

Cd at (0.441, 0.627, 0.626)

Figure 3 Theoretical migration energy of Cd atoms in CIS calculated with LST/QST and NEB methods.

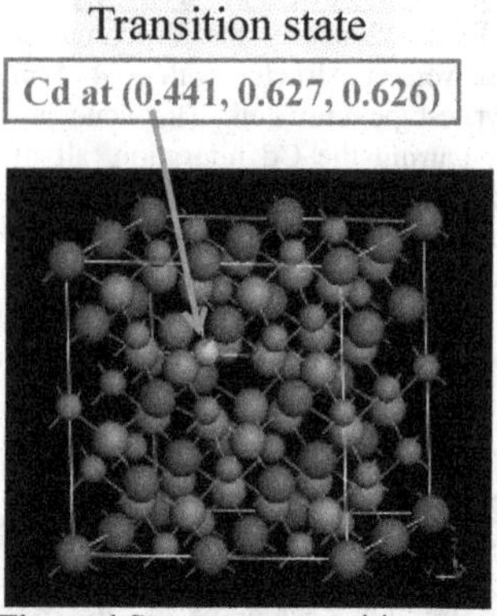

Figure 4 Structure at transition state of Cd migration determined by NEB method.

The chemical reactions can advance when the activation energy E_a of the reaction is 10, 20, or 30 times as large as $k_B T$, where k_B is Boltzmann constant and T is absolute temperature [17]. The $E_a/k_B T$ ratios of Cd and Cu migration in CIS at room temperature (T=25°C) and usual annealing temperature in fabricating CIGS solar cells, (T=100, 200, and 300°C) are listed in Table I. The theoretical $E_a/k_B T$ ratios of Cd migration at 25, 200, 250, and 300°C are 38.5, 30.8, 24.3 and 20.0, respectively. Those of Cu migration at 25, 200, 250, and 300°C are 41.2, 33.0, 26.0 and 21.5, respectively. These values at above 200°C are less than 30. Therefore, at the annealing temperatures of CIS solar cells above 200°C, diffusions of Cd atom in CIS easily occur as is the case for Cu diffusion.

Table I Theoretical and experimental migration energies of Cd and Cu atom in CIS. $E_a/k_B T$ ratios at 25, 100, 200, 300°C are shown.

		Migration energy (eV)	$E_a/k_B T$ ratio			
			25 °C	100 °C	200 °C	300 °C
Cd migration	Theoretical	0.99	38.5	30.8	24.3	20.0
	Experimental [16]	1.04	40.5	32.3	25.5	21.0
Cu migration	Theoretical [5]	1.06	41.2	33.0	26.0	21.5
	Experimental [18]	1.05	40.9	32.7	25.8	21.3

Diffusion coefficient of Cd atom in CIS

The diffusion coefficient of Cd atom in CIS was calculated as follows [18]:

$$D = D_0 \exp(-\frac{E_a}{k_B T}), \qquad D_0 = \frac{1}{6}Z v_0 c_v d^2.$$

Here, D is the diffusion coefficient. D_0, E_a, k_B, and T show the pre-exponential factor, activation energy, Boltzmann constant, and absolute temperature, respectively. E_a is the calculated migration energy of Cd atom in CIS. The pre-exponential factor was calculated from the number of nearest neighbors (Z=4), total jump frequency (ν_0 is about 10^{13} s^{-1}), concentration of Cu vacancy (c_v=18% [19]), and jump distance (Cu$\rightarrow$V$_{Cu}$). The calculated diffusion coefficient of Cd atom in CIS was D_{Cd} = 2.0×10^{-3} exp(-0.99 eV/k_BT) cm^{2}s^{-1}. The diffusion of Cd atom in the CIGS layer was examined by the radiotracer technique, and diffusion coefficient of Cd was determined by the Arrhenius equation D_{Cd} = 4.8 $\times10^{-4}$ exp (-1.04 eV/k_BT) cm^{2}s^{-1} [16]. The theoretical diffusion coefficient of Cd atom in CIS agrees with the experimental one. The diffusion coefficient of Cd atom in CIS at 100 °C is high enough to advance the diffusion. Therefore, Cd diffusion can advance easily at a low temperature. After the CdS layer was deposited on the Cu(In,Ga)Se$_2$ film by CBD, Cd diffusion was observed near the surface of the CIGS film in the SIMS depth profile [1], even though CBD is a thin-film deposition method at low temperatures (usually about 60-80 °C).

ACKNOWLEDGMENTS

This work was supported by the Incorporated Administrative Agency New Energy and Industrial Technology Development Organization (NEDO) under the Ministry of Economy, Trade and Industry (METI).

REFERENCES
1. T. Wada, S. Hayashi, Y. Hashimoto, S. Nishiwaki, T. Sato, T. Negami and M. Nishitani, Proc. 2nd World Conf. Photovoltaic Engineering Conversion, 1998, p. 403.
2. K. Ramanathern, H. Wiesner, S. Asher, D. Niles, R. N. Bhattacharya, J. Keane, M. A. Contreras, and R. Noufi: Proc. 2nd World Conf. Photovoltaic Engineering Conversion, 1998, p. 477.
3. T. Nakada and A. Kunioka, Appl. Phys. Lett. 74, 26 (1999).
4. T. Maeda and T. Wada, Jpn. J. Appl. Phys., accepted.
5. S. Nakamura, T. Maeda, and T. Wada, Jpn. J. Appl. Phys. 52, 04CR01 (2013).
6. B. Delley, J. Chem. Phys. 92, 508 (1990).
7. B. Delley, J. Phys. Chem. 100, 6107 (1996).
8. B. Delley, J. Chem. Phys. 113, 7756 (2000).
9. B. Delley, Int. J. Quantum Chem. 69, 423 (1998).
10. J. P. Perdew, J. A. Chevary, S. H. Vosko, K. A. Jackson, M. R. Pederson, D. J. Singh, and C. Fiolhais, Phys. Rev. B 46, 6671 (1992).
11. J. P. Perdew, K. Burke, and M. Ernzerhof, Phys. Rev. Lett. 77, 3865 (1996).
12. H. J. Monkhorst and J. D. Pack, Phys. Rev. B 13, 1588 (1976).
13. T. A. Halgren and W. N. Lipscomb, Chem. Phys. Lett. 49, 225 (1977).
14. N. Govind, M. Petersen, G. Fitzgerald, D. King-Smith, and J. Andzelm, Comput. Mater. Sci. 28, 250 (2003).
15. G. Henkelman and H. Jonsson, J. Chem. Phys. 113, 9978 (2000).
16. K. Hiepko, J. Bastek, R. Schlesiger, G. Schmitz, R. Wuerz, and N. A. Stolwijk, Appl. Phys. Lett. 99, (2011) 234101.
17. J. Ogborn, R. Marshall, and I. Lawrence, Advancing Physics: A2 Student Book Second Edition, (Oxford University Press, Oxford, 2008).
18. J. Pohl and K. Albe, J. Appl. Phys. 108, 023509 (2010).
19. S. Yamazoe, H. Kou, and T. Wada, J. Mater. Res. 26, 1504 (2011).

Mater. Res. Soc. Symp. Proc. Vol. 1538 © 2013 Materials Research Society
DOI: 10.1557/opl.2013.981

Characterization of Electron-Induced Defects in Cu (In, Ga) Se$_2$ Thin-Film Solar Cells using Electroluminescence

Shirou Kawakita[1], Mitsuru Imaizumi[1], Shogo Ishizuka[2], Hajime Shibata[2], Shigeru Niki[2], Shuichi Okuda[3] and Hiroaki Kusawake[1]

[1]Japan Aerospace Exploration Agency (JAXA), Tsukuba, Ibaraki, 305-8505 Japan
[2]Insititute of Advanced Industrial Science and Technology (AIST), Tsukuba, Ibaraki, 305-8568 Japan
[3]Osaka Prefecture University (OPU), Sakai, Osaka, 599-8570 Japan

ABSTRACT

CIGS solar cells were irradiated with 250 keV electrons, which can create only Cu-related defects in the cell, to reveal the radiation defect. The EL image of CIGS solar cells before electron irradiation at 120 K described small grains, thought to be those of the CIGS. After 250 keV electron irradiation of the CIGS cell, the cell was uniformly illuminated compared to before the electron irradiation and the observed grains were unclear. In addition, the EL intensity rose with increasing electron fluence, meaning the change in EL efficiency may be attributable to the decreased likelihood of non-irradiative recombination in intrinsic defects due to electron-induced defects. Since the light soaking effect for CIGS solar cells is reported the same phenomena, the 250 keV electron radiation effects for CIGS solar cells might be equivalent to the effect.

INTRODUCTION

CIGS solar cells have high attractive solar cells for space applications, since the cells have the highest efficiency among all thin-film solar cells [1], are lightweight and flexible with film substrates [2, 3, 4], and have excellent radiation tolerance in a space environment [5]. In particular, their radiation tolerance has been proved; not only in ground-based radiation irradiation tests but also demonstrations with small satellites in space [6].

CIGS solar cells have excellent radiation tolerance, which means their electrical properties are not degraded by 1MeV electrons. Conversely, cell performance is impaired with exposure to proton irradiation, similar to other solar cell types. The radiation damage to the cells caused by proton irradiation gradually recovers when the irradiated cells are kept even at room temperature and the recovery rate is temperature-dependent [7]. The radiation defect in CIGS solar cells, which impair their performance, was reported as an In antisite defect [8]. However, it remains unclear whether the other types of defects, namely Cu, Ga and Se Frenkel-pairs in CIGS, which are simultaneously generated by radiation, degrade cell performance or not. Therefore, we investigated these defects in CIGS solar cells induced by low energy electrons, enabling the type of radiation defect in the solar cells to be selected.

The electrical output performance of CIGS solar cells was not degraded by 250 keV electron irradiation, which can generate Cu-related defects in CIGS [9] and the roll–over behavior featured in the current-voltage characteristic under light illumination was reduced by irradiation. The increased carrier density produced by 250 keV electrons was found to reduce the roll-over. These results suggest that Cu-related defects induced by 250 keV electrons differ from those generated during 1 MeV electron irradiation, which degrades the electrical performance of CIGS solar cells. However, the defect induced by 250 keV electrons is not revealed.

Electroluminescence (EL) is a powerful tool to analyze semiconductor defects. Using EL from the CIGS solar cells, metastable defects were investigated [10]. We analyzed the defects induced by 250keV-electrons in CIGS solar cells with the EL.

EXPERIMENTS

CIGS solar cells were fabricated on glass substrates with co-evaporation [11]. The [Ga]/([Ga]+[In]) composition ratio of the CIGS layers was about 0.4 and the solar cells were of the bare type without an anti-reflective coating. Specifically, the short circuit density Jsc, the open circuit voltage Voc, the fill factor FF, and efficiency under AM0 condition were 38.1 mA/cm^2, 702 mV, 0.721, and 14.29 %, respectively.

The electron irradiation tests were conducted using a Cockcroft Walton electron accelerator at Osaka Prefecture University (OPU). The cells were irradiated in a vacuum with electron energy at 250 keV. Since the electrons heated the cells during the tests, the cells were cooled to less than 150 K using LN$_2$ to prevent the thermal annealing effect. EL images of the cells were observed with a cooled InGaAs CCD Camera. The EL spectrum was measured by a small monochromator with a cooled Si array detector.

RESULTS and DISCUSSION

EL images of a CIGS solar cell before electron irradiation are shown in Fig. 1. The EL image at R.T. had uniform luminescence. In contrast, the image at low temperature appeared grainy, seemingly indicating CIGS grain boundaries. However, the images must be evaluated in detail to clarify the origin of the grains.

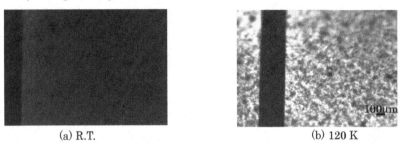

(a) R.T.   (b) 120 K

Figure 1. EL images of CIGS solar cells before electron irradiation at (a) room temperature and (b) 120 K .

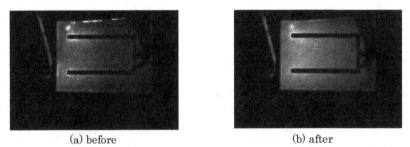

(a) before   (b) after

Figure 2. EL images of CIGS solar cells (a) before and (b) after electron irradiation at 120 K. The electron fluence was 3×10^{15} cm^{-2} .

The EL image before electron irradiation at 120K displayed grains in Fig. 2 (a) equivalent to Fig. 1 (b), and local luminescence from the CIGS. Meanwhile, the luminescence from the CIGS cell-irradiated electrons in Fig. 2 (b) expanded uniformly and showed indefinite grains compared to that before electron irradiation. Fig. 3 shows the EL total intensity of the CIGS cell irradiated with electrons, with intensities estimated from the EL images. The intensity rose with increasing electron fluence. The EL spectrum in Fig. 4 indicates that near band-edge emission from CIGS layer was 1.24 eV and the intensity of the near band-edge emission also rose with increasing electron fluence. This result explains how the radiation recombination at the near band-edge rose with increasing electron fluence, as in Fig. 3. The change in EL efficiency may be attributable to the decreased likelihood of non-irradiative recombination in intrinsic defects due to electron-induced defects. These results correspond to an improvement in the roll-over behavior in current-voltage characteristics under light condition because of the increase the carrier density in the cell by 250 keV electrons.

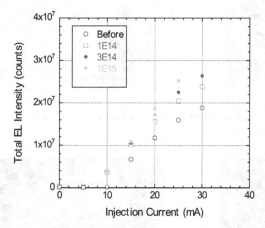

Figure 3. Total area EL intensities as a function of the injection current in the CIGS solar cell irradiated with electrons (Note that 1E14, 3E14, 1E15 and 3E15 denote irradiations to fluencies of 1×10^{14}, 3×10^{14}, 1×10^{15} and 3×10^{15} cm^{-2}, respectively.) as estimated by the EL images at 120 K.

The increased intensity of EL and carrier density characterized the light-soaking effect for CIGS solar cells [10]. This effect is said to be due to the change in charge in the V_{Cu}-V_{Se} metastable defect in CIGS solar cells by electrons generated by light. These results suggest that the 250 keV electron radiation effects for CIGS solar cells might be equivalent to the light soaking effect.

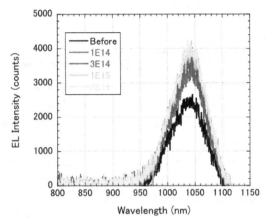

Figure 4. EL spectrum measured at 120 K in CIGS solar cells irradiated with electrons.

CONCLUSION

The effect of low-energy electrons on CIGS solar cells was evaluated to solve the mechanism of radiation-induced defects in cells by measuring EL. Electrons with 250 keV can create Cu-related defects in CIGS cells, which were detected the increased EL intensity for the cell. This result differs from the decreasing EL intensity for CIGS solar cells irradiated with proton and electrons with over 1 MeV. An increase in carrier concentration and improved roll-over behavior featured in current-voltage characteristic under light condition were reported for CIGS solar cells irradiated with 250 keV electrons. The mechanism is deduced to be equivalent to that of the light-soaking effect for CIGS solar cells, since these phenomena showed this effect. Therefore, Cu-related defects in CIGS may not be radiation capable of impairing cell performance in CIGS solar cells.

ACKNOWLEDGEMENTS

We would like to thank Messrs. Takashi Oka of OPU and Mitsunobu Sugai of Advanced Engineering Services for their extensive and kind cooperation in the electron irradiation tests and electrical measurements.

REFERENCES

1 . P. Jackson, D. Hariskos, E. Lotter, S. Paetel, R. Wuerz, R. Menner, W. Wischmann and M. Powalla, *Progress in Photovoltics*, **19**, 894 (2011).

2 . A. Chirila, P. Bloesch, A. Uhl, S. Seyrling, F. Pianezzi, S. Buecheler, C. Fella, S. Nishiwaki, Y. E. Romanyuk and A. N. Tiwari, *Proceeding of the 5th World Conference on Photovoltaic Energy Conversion*, Valencia, 2010, pp. 3403-3405.

3 . S. Ishizuka, T. Yoshiyama, K. *Mizukoshi, A. Yamada and S. Niki, Solar Energy Materials and Solar Cells* **94**, 2052 (2010).

4 . K. Moriwaki, M. Nangu, S. Yuuya, S. Ishizuka and S. Niki, *Proceeding of the 5th World Conference on Photovoltaic Energy Conversion*, Valencia, 2010, pp. 2858-2861.

5 . T. Hisamatsu, T. Aburaya and S. Matsuda, *Proceeding of the 2nd World Conference on Photovoltaic Energy Conversion*, Vienna, 1998, pp.3568 – 3571.

6 . S. Kawakita, M. Imaizumi and M. Takahashi, *Proceeding of the 26th European Photovoltaic Solar Energy Conference*, Hamburg (2011) pp. 210-213.

7 . S. Kawakita, M. Imaizumi, M. Yamaguchi, K. Kushiya, T. Ohshima, H. Itoh and S. Matsuda, *Jpn. J. Appl. Phys.* **41**, L797 (2002).

8 . J. F. Guillemoles, L. Kronik, D. Cahen, U. Rau, A. Jasenek and H. W. Schock, *J. Phys. Chem.* B. **104**, 4849 (2000).

9 . S. Kawakita, M. Imaizumi, S. Ishizuka, S. Niki, S. Okuda and H. Kusawake. *Thin Solid Films*, in press.

1 0 . M. Igalson, P. Zabierowski, D. Przado, A. Urbaniak, M. Edoff and W. N. Shafarman, *Solar Energy Materials and Solar Cells* **93**, 1290 (2009).

1 1 . S. Ishizuka, K. Sakurai, A. Yamada, H. Shibata, K. Matsubara, M. Yonemura, S. Nakamura, H. Nakanishi, T. Kojima and S. Niki, *Jpn. J. Appl. Phys.* **44** L679 (2005).

Mater. Res. Soc. Symp. Proc. Vol. 1538 © 2013 Materials Research Society
DOI: 10.1557/opl.2013.1044

Impact of maximum copper content during the 3-stage process on CdS thickness tolerance in Cu(In,Ga)Se$_2$-based solar cell

Thomas Lepetit[1], Ludovic Arzel[1], Nicolas Barreau[1]

[1]Institut des Matériaux Jean Rouxel (IMN), Université de Nantes, CNRS – UMR 6502
2 rue de la Houssinière, BP 32229, 44322 Nantes cedex 3, France

ABSTRACT

The tolerance of photovoltaic performances of Cu(In,Ga)Se$_2$-based (CIGSe) solar cells prepared from 3-stage grown absorbers to cadmium sulfide (CdS) buffer layer thickness was investigated. We focus on the influence of the maximum Cu content y = [Cu]/([In]+[Ga]) reached during the co-evaporation process on this tolerance. By increasing the duration of the 2nd stage we varied y$_{max}$ from 0.93±0.11 up to 1.06±0.12. Although final Cu content and CIGSe surface morphology seem to be similar for all absorbers, the photovoltaic performance of cells with higher maximum Cu content are better; moreover they tolerate much thinner CdS buffers (down to 10 nm-thick) without open circuit voltage or fill factor loss. Cells with lower y$_{max}$ exhibit more erratic performance and J(V,T) measurements show a specific voltage distribution for thin CdS. From these results it appears possible to decrease the CdS buffer layer thickness if it is deposited on adapted absorbers.

INTRODUCTION

Laboratory scale thin film solar cells based on co-evaporated Cu(In,Ga)Se$_2$ (CIGSe) absorber and chemical bath deposited (CBD) CdS buffer have reached conversion efficiency beyond 20 % when the CIGSe is grown following the 3-stage process [1]. Reaching such very high efficiency requires photovoltaic grade CIGSe which tolerates very thin buffer layer. Indeed, this condition is necessary to lower the absorption of short wavelength photons by CdS and consequently increase the short circuit current density (J$_{sc}$). Unfortunately, lowering the CdS thickness usually impacts long wavelength photogenerated carrier collection and lowers the open circuit voltage (V$_{oc}$) and the fill factor (FF) [2]. Contradictory conclusions concerning the optimal absorber Cu content y = [Cu]/([In]+[Ga]) have been reported in the literature [1,3]. In the present study we highlight that the final Cu content is not the only parameter to take into account and that the Cu content at the end of the second stage is crucial to obtain an absorber with the required properties.

EXPERIMENT

Thin films growth

CIGSe-based solar cells consisting of a Al/ZnO:Al/i-ZnO/CdS/CIGSe/Mo/SLG stack have been synthesized following the baseline of the IMN. We first covered soda-lime glass (SLG) substrates with a 500 nm Mo back contact layer by DC-sputtering. We then co-evaporated the 2 μm thick CIGSe absorber film following the three-stage process. By varying the duration of the 2nd stage we were able to change the maximum of Cu content of the absorber during the deposition. Figure 1 shows the End Point Detection (EPD) signal [4] for the 3 different absorbers grown for this study. The corresponding Cu contents at the end of the 2nd stage (y_{max}) and at the end of the deposition process (y_{final}) are reported in table I. During the deposition, copper and elements III selection is made by shutters. As the opening of these metal source shutters might influence the evaporation rates, the substrates were hidden for 2 minutes between the end of the 2nd and the beginning 3rd stage so that indium and gallium sources retrieved their stability. Fluxes should therefore be the same during the 1st and 3rd stage. CdS buffer layers were then grown on the absorber by chemical bath deposition. By varying the dipping time in the bath containing cadmium acetate, thiourea and ammonia at 60°C, we varied the thickness of this layer. Table I shows the three different optical equivalent thickness used for this study. To complete cells we finally RF-sputtered a 50 nm-thick undoped zinc oxide / 300 nm-thick aluminum-doped zinc oxide bilayer and we deposited aluminum front contacts using the electron beam technique. It is important to notice that at the exception of absorber and buffer layers, all the films are similar from one sample to another as they were grown in the same conditions.

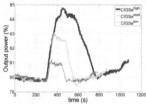

Figure 1 – End Point Detection signals illustrating how much copper was brought at the end of the 2nd stage. CIGSe^{high}, CIGSe^{med} and CIGSe^{low} correspond to a high, medium and low excess of copper, respectively

Characterizations

We performed material analysis on SLG/Mo/CIGSe stack. The composition of the absorbers has been determined by Energy Dispersive X-raw spectroscopy (EDX) using a Jeol JSM 5800LV scanning electron microscope (SEM) equipped with a dispersive energy spectrometer PGT IMIX germanium. Although composition throughout the absorber is inhomogeneous we measured final Cu content y_{final} and we deduced the maximum one y_{max} at the end of the 2nd stage using equation 1 where $y = [Cu]/([In]+[Ga])$ and t_1 and t_3 correspond to the duration of stage 1 and 3 during the 3-stage process, respectively.

$$y_{max} = \frac{t_1 + t_3}{t_1} y_{final} \tag{1}$$

The surface morphology of absorbers was analyzed with a JPK Nanowizard 3 atomic force microscope (AFM) equipped with Nanosensors PPP-NCHR-50 tips. Buffer layers were too thin to be viewed with SEM so we estimated their thicknesses with transmittance spectra of CdS deposited on glass as explained in [5].We studied the transport mechanisms by measuring J=f(V) responses of the cells cooled down with liquid nitrogen at temperatures from 120 K to 300 K in a Janis cryostat equipped with glasses transparent to wavelength from 280 up to 2000 nm.

DISCUSSION

Material properties

Table I summarizes compositional and morphological information about synthesized films extracted from EDX (y), AFM (roughness), SEM and transmittance spectroscopy (thickness). The final Cu content y_{final} is 0.87 ± 0.03 for all absorbers. However, Cu content has evolved during the process to reach a different maximum value y_{max} at the end of the 2nd stage. Amongst the 3 absorbers chosen for this study one has reached a composition above the stoichiometry ($\mathbf{CIGSe^{high}}$, $y_{max}>1$), another just reached it ($\mathbf{CIGSe^{med}}$, $y_{max}\approx1$) and the last one stayed below ($\mathbf{CIGSe^{low}}$, $y_{max}<1$). The range of the root mean square (RMS) roughness, determined by AFM measurements on absorber surfaces, is narrow enough to assume this difference is not the origin of such different electrical behaviors and cannot explain that one absorber can tolerate thinner buffer layer than the others, as we will see in the following section.

Table I – Compositional and morphological information about the films

Film designation	Thickness [nm]	Roughness (RMS) [nm]	y_{max}	y_{final}
CIGS^{high}	1950	58.3	1.06 ± 0.12	0.90 ± 0.10
CIGS^{med}	1800	50.9	1.01 ± 0.11	0.89 ± 0.10
CIGS^{low}	1750	49.4	0.93 ± 0.11	0.84 ± 0.10
CdS^{50}	51 (fitted)			
CdS^{30}	32 (fitted)			
CdS^{10}	9 (fitted)			

SEM cross sections of the 3 absorbers are shown in figure 2. There appears to be a large grain size distribution whatever the maximum copper content reached during the deposition. No clear difference can be noticed between the 3 absorbers.

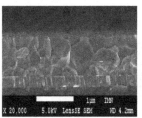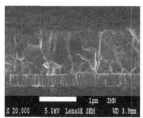

Figure 2 – SEM cross sections of absorbers with ymax varying from 0.93 ± 0.11 (left) to 1.06 ± 0.12 (right)

Photovoltaic performance

External quantum efficiencies, plotted in figure 3 for CIGSe^{high} and CIGSe^{low} absorbers show that each CdS thickness leads to the same short wavelength photons absorption on all absorbers. Although the CdS thickness was controlled through the dipping time duration in the bath, each buffer layer thickness (10, 30 or 50 nm) is similar from one absorber to another.

Surface of the absorbers had no impact on the growth speed of the CdS. Short circuit current densities are close for all samples. Long wavelength photogenerated carrier collection loss is higher as the CdS buffer layer becomes thinner and this phenomenon is accentuated on absorber whose y_{max} is low.

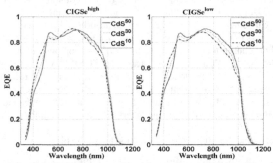

Figure 3 – External Quantum Efficiency performed on cells with high and low Cu content y_{max} during the process, for various CdS thickness

For each absorber/buffer combination, 9 in total, 28 cells with an area of about 0.5cm² have been synthesized and tested under AM1.5 solar spectrum. Figure 4 gives statistics on V_{oc} and FF for all the 252 cells. First it appears that the higher y_{max} the better performance whatever the CdS thickness. Moreover, the performance of cells based on CIGSe^{high} absorber appears much less affected by the decrease of CdS thickness. Relative losses between cells exhibiting thick and thin buffer are very low compared to those on other absorbers. One can also notice the spread in the performance of cells based on CIGSe^{low} absorber.

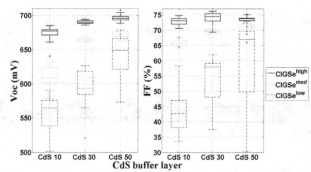

Figure 4 – Each box gives statistical results on 28 cells for open circuit voltage and fill factor. The box has lines at the lower quartile, median and upper quartile values. Lines extending from each end of the box show the extent of the rest of the data

Transport

J=f(V) responses of samples with CIGSe^{high} and CIGSe^{low} absorber and thin (CdS^{10}) and thick (CdS^{50}) buffer are plotted in figure 5, for temperature varying from 120 K to 260 K. No clear difference can be noticed in the electrical behavior of cells grown on CIGSe^{high} absorber when we decrease the CdS thickness. The electronic transport is similar with both thin and thick buffer. In contrast, the transport depends on the thickness of CdS in cells grown on CIGSe^{low} absorber, probably reflecting a change in the energy band "structure" when we reduce the buffer layer thickness. Voc is blocked around 675 mV and a thermally activated barrier for majority carriers appears below 220 K.

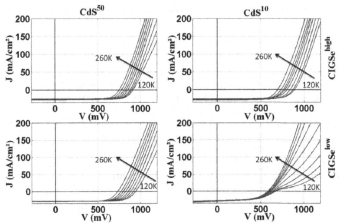

Figure 5 – I(V,T) measurements performed on cells exhibiting extreme value for y$_{max}$ and CdS thickness, from 120 K (best I(V)) to 260 K

CONCLUSIONS

According to these results it appears possible to keep V$_{oc}$ and FF high while decreasing the CdS thickness if the absorber has reached a copper content higher than one during the co-evaporation process. On such absorber, the carrier transport seems not to be related to the buffer layer thickness. In fact, collection depth seems to remain the same for all samples grown on this absorber, as long wavelength photo-generated carrier collection is quite similar on EQE measurements whatever the CdS thickness. Moreover J(V,T) measurements reveal no difference in electrical behavior on such absorber when we decrease the CdS thickness down to 10 nm. In contrast, performance becomes erratic on absorber whose copper content stayed below the stoichiometry during the deposition and fall off quickly as we decrease the buffer layer thickness. Growing very thin buffer layer on this CIGSe layer leads to an additional barrier for majority carrier and a V$_{oc}$ limitation.

This CdS thickness tolerance on photovoltaic performance seems not to be purely related to morphological issue at the interface between buffer and absorber, as differences revealed by AFM characterization of absorber surfaces may be too thin to explain such differences in electrical behavior. Reaching different maximum copper content y_{max} during the deposition process may rather influence material and/or electronic properties of the absorber sub-surface. Intermixing of elements at the absorber/buffer interface, reported in [6], could therefore be enhanced or reduced if the composition or the structure of the CIGSe sub-surface was changed, leading to a different sensitivity to the CdS thickness. The increase of CIGSe grain size with y_{max} could also change the density of electronic defects located at grain boundaries or within the grains. High y_{max} could then passivate detrimental electronic defects or create beneficial ones within the bulk and lower the influence of those located at the buffer interface.

To conclude, thinning the CdS buffer layer in high efficiency Cu(In,Ga)Se$_2$-based solar cell appears to be possible if the absorber has been synthesized following the 3-stage process and has reached a copper content higher than one at the end of the second stage. In fact, such absorber exhibits properties that lower the influence of the CdS buffer as the electronic transport may be independent from its thickness.

REFERENCES

1.   P. Jackson, D. Hariskos, E. Lotter, S. Paetel, R. Wuerz, R. Menner, W. Wischmann, and M. Powall, *Progress in Photovoltaics: Research and Applications* **19**, 894-897 (2011)
2.   M. A. Contreras, M. J. Romero, B. To, F. Hasoon, R. Noufi, S. Ward, and K. Ramanathan, *Thin Solid Films,* **403**, 204-211 (2002)
3.   M. A. Contreras, M. J. Romero and R. Noufi, *Thin Solid Films* **511**, 51-54 (2006)
4.   J. Kessler, J. Scholdstrom, L. Stolt, *Photovoltaic Specialists Conference, 2000. Conference Record of the Twenty-Eighth IEEE,* 509-512, (2000)
5.   F. Couzinie-Devy, L. Arzel, N. Barreau, C. Guillot-Deudon, S. Harel, A. Lafond and J. Kessler, *Journal of Crystal Growth,* **312.4**, 502-506, (2010)
6.   C. Heske, D. Eich, R. Fink, E. Umbach, T. Van Buuren, C. Bostedt, and F. Karg, *Applied physics letters,* **74**, 1451-1453, (1999)

Mater. Res. Soc. Symp. Proc. Vol. 1538 © 2013 Materials Research Society
DOI: 10.1557/opl.2013.976

Effects of additives on the improved growth rate and morphology of Chemical Bath Deposited Zn(S,O,OH) buffer layer for Cu(In,Ga)Se₂- based solar cells

Thibaud Hildebrandt[1], Nicolas Loones[1], Nathanaelle Schneider[1], Muriel Bouttemy[2], Jackie Vigneron[2], Arnaud Etcheberry[2], Daniel Lincot[1] and Negar Naghavi[1]

[1] Institute of Research & Development on Photovoltaic Energy (IRDEP), EDF/CNRS/Chimie-ParisTech, UMR 7174, 6 quai Watier, 78400 Chatou, France
[2] Institute Lavoisier of Versailles (ILV), CNRS/UVSQ, UMR 8180, 45 av. des Etats-Unis, 78035 Versailles, France

ABSTRACT

Zn(S,O,OH) Chemical Bath Deposited (CBD) remains one of the most studied Cd-free buffer layer for replacing the CBD-CdS buffer layer in a Cu(In,Ga)Se₂-based (CIGSe) solar cells and has already demonstrated its potential to lead to high-efficiencies. However, in order to further increase the deposition rate of the Zn(S,O,OH) layer during the CBD, the inclusion of additives can be a reasonable strategy, as long as the efficiencies of solar cells are maintained. The aim of this work is to understand the effect of the introduction of additives such as hydrogen peroxide (H₂O₂), H₂O₂+ethanolamine (C₂H₇NO) and H₂O₂+tri-sodium citrate (Na₃C₆H₅O₇) during CBD on the deposition mechanism, the growth rate and the quality of the buffer layer. It has been shown that the combined use of H₂O₂ and citrate in the bath formulation allows the deposition of Zn(S,O,OH) via a mix of "ion-by-ion" and "cluster-by-cluster" mechanisms that have good properties as buffer layers leading to high efficiency solar cells.

INTRODUCTION

In order to achieve high-efficiency Cu(In,Ga)Se₂ (CIGSe)-based solar cells, the use of cadmium sulfide (CdS) buffer layer is needed. However, because of cadmium toxicity and the eventual gain in solar cell performances with the use of a wider bandgap material, many Cd-free materials have been studied [1]. Among them, Chemical Bath Deposited (CBD) Zn(S,O,OH) buffer layer remains one of the most studied to replace CdS in CIGSe-based solar cells, and has already demonstrated its potential to achieve high-efficiency solar cells and modules. Indeed, Zn(S,O,OH) material has previously allowed the achievement of conversion efficiencies up to 19.4% [2]. However, a key issue to implement CBD-Zn(S,O,OH) process in CIGSe production line is both the deposition time and the deposition homogeneity, which depend on the growth mechanism of the buffer layer. A fast Zn(S,O,OH) CBD process with the use of hydrogen peroxide H₂O₂ has been previously reported [3]. However, in that case, the deposition process is not optimal as it leads to a "cluster-by-cluster" growth mechanism which is not well adapted for large scale deposition. In order to have a more adapted growth mechanism for large scale deposition while increasing the deposition rate, complexing agents such as ethanolamine (C₂H₇NO) and tri-sodium citrate (Na₃C₆H₅O₇) have been introduced in the deposition bath. In this paper, we present the influence of these additives mixed to H₂O₂ on the growth mechanisms and on the properties of the layers deposited. These new buffer layers will be tested on CIGSe-based solar cells and the photovoltaic properties will be presented.

EXPERIMENT

Film synthesis

Glass/Mo/CIGSe absorber layers are provided by Manz. Zn(S,O,OH) buffer layers are grown by Chemical Bath Deposition (CBD) using zinc sulfate (ZnSO$_4$) as zinc salt, thiourea (SC(NH$_2$)$_2$) as sulfur precursor and ammonia (NH$_3$) as complexing agent. For a "classical" CBD-Zn(S,O,OH) buffer layer, the reactants are introduced as follow : [ZnSO$_4$]=0.15M, [SC(NH$_2$)$_2$]=0.65M, [NH$_3$]=2M and deposition temperature set at T$_{deposition}$ = 80°C [4]. The different chemical bath compositions when additives are used are presented on Table 1. For Quartz Crystal Microgravimetry (QCM) analysis and the determination of growth mechanisms, the deposition temperature has been varied between 50°C and 80°C. After Zn(S,O,OH) deposition, the samples are rinsed in NH$_3$ solution followed by de-ionised water in order to avoid Zn(OH)$_2$ post-precipitation at the surface of the films. Solar cells are finally completed with sputtered window layers ZnMgO and ZnO:Al. In order to reduce metastability of the cells, samples are annealed in air for 10 min at 200°C and light-soaked for 1 hour.

Film characterization

Growth mechanisms are determined via in situ measurements carried out with Quartz Crystal Microgravimetry (QCM). Both surface and cross section of the films have been observed using a Zeiss SUPRA 55 Scanning Electronic Microscope (SEM). The atomic composition of the films are confirmed by the use of Thermo Fisher k-alpha X-ray Photoelectron Spectroscopy (XPS). The current-voltage J(V) characterization has been performed at 25°C under AMG1.5 illumination normalized to 1000W/m².

RESULTS AND DISCUSSION

Influence of the bath composition on the deposition rate and growth mechanism

Previous studies have shown the relevance of the use of hydrogen peroxide in CBD-Zn(S,O,OH). In Figure 1 the QCM growth curve of Zn(S,O,OH) films has been compared as a function of the composition of deposition bath (Table 1): a bath without additive and baths with hydrogen peroxide (H$_2$O$_2$), H$_2$O$_2$+ethanolamine (C$_2$H$_7$NO) and H$_2$O$_2$+tri-sodium citrate (Na$_3$C$_6$H$_5$O$_7$). The correspondence between mass uptake per surface unit, noted m (in g cm^{-2}), and the equivalent thickness of the layer, e (cm), is straightforward: m = ρe , where ρ is the masse density.
Indeed, as observed on Figure 1, it appears that H$_2$O$_2$ not only dramatically increases the deposition rate of Zn(S,O,OH), but also reduces both the induction time and the reactant concentrations (Table 1). The combined use of hydrogen peroxide and ethanolamine or tri-sodium citrate leads to a slightly lower deposition rate, but still higher than the "classic" deposition without additive. With the use of these additives it is then possible to highly reduce

the deposition time and the concentration of the reactants in the bath for the same material thickness deposited, as shown on Figure 1.

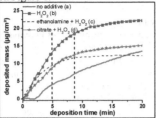

Figure 1: influence of H_2O_2 (b), ethanolamine + H_2O_2 (c), citrate + H_2O_2 (d) in comparison with a "classical" deposition with no additive (a) on the deposition rate

	Thiourea (M)	ZnSO₄ (M)	NH₃ (M)	H₂O₂ (M)	etha. (M)	Citrate (M)
a	0.65	0.15	2	-	-	-
b	0.4	0.1	2	0.22	-	-
c	0.4	0.1	1	0.22	0.1	-
d	0.4	0.1	1	0.22	-	0.1

Table 1: chemical composition of the different deposition bath depending on the additives introduced

Generally CBD involves two main mechanisms: one, corresponding to a "low" activation energy (Ea) process (Ea < 20 kJ.mol⁻¹), is called "cluster-by-cluster" [5] and involves a reaction in solution, with the formation of $Zn(OH)_2$, that diffuses to the substrate, and then reacts with S^{2-} :
$$Zn(OH)_2 + S^{2-} -> ZnS + 2\ OH^-.$$
For this mechanism as the growth of the thin film is controlled by diffusion it is more difficult to control a homogenous growth on large areas.
The other mechanism referred as "ion-by-ion" corresponds to a "high" Ea process (Ea > 40 kJ.mol⁻¹). In that case, the deposition takes place at the surface of the substrate, following :
$$Zn^{2+} + S^{2-} -> ZnS$$
This mechanism is known to be the best for large scale deposition as it offers a better control of the film growth, leading to more homogenous deposition on large surfaces [5].
In the case of an intermediate activation energy, there is a mix of the two mechanisms previously described (20 kJ.mol⁻¹ < Ea < 40 kJ.mol⁻¹). These mechanisms involved during deposition depend on the temperature.

In order to have a better understanding of the deposition mechanisms of our films, in situ growth deposition studies are carried out using QCM, as a function of deposition temperature.
It is then possible, based on the Arrhenius Law:

$$v = A\exp(-Ea/RT)$$

with v the deposition rate, A a constant, Ea the activation energy, R the gas constant and T the deposition temperature, to determine the activation energy Ea of the different deposition bath compositions, and to deduce the deposition mechanism associated by plotting log v as a function of 1000/T, as presented on Figure 2.

As observed when no additive is introduced in the bath, the deposition rate remains quite low and, as shown on Table 2, the deposition rate depends on the temperature of deposition leading to an Ea=45 kJ.mol^{-1} which is typical of an "ion-by-ion" mechanism. When hydrogen peroxide is introduced, a low activation energy Ea = 16 kJ.mol^{-1} is determined, which is characteristic of a cluster-by-cluster mechanism. However, as "ion-by-ion" is known to lead to more homogenous and high-quality films [5], complexing agents have been introduced in the bath and coupled with H_2O_2: ethanolamine (C_2H_7NO) and tri-sodium citrate ($Na_3C_6H_5O_7$). In the case of the H_2O_2/ethanolamine mix, the deposition is still characterized by a low activation energy Ea = 9 kJ.mol^{-1} and thus a "cluster-by-cluster" mechanism. However, in the case of the hydrogen peroxide/tri-sodium citrate, a slight increase of the activation energy is observed, reaching 20 kJ.mol^{-1}. This Ea is characteristic of a "cluster-by-cluster" mixed to "ion-by-ion" mechanism.

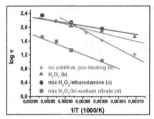

Chemical composition	Ea (kJ.mol^{-1})
no additive (a)	45 (ion-by-ion)
H_2O_2 (b)	16 (cluster-by-cluster)
ethanolamine + H_2O_2 (c)	9 (cluster-by-cluster)
citrate + H_2O_2 (d)	20 (mix)

Figure 2 : evolution of log v as a function of 1/T for each composition of the deposition bath : a) no additive, b)H_2O_2, c) mix of H_2O_2/ethanolamine and d) mix of H_2O_2/tri-sodium citrate

Table 2 : activation energy and growth mechanism associated of the deposition depending on the chemical composition of the bath

Influence of the bath composition on the morphology

The morphology of the films has been investigated by Scanning Electron Microscopy (SEM) and is presented on Figure 3.

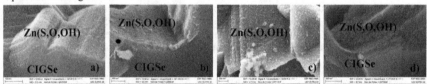

Figure 3 : SEM images of a) Zn(S,O,OH) deposited without additives ($t_{deposition}$ = 8minutes), b) Zn(S,O,OH) deposited with H_2O_2 ($t_{deposition}$ = 5min), c) Zn(S,O,OH) deposited with a mix H_2O_2/ethanolamine ($t_{deposition}$ =5min) and d) Zn(S,O,OH) deposited with a mix H_2O_2/tri-sodium citrate ($t_{deposition}$ = 5min). The films are deposited on CIGSe substrates

The SEM images (Figure 3) show that the "classic" CBD-Zn(S,O,OH) with no additives leads to films which are dense, covering and highly homogeneous. When hydrogen peroxide is introduced, the film remains dense and covering. However when H_2O_2 is mixed to ethanolamine, the film becomes unhomogeneous with apparition of clusters at the surface of the films, as shown on Figure 3 c). The association of tri-sodium citrate to hydrogen peroxide leads to a smoother surface, dense, covering and thick layer, as shown on Figure 3d).

Photovoltaic properties

The use of new Cd-free buffer layers is relevant only if the solar cells associated properties are good. Hence, the four different chemical bath compositions for CBD-Zn(S,O,OH) deposition have been tested on CIGSe-based solar cell and compared to standard CdS-based solar cell. The electrical properties of the cells are presented on Figure 4 and Table 3. From this study it appears that all different CBD-Zn(S,O,OH)-based solar have good properties although cells with H_2O_2 or a H_2O_2/ethanolamine mix present lower efficiencies than the CdS references, especially due to a lower FF. Cells without additive or with a H_2O_2/citrate mix present efficiency similar or even better than the CdS buffer layer. The Zn(S,O,OH) growth mechanism seems to influence the FF of the cells : indeed, a cluster-by-cluster mechanism leads to a low FF while an ion-by-ion mechanism promotes an improvement of the FF .

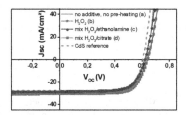

Figure 4: current-voltage curves of Mo/CIGSe/buffer layer/ZnMgO or i-ZnO/ZnO:Al after post-treatment (10 min of air annealing and 1 hour light-soaking)

Buffer Layer	Eff. (%)	Jsc (mA/cm²)	Voc (V)	FF (%)
No additive (a)	13.4	29.4	0.627	72
H_2O_2 (b)	12.9	28.7	0.638	68
H_2O_2/ethanolamine (c)	12.1	28.3	0.62	69.4
H_2O_2/citrate (d)	13.5	30.1	0.635	70.5
CdS-based	13.3	27.9	0.628	76.5

Table 3: photovoltaic properties of glass/Mo/CIGSe/Buffer layer/i-ZnO or ZnMgO/ZnO:Al depending on the Buffer Layer composition (average values after 10min of air annealing and 1 hour light-soaking)

CONCLUSIONS

Fast CBD-Zn(S,O,OH) deposition has been previously developed with the use of hydrogen peroxide as an additive and associated to a "cluster-by-cluster" growth mechanism. However, despite the good quality of the buffer layer thus deposited, it has not been possible to outperform CdS-based properties. In order to promote an "ion-by-ion" growth mechanism; which is known to be more adapted to large scale deposition; while keeping the high deposition rate, new complexing agents have been mixed to hydrogen peroxide based CBD bath: ethanolamine and tri-sodium citrate. In these mixes, hydrogen peroxide allows an increase of the deposition rate, and thus a shorter deposition time and less reactants. The addition of ethanolamine or tri-sodium citrate allows a better complexation of the zinc ions. In situ measurements have shown that tri-sodium citrate leads to a combination of "cluster-by-cluster" and "ion-by-ion" mechanisms, contrary to ethanolamine, which leads to a "cluster-cluster" mechanism. Based on this study, the Zn(S,O,OH) growth mechanism seems to influence the FF of the cells : indeed, a cluster-by-cluster mechanism leads to a low FF while an ion-by-ion mechanism promotes an improvement of the FF . Finally, the combination of hydrogen peroxide and tri-sodium citrate leads to good photovoltaic properties similar or even slightly better than CdS-based reference. This work offers a better understanding of the role of additives for the improvement of Zn(S,O,OH) buffer layer in order to outperform significantly CdS solar cells records.

ACKNOWLEDGMENTS

The authors would like to thank Christian Cossange from EDF for the SEM images and Manz for providing CIGSe absorbers.

REFERENCES

[1] N. Naghavi, D. Abou-Ras, N. Allsop, N. Barreau, S. Bücheler, A. Ennaoui, C. H. Fischer, C. Guillen, D. Hariskos, J. Herrero, et others, « Buffer layers and transparent conducting oxides for chalcopyrite Cu(In,Ga)(S,Se)$_2$ based thin film photovoltaics: present status and current developments », *Prog. Photovoltaics Res. Appl.*, vol. 18, n° 6, p. 411–433, 2010.

[2] D. Hariskos, R. Menner, P. Jackson, S. Paetel, W. Witte, W. Wischmann, M. Powalla, L. Bürkert, T. Kolb, M. Oertel, B. Dimmler, et B. Fuchs, « New reaction kinetics for a high-rate chemical bath deposition of the Zn(S,O) buffer layer for Cu(In,Ga)Se$_2$-based solar cells », *Prog. Photovoltaics Res. Appl.*, vol. 20, n° 5, p. 534-542, 2012.

[3] M. Buffière, S. Harel, L. Arzel, C. Deudon, N. Barreau, et J. Kessler, « Fast chemical bath deposition of Zn(O,S) buffer layers for Cu(In,Ga)Se$_2$ solar cells », *Thin Solid Films*, vol. 519, n° 21, p. 7575-7578, 2011.

[4] C. Hubert, N. Naghavi, A. Etcheberry, O. Roussel, D. Hariskos, M. Powalla, O. Kerrec, et D. Lincot, « A better understanding of the growth mechanism of Zn(S,O,OH) chemical bath deposited buffer layers for high efficiency Cu(In,Ga)(S,Se)$_2$ solar cells », *Phys. Status Solidi*, vol. 205, n° 10, p. 2335-2339, 2008.

[5] G. Hodes, *Chemical Solution Deposition Of Semiconductor Films*. New York: Marcel Dekker, 2003.

Mater. Res. Soc. Symp. Proc. Vol. 1538 © 2013 Materials Research Society
DOI: 10.1557/opl.2013.1008

A low temperature, single step, pulsed d.c magnetron sputtering technique for copper indium gallium diselenide photovoltaic absorber layers

Sreejith Karthikeyan[1], Kushagra Nagaich[1], Arthur E Hill[2], Richard D Pilkington[2], and Stephen A Campbell[1]

[1]Department of Electrical and Computer Engineering, University of Minnesota, Minneapolis, MN -55414, USA
[2]Materials and Physics Research Centre, University of Salford, Salford, M5 4WT, UK

ABSTRACT

Pulsed d.c Magnetron Sputtering (PdcMS) has been investigated for the first time to study the deposition of copper indium gallium diselenide (CIGS) thin films for photovoltaic applications. Pulsing the d.c. in the mid frequency region enhances the ion intensity and enables long term arc-free operation for the deposition of high resistivity materials such as CIGS. It has the potential to produce films with good crystallinity, even at low substrate temperatures. However, the technique has not generally been applied to the absorber layers for photovoltaic applications. The growth of stoichiometric p-type CIGS with the desired electro-optical properties has always been a challenge, particularly over large areas, and has involved multiple steps often including a dangerous selenization process to compensate for selenium vacancies. The films deposited by PdcMS had a nearly ideal composition ($Cu_{0.75}In_{0.88}Ga_{0.12}Se_2$) as deposited at substrate temperatures ranging from no intentional heating to 400 ^{0}C. The films were found to be very dense and pin-hole free. The stoichiometry was independent of heating during the deposition, but the grain size increased with substrate temperature, reaching about ~ 150 nm at 400 ^{0}C. Hot probe analysis showed that the layers were p-type. The physical, structural and optical properties of these films were analyzed using SEM, EDX, XRD, and UV-VIS-NIR spectroscopy. The material characteristics suggest that these films can be used for solar cell applications. This novel ion enhanced single step low temperature deposition technique may have a critical role in flexible and tandem solar cell applications compared to other conventional techniques which require higher temperatures.

INTRODUCTION

Copper Indium Gallium diSelenide ($CuInGaSe_2$ – CIGS) is a well-known absorber layer in the thin film solar cell industry. Recently researchers reported a laboratory scale efficiency of 20.3%[1]. Various commercial industries such as Shell Solar, Würth Solar, Nanosolar, Miasolé, and Ascent Solar make use of CIGS based solar cells. CIGS layers are extensively studied by researchers worldwide because of their solar cell application potential and their stability with respect to radiation damage. Various physical and chemical methods have been employed for the formation of this layer. These techniques typically require multi step processes in order to achieve the stoichiometry required for solar cell applications. The majority of these techniques produce films with low Se content. Unless the film is deposited in a very Se-rich ambient, post deposition selenization is usually required to prevent Se vacancies which act as donors, compensating the absorber. The use of H_2Se during selenization requires extensive health and safety precautions and can introduce secondary phases such as In_2Se and $CuSe$ which degrade the properties of the CIS/CIGS films [2]. Another disadvantage of these methods is the requirement for long high temperature heat treatments, typically > 500 °C for the crystallization of CIGS. This is very close to the softening point of the soda lime glass commonly used as a substrate [3]. The need for high temperatures limits the use of this material for flexible solar cells, tandem solar cell applications and roll to roll production of this material. The complexity of the manufacturing process also leads to a large gap in efficiency between lab scale and the industrial level. These factors suggest that there is a significant role for a single step, low temperature manufacturing process for the deposition of CIGS absorber layer. This paper reports the material properties of CIGS absorber layers deposited by ion enhanced pulsed d.c magnetron sputtering (PdcMS) from a CIGS powder target as a function of the substrate temperature. PdcMS is commonly used in the reactive sputtering of dielectric materials. Pulsing the D.C. voltage in the mid frequency region (10 to 350 kHz) suppresses arc formation at the target and hence enables the deposition of dielectric materials in a long-term arc-free environment [4, 5]. PdcMS can also use the enhanced ion flux [6] to help to crystallise the film at low substrate temperatures. The use of a powder target can cut down the material wastage associated with the "race track" effect.

EXPERIMENT

The sputtering system was custom designed, with a 2" unbalanced magnetron capable of using a powder sputter target in a sputter-up configuration [7, 8]. $CuInGaSe_2$ powder from Testbourne Ltd with 22% Cu, 23% In, 3% Ga and 52% Se (atomic %) was used as the target. The system was pumped to obtain a base pressure of 5.5×10^{-6} mbar. Argon was used as sputtering gas at 7.5 $\times 10^{-5}$ mbar. An Advanced Energy Pinnacle Plus pulsed d.c. power supply drove the cathode. Sputtering was done for 1 hour in a constant power mode (70W) at a frequency of 130 kHz and a pulse off-time of 1 μs. Films were deposited on Mo coated soda lime glass (SLG) wafers and bare SLG wafers that had been cleaned with isopropyl alcohol and acetone in an ultrasonic bath prior to sputtering. Substrate temperatures from no intentional heating (~ 67 °C) to 400 °C were used. X-ray diffraction analysis was carried out using a Bruker-AXS Microdiffractometer with Cu Kα radiation. The morphology of the bilayer films was studied with a JEOL 6500 SEM. The compositional analysis was carried out by Energy-dispersive X-ray spectroscopy using an Oxford Instruments Inca X-Act EDS. The thickness of the deposited films was measured with a KLA Tencor P-16 stylus profilometer. The optical properties were analyzed using an Aquila NKD 8000 UV-VIS-NIR spectrophotometer.

DISCUSSION

Sample thicknesses and compositions are summarized in Table 1. The films were analyzed using a hot probe analysis. All were found to be p-type. The thickness of the films reduced with increasing substrate temperature. The deposited films were found to be very close to the starting powder composition within the error of the EDX measurement. This confirms the ability of PdcMS to prepare CIGS films in a single step process which maintains the stoichiometry for solar cell applications. Similar results were observed for $CuInSe_2$ from various starting composition of the powder from our preliminary PdcMS study [7]. This effect has not been explained in the literature; however we believe a simple model which invokes standard BCF growth theory may be able to explain the effect [9]. Adatoms on the surface of the film diffuse laterally until multiple chemical bonds are formed. Prior to this they are weakly bound to the surface and so liable to removal by sputtering from the next PdcMS pulse or, if they remain ions, by electrostatic force during the reverse PdcMS pulse. The existence of a large-fluence, low-energy ion stream distinguishes this process from conventional sputtering since the flood of low energy ions may be quite effective in removing excess adatoms. Atoms that bond quickly will be selectively incorporated into the film. This favors both the majority phase material on the growing film and materials with the highest bond energy since these bonds tend to have the highest kinetic rate coefficients. This effect may increase process latitude for forming the most energetically favored material. Various research groups reported enhanced ionization in PdcMS plasma [6, 10].

Table 1 Thickness and composition of samples deposited at various substrate temperatures.

Sample Name	Substrate Temperature °C	Thickness μm	%Cu	%In	%Ga	%Se
S1	67	2.356	21.21	24.09	3.90	50.80
S2	200	2.254	20.34	24.53	4.33	50.80
S3	300	2.173	19.9	26.17	4.40	49.53
S4	400	2.032	21.99	25.80	3.78	48.43

The structural analysis revealed that films prefer (112) orientation even without substrate heating. However, observed a (110) secondary peak from $(InGa)_2Se_3$ (JCPDS 78-1745) phase next to CIGS (112) which disappears for films deposited at 300°C and above. Wang et al studied the effect of substrate temperature on the growth properties of CIGS films by a single stage co-evaporation technique [3]. They observed the same secondary phases for films deposited at low substrate temperature and they also observed the disappearance of these peaks > 450 °C [3]. Zang et al also observed these secondary peaks along with CuSe secondary peaks for films deposited using a three stage co evaporation process and these peaks disappeared > 500°C [11]. Sputtering also resulted in secondary phases at low substrate temperatures [12]. In our case the enhanced ionization of the sputtering species resulted in a shift to the low temperature regime with none of the additional selenization steps associated with conventional growth techniques. The intensity of the (112) peak was found to be increased with increase in substrate temperature up to 300 °C and reduced for the sample deposited at 400 °C.

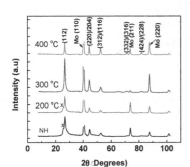

Figure 1 X-dray diffraction spectra of CIGS films deposited at various substrate temperatures.

Electron micrographs indicate (Figure 2) that the films consisted of very dense clusters of particles with sharp edges. The grain size increased with increased temperature. The cross-section image of the sample deposited at 300 °C (Figure 2 (c)) shows that films consisted of very dense columns aligned to the Mo back contact at bottom. At the top they form larger grains ~ 250 nm. The reported grain size of CIGS films obtained by PdcMS is comparable to other growth techniques with respect to their growth temperatures [3, 11]. Wang reported ~ 450nm for films deposited at 350 °C for films deposited by a single stage co-evaporation method [3]. Halbe *et al* studied the effect of another ion enhanced sputtering technique called high power impulse magnetron sputtering (HiPIMS) for the growth of CIGS. They deposited metallic layers using HiPIMS and observed higher density CIGS films compared to DC sputtered films. This improved the shunt resistance, hence resulted in higher efficiency (13.1% compared with 10.2% efficiency of CIGS cells deposited using DC sputtering) [13]. Figure 2c clearly shows small, tightly packed grains with few voids, in agreement with the observations of Halbe *et al*. We also found that these films adhered well to the Mo bilayer back contact.

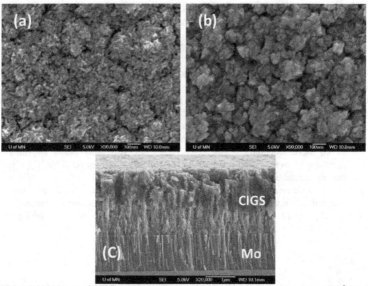

Figure 2 SEM image of CIGS samples (a) deposited with no heating (b) deposited at 300 °C and (c) cross-section image of the sample deposited at 300 °C.

The transmission and reflectance (not shown) spectra of these samples were recorded from 800 nm to 1800 nm. The transmittance improved with increase in substrate temperature. The absorption coefficient α was calculated using the following equation [14];

$$\alpha = \frac{1}{d} \cdot \ln \left(\frac{\left((1-R)^2 \right)}{2 \cdot T} + \sqrt{\left(\frac{(1-R^4)}{4T^2} + R^2 \right)} \right) \qquad (1)$$

where T is the transmittance, R is the reflectance and d is the thickness of the film. The calculated value of α is used to plot $(\alpha h\nu)^2$ versus hν. The intercept of the straight-line portion of the graph on the hν axis gives the optical band gap energy. The band gap increased from 1.04 eV for the sample deposited with no intentional heating to 1.11 eV for the sample deposited at 300 °C and decreased slightly to 1.09 eV for the sample deposited at 400 °C. The Ga was found to be close to the starting material composition. It is expected that using different starting compositions of Ga content and substrate bias will control the Ga % in the films to produce higher band gap materials.

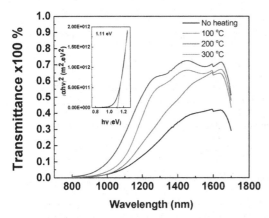

Figure 3 Transmission spectra of CIGS films deposited at various temperatures (inset - band gap analysis of 300°C).

CONCLUSION

CIGS films were deposited by an ion enhanced PdcMS technique. It was found that these films could be deposited in single phase at 300 °C. Films deposited at 300 °C were tightly packed and showed an average grain size of about ~250 nm. Optical analysis revealed that the band gap is around 1.10 eV, close to the operating band gap of solar cells. The deposition of single phase CIGS films without the need of any additional selenization process at low temperatures favors their application for flexible and tandem solar cell applications.

ACKNOWLEDGEMENT

Sreejith Karthikeyan acknowledges the support of Minnesota Institute for Renewable Energy and the Environment (IREE). This work was conducted in Nanofabrication Center (NFC), University of Minnesota, an NNIN-supported facility. Parts of this work were carried out in the University of Minnesota I.T. Characterization Facility, which receives partial support from NSF through the MRSEC program.

REFERENCES

1   P. Jackson, D. Hariskos, E. Lotter, S. Paetel, R. Wuerz, R. Menner, W. Wischmann, and M. Powalla, *Progress in Photovoltaics: Research and Applications*, doi: 10.1002/pip.1078 (2011).

2   S. Deok Kim, H. J. Kim, K. Hoon Yoon, and J. Song, *Sol. Energy Mater. Sol. Cells* **62**, 357 (2000).

3   H. Wang, Y. Zhang, X. L. Kou, Y. A. Cai, W. Liu, T. Yu, J. B. Pang, C. J. Li, and Y. Sun, *Semicond. Sci. Technol.* **25**, 055007 (2010).

4   P. J. Kelly, J. Hisek, Y. Zhou, R. D. Pilkington, and R. D. Arnell, *Surface Engineering* **20**, 157 (2004).

5   P. J. Kelly and R. D. Arnell, *Vacuum* **56**, 159 (2000).

6   J. W. Bradley, H. Bäcker, P. J. Kelly, and R. D. Arnell, *Surf. Coat. Technol.* **135**, 221 (2001).

7   S. Karthikeyan, A. E. Hill, R. D. Pilkington, J. S. Cowpe, J. Hisek, and D. M. Bagnall, *Thin Solid Films* **519**, 3107 (2011).

8   S. Karthikeyan, A. E. Hill, J. S. Cowpe, and R. D. Pilkington, *Vacuum* **85**, 634 (2010).

9   W. Burton, N. Cabrera, and F. Frank, *Philosophical Transactions of the Royal Society of London. Series A, Mathematical and Physical Sciences*, 299 (1951).

10   J. W. Bradley, H. Bäcker, Y. Aranda-Gonzalvo, P. J. Kelly, and R. D. Arnell, *Plasma Sources Science and Technology* **11**, 165 (2002).

11   L. Zhang, Q. He, W.-L. Jiang, F.-F. Liu, C.-J. Li, and Y. Sun, *Sol. Energy Mater. Sol. Cells* **93**, 114 (2009).

12   G. S. Chen, J. C. Yang, Y. C. Chan, L. C. Yang, and W. Huang, *Sol. Energy Mater. Sol. Cells* **93**, 1351 (2009).

13   A. Halbe, P. Johnson, S. Jackson, R. Weiss, U. Avachat, A. Welsh, and A. Ehiasarian, in *Photovoltaic Materials and Manufacturing Issues II - Materials Research Society Symposium Proceedings* (Materials Research Society, 2009), Vol. 1210, p. 179.

14   H. Neumann, P. A. Jones, H. Sobotta, W. Hörig, R. D. Tomlinson, and M. V. Yakushev, *Cryst. Res. Technol.* **31**, 63 (1996).

Mater. Res. Soc. Symp. Proc. Vol. 1538 © 2013 Materials Research Society
DOI: 10.1557/opl.2013.1053

Effect of Location of Sodium Precursor on the Morphological and Device Properties of CIGS Solar Cells

Neelkanth G. Dhere[1], Ashwani Kaul[1] and Helio Moutinho[2]

[1]Florida Solar Energy Center, 1679 Clearlake Road, Cocoa, FL 32922, USA.

[2]National Renewable Energy Laboratory, Golden CO, USA.

ABSTRACT

Sodium plays an important role in the development of device quality CIGS (Cu-In-Ga-Se) and CIGSeS (Cu-In-Ga-Se-S) chalcopyrite thin film solar cells. In this study the effect of location of sodium precursor on the device properties of CIGS solar cells was studied. Reduction in the surface roughness and improvement in the crystallinity and morphology of the absorber films was observed with increase in sodium quantity from 0 Å to 40 Å and to 80 Å NaF. It was found that absorber films with 40 Å and 80 Å NaF in the front of the metallic precursors formed better devices compared to those with sodium at the back. Higher open circuit voltages and short circuit current values were achieved for devices made with these absorber films as well.

INTRODUCTION

Sodium plays a useful role in the growth and doping of CIGS thin film solar cells. Several beneficial effects of sodium in copper chalcopyrite thin-film solar cells are reported in the literature. It has been shown that the presence of sodium plays a very critical role during the growth of CIGS absorber layer and is beneficial for the device performance. Several sodium precursors have been explored to determine their effect on the growth of CIGS solar cells. Among those studied, sodium fluoride, NaF has been found to be the best choice. This is because NaF is non-hygroscopic, stable in air and evaporates stoichiometrically [1]. It has been shown that selenization of the film containing sodium results in the formation of $NaSe_x$ compounds that delay the growth kinetics of the CIGS phase for better incorporation of selenium in the film [2]. Sodium has a tendency to reduce detrimental point defects. It reduces compensating donors by substituting selenium vacancies V_{se} and, therefore, increases the p-type doping. Sodium also replaces In_{Cu} anti-site defects further reducing the compensating donors [3]. Apart from reducing the compensating donor, sodium also replaces copper vacancies thereby minimizing the formation of ordered defect compounds (ODC) and thus favors widening the α-phase region. Sodium also promotes increase in grain size and preferred (112) orientation of CIGS films [4-6]. Sodium has also been shown to passivate the surface and grain boundaries of CIGS films by promoting oxygen incorporation [2]. The overall effect of sodium on the device performance is noted by an increase in efficiency by improvement in fill factor and open circuit voltage.

Addition of sodium helps in fabricating Cu-poor films with higher device efficiencies without a KCN treatment. It has been shown that the optimum amount of sodium for better device performance is determined by the process used to prepare the absorbers films [6]. Therefore, films prepared by rapid thermal annealing (RTA) will need higher amount of sodium compared to films synthesized in conventional furnaces with slower temperature ramp rates.

The highest efficiencies of CIGS thin-film solar cells have been achieved by using sodalime glass as substrate material. Sodalime glass contains significant amounts of sodium in the form of 13-15% Na_2O. The sodium from the substrate diffuses through the Mo back contact layer into the absorber. Nonuniform sodium diffusion can create defects at a few locations in the CIGS layer. In order to ensure uniformity in sodium incorporation, the preferred technique is to use alkali barriers between the sodalime glass and molybdenum back contact and deposit a known quantity of sodium, traditionally, by thermal evaporation of sodium precursor on molybdenum before the absorber growth [7-8]. However, for large area substrates the uniformity in Na deposition will still be compromised and this method of sodium incorporation adds another step in the process rendering it unsuitable for large volume production. Various methods have been tried to minimize this problem. In a recent study by Shogo et al, soda-lime glass thin-film deposited by rf-magnetron sputtering was used as sodium source for controlled incorporation of Na in CIGS films grown on flexible substrates [9]. The large area uniformity of deposited sodium is the noted advantage of this technique compared to the sodium deposited by traditional thermal evaporation.

In recent years an alternative technique has been developed wherein sodium doped molybdenum back contact has been explored. In fact, efficiencies of 13.7% have been achieved on flexible titanium foils using this method [10]. It is possible to control the optimum amount of sodium by sputtering a layer of know thickness from such a target. In an in-line manufacturing set-up, this technique would eliminate the need for additional deposition for sodium precursor thus contributing to savings in terms of time and cost for the additional equipments. The benefits of this method are being realized to the point that some companies are already producing standard (3, 5, and 10 at. % Na) and custom made sodium doped molybdenum targets know commercially as MoNa. This technique is also suitable on a production scale for large areas and would eliminate the additional step for deposition of sodium precursor. However, this can only provide sodium at the back. In the prior work carried out at FSEC, the effect of NaF concentration on the absorber quality and device performance of CIGSeS thin film solar cells has been studied. Normally, the sodium precursor is deposited over the molybdenum back contact layer before the deposition of CuGa-In metallic precursor in a two-stage process. In a recent study, record efficiency of 20.4 % was obtained for laboratory scale CIGS solar cells on polyimide sheet by coevaporation [11]. Since not much work has been carried out to investigate the effect of location of sodium precursor on the properties of the absorber films prepared by a two stage process, an attempt was made to investigate this effect on the properties of the absorber films prepared by selenization-sulfurization of metallic precursors. Initially, optimization of process parameters was carried out for the preparation of absorber films. Using the optimized parameters, absorber films were prepared with various amounts of NaF in the front and at the back of the CuInGa metallic precursors. Solar cell devices were also completed on the as-prepared absorber films.

EXPERIMENTAL

At first, molybdenum back-contact layer was deposited on sodalime glass substrates having an alkali diffusion barrier. A two-stage process was used for the preparation of the absorber layer. The first step consists of the deposition of the CuGa-In metallic precursors by DC magnetron sputtering. Initial experimentation was carried out to optimize the sodium quantity for preparation of device quality absorbers. NaF was deposited by high vacuum thermal evaporation. The Cu-Ga and In precursors were selenized at 500 °C for 60 minutes using diluted diethyl selenide (DESe) vapor as selenium source with 40 Å and 80 Å NaF at the back of metallic precursor as well as in the front of the metallic precursor. Later the two best devices were selenized at 500 °C for 40 min and sulfurized for 5 minutes to improve Voc. The as-prepared absorbers were characterized using scanning electron microscopy (SEM), x-ray diffraction (XRD), and atomic force microscopy (AFM) to study the morphology and crystal structure. I-V measurements were carried out to study the device performance of absorber films.

RESULTS AND DISCUSSIONS

In the beginning, the effect of NaF on the morphology and growth of the absorber films was studied for films with no NaF, 40 Å and 80 Å NaF at the back of the metallic precursor films. Figure 1 shows the SEM micrographs of the resulting films. It is evident from the micrographs that the presence of sodium improves the crystallinity and grain size of the films. Also, the XRD analysis (figure 2) showed a slight improvement in the (112) preferred orientation as evident from Table I.

Table I. Effect of NaF on the degree of preferred orientation for CIGSe thin film

Orientation	No NaF	40 Å	80 Å
112	0.958	1.032	1.041
220/204	0.649	0.635	0.599
312/116	0.970	0.898	0.857

In order to understand whether the location of sodium precursor has any impact on the structure and morphology and on device performance of the absorber films, it was decided to deposit 40 Å and 80 Å NaF on top of the metallic precursor films. The resulting films were selenized at 500 °C for 60 minutes in diluted DESe vapor.

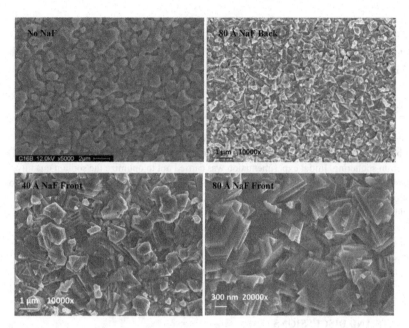

Figure 1. SEM micrographs of as-prepared CIGSe absorber films with various quantities of NaF

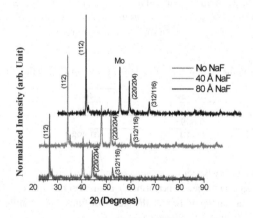

Figure 2. XRD spectra of CIGSe films with various quantities of NaF

From the SEM micrograph in figure 1, it is clear that the presence of sodium improved the morphology and grain size of the absorber films. In fact films with sodium in the front showed much better crystallinity with well faceted, compact and continuous grains in comparison to absorbers films with sodium at the back. Even with 80 Å NaF at the back of the absorber, the grains are much smaller than and not as compact as the grains in the absorber with 40 Å NaF at the front which shows large, compact and faceted grains. Addition of sodium improves the surface morphology and favors homogenization of the films, thereby minimizing and even eliminating the hillock like growth on the film surface and reducing defects. Sodium also has a high mobility and has the ability to act as a fluxing agent by imparting higher mobility to the other atomic species during the growth of the absorber film. Higher mobility of atoms results in improved grain growth as evident from SEM micrographs in figure 1. Atomic force microscopy (AFM) was carried out in order to study the impact of sodium on surface roughness of the absorber films. Comparison of average surface roughness values (figure 3) obtained with AFM indicates that sodium enhances surface diffusivity of the atomic species and results in smoothening of the absorber surface with a reduction in the surface roughness.

Devices were completed on all the absorber films with various amounts of NaF in the front and at the back of the absorber films. In order to improve the performance of the CIGSe devices it was planned to carry out sulfurization following selenization process. To begin with, precursors with 40 Å and 80 Å NaF in the front were selenized and sulfurized for 40 minutes and 5 minutes respectively. The selenization time was reduced so that the films are slightly selenium deficient and that the sulfur would substitute for the selenium vacancies. The results obtained with illuminated I-V measurements are presented in Table II.

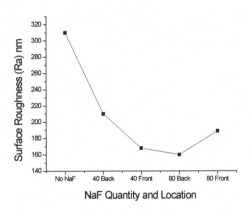

Figure 3. Variation of surface roughness with quantity and location of NaF precursor for CIGSe and CIGSeS thin film absorber

Table II. Current voltage data for the absorber films prepared with various quantities of NaF

Parameter	Voc (V)	Jsc (mA/cm^2)	FF (%)	η (%)	R$_s$ (Ω-cm^2)	R$_{sh}$ (Ω-cm^2)
40 Å Back, CIGSe	0.38	30.10	54.4	6.29	2.3	455
80 Å Back, CIGSe	0.42	38.50	58.7	9.60	1.8	500
40 Å Front, CIGSe	0.46	35.25	59.2	9.67	3.2	1050
80 Å Front, CIGSe	0.46	38.00	66.0	11.80	1.5	1226
40 Å Front, CIGSeS	0.51	38.24	61.4	11.85	3.0	1175
80 Å Front, CIGSeS	0.49	37.51	66.8	12.20	2.3	819

Table III. Elemental composition of absorber films determined using EPMA at 10kV

Element	40 Å Front, CIGSe	40 Å Front, CIGSeS	80 Å Front, CIGSeS
Cu	20.72	19.58	21.72
In	28.75	29.59	28.17
Ga	1.00	1.13	0.86
Se	49.5	45.78	45.29
S	NA	3.92	3.95
Cu/(In+Ga)	0.70	0.64	0.75
Ga/(In+Ga)	0.03	0.04	0.03
S/(S+Se)	NA	0.08	0.08

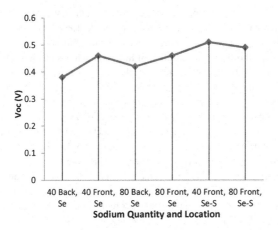

Figure 4. Variation of Voc with quantity and location of NaF precursor for CIGSe and CIGSeS thin film absorber

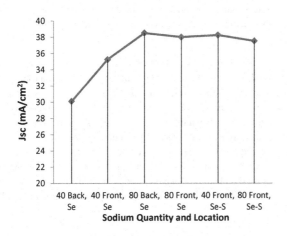

Figure 5. Variation of Jsc with quantity and location of NaF precursor for CIGSe and CIGSeS thin film absorber

Improvement in Jsc, Voc and fill-factor is clearly evident from the I-V data in Table II and figures 4 and 5 with increase in the NaF content from 40 Å to 80 Å. Table III provides the elemental composition of as-prepared absorber films. The effect of sulfur on increase in the Voc values is evident. It has been shown that selenization of films containing sodium results in the formation of $NaSe_x$ compounds that retard the growth of the CIGSe phase and thereby facilitates better incorporation of selenium into the film [2]. Presence of sodium improved the quality of the absorber with enhancement in device characteristics for increase in quantity from 40 Å to 80 Å. With sodium readily available in the front of the metallic precursor this process of selenium incorporation through the formation of $NaSe_x$ compounds is much faster. Sodium has a tendency to reduce detrimental point defects. It reduces compensating donors by substituting selenium vacancies V_{se} and therefore increases the p-type conductivity. Therefore, higher p-type doped material with minimum defect results. This improvement is seen with improved photon absorption resulting in higher Jsc values. The presence of NaF on the front of the precursor is improving the electronic properties of material in the upper region, thereby, helping to passivate the defects near the CIGS/CdS heterojunction. This effect has also been observed elsewhere [12]. This accounts for an increase in Voc which is evident in absorbers that had NaF in the front of the precursor. NREL certified device efficiency of 12.2 % was obtained for absorber with 80 Å NaF in the front. Earlier, NREL certified efficiencies of 13.73% and 12.78% have been obtained at this laboratory using an alkali barrier on sodalime glass substrate in a conventional furnace and in a rapid thermal processing system respectively [13-15]. The supply of sodium was obtained with vacuum evaporated NaF behind the CuGaIn metallic precursor followed by selenization-sulfurization of the precursors in diluted DESe vapor in a conventional furnace and in a rapid thermal processing system using vacuum evaporated selenium layer on the metallic precursor.

From the above study it can be concluded that better CIGSeS absorbers can also be prepared with deposition of NaF on top of the metallic precursor films. This is a useful finding both from a scientific understanding and also from a manufacturing view-point. What is implied from this study is that irrespective of the sodium location, it is possible to obtain devices with reasonable efficiencies. Moreover, it is shown that sodium on top of the precursor provided comparatively better devices than sodium at the back. This finding can be effectively utilized for a useful application so as to speed up the deposition process for fabrication of CIGS solar cells, which is one of the goals of this study. Similar to using sodium doped molybdenum targets for deposition of back contact, it is possible to dope a controlled amount of sodium in the metallic precursor targets and thereby eliminate the need for a separate step for the deposition of NaF.

CONCLUSIONS

Optimization of sodium precursor was carried out in this research work and it was found that good quality CIGSeS absorbers by a two stage process can also be prepared with deposition of NaF in the front of the metallic precursor films. It was shown that sodium in the front of the precursor provides comparatively better devices than sodium at the back. Improvement in the absorber crystallinity and grain size and reduction in surface roughness of the absorber films was observed with increase in NaF quantity from 40 Å to 80 Å. Improved crystal structure and morphology was seen for films with sodium in the front. Sulfurization of absorber films resulted in improvement in the open-circuit voltages. NREL certified device efficiency of 12.2 % was obtained for CIGSeS device with 80 Å NaF in the front. The results indicate that one could incorporate an optimum quantity of sodium within the metallic target and thereby eliminate the need for any additional step for sodium deposition thus saving considerable time in manufacturing line, similar to using sodium doped molybdenum targets for deposition of back contact. This is a useful finding both from a scientific understanding and also from a manufacturing view point.

REFERENCES

1. M.A. Contreras, B. Egaas, P. Dippo, J. Webb, J. Granata, K. Ramnathan, S. Asher, A. Swartzlander and R. Noufi, "On the Role of Na and Modifications to Cu(In,Ga)Se$_2$ Absorber Materials Using Thin-MF (M=Na, K, Cs) Precursor Layers," conference proceedings of 26th IEEE Photovoltaic Specialists Conference, Anaheim, IEEE Press, Piscataway, pp. 359-362, 1997.
2. D. Braunger, S. Zweigart and H.W. Schock, "The Influence of Na and Ga on the Incorporation of the Chalcogen in Polycrystalline Cu(In,Ga)(S,Se)2 Thin-Films for photovoltaic Applications," Proceedings of 2nd World Conference on Photovoltaic Solar Energy Conversion, Vienna, pp. 1113, 1998.
3. S.H. Wei, S.B. Zhang and A. Zunger, "Effects of Na on the Electrical and Structural Properties of CuInSe$_2$," Journal of Applied Physics, vol. 85, pp. 7214-7218, 1999.
4. V. V. Hadagali, "Study of the effects of sodium and absorber microstructure for the development of CuIn$_{1-x}$Ga$_x$Se$_{2-y}$S$_y$ thin film solar cells using an alternative selenium precursor," Ph.D. Dissertation, University of Central Florida, Spring 2009.
5. P. S. Vasekar, "Effect of sodium and absorber thickness on CIGS2 thin film solar cells," Ph.D. Dissertation, University of Central Florida, Spring 2009.
6. S. A. Pethe, "Optimization of process parameters for reduced thickness CIGSeS thin film solar cells," Ph.D. Dissertation, University of Central Florida, Fall 2010.
7. J. Palm, V. Probst, F. H. Karg, "Second generation CIS solar modules," *Solar Energy*, vol. 77, pp. 757–765, 2004.
8. K. Granath, M. Bodegard, and L. Stolt, "The effect of NaF on Cu(In,Ga)Se$_2$ thin film solar cells," *Solar Energy Materials and Solar Cells*, vol. 60, pp. 279-293, 2000.
9. S. Ishizuka, A. Yamada, and S. Niki, "Efficiency enhancement of flexible CIGS solar cells using alkali-silicate glass thin layers as an alkali source material," *IEEE Photovoltaic Specialists Conference*, pp. 2349-2353, 2009.

10. R. Wuerz, A. Eicke, F. Kessler, P. Rogin, O. Yazdani-Assl, "Alternative sodium sources for Cu(In,Ga)Se$_2$ thin-film solar cells on flexible substrates," Thin Solid Films, pp. 7268-7271, 2011.

11. F. Pianezzi, A. Chirila, P. Reinhard, J. Perrenoud, S Nishiwaki, S. Buecheler and A. Tiwari, "A novel surface treatment for Cu(In,Ga)Se2 thin films for highly efficient solar cells," MRS Spring Meeting, San Francisco, CA, 2013.

12. P. T. Erslev, J. W. Lee, W. N. Shafarman, J. D. Cohen, " The influence of Na on metastable defect kinetics in CIGS materials," Thin Solid Films, pp. 2277-2281, 2009.

13. A. A. Kadam and N.G. Dhere, "Highly efficient CuIn$_{1-x}$Ga$_x$Se$_{2-y}$S$_y$/CdS thin-film solar cells by using diethylselenide as selenium precursor," Solar Energy Materials and Solar Cells, Vol. 94, Issue 5, pp. 738-743, 2010.

14. S.S. Kulkarni, G.T. Koishiyev, H. Moutinho, and N.G Dhere, "Preparation and characterization of CuIn$_{1-x}$Ga$_x$Se$_{2-y}$S$_y$ thin film solar cells by rapid thermal processing," Thin Solid Films, Vol. 517, Issue 7, pp. 2121-2124, 2009.

15. N. G. Dhere and A. A. Kadam, U.S. Patent, "CuIn$_{1-x}$Ga$_x$Se$_{2-y}$S$_y$ (CIGSS) thin film solar cells prepared by selenization /sulfurization in a conventional furnace using a new precursor," U.S. Patent 7632701.

Mater. Res. Soc. Symp. Proc. Vol. 1538 © 2013 Materials Research Society
DOI: 10.1557/opl.2013.1011

Observation of Sodium Diffusion in CIGS Solar Cells with Mo/TCO/Mo Hybrid Back Contacts

Yukiko Kamikawa[1], Hironori Komaki[1], Shigenori Furue[1], Akimasa Yamada[1], Shogo Ishizuka[1], Koji Matsubara[1], Hajime Shibata[1], Shigeru Niki[1]

[1]Research Center for Photovoltaic Technologies, National Institute of Advanced Industrial Science and Technology, 1-1-1 Umezono, Tsukuba, Ibaraki 305-8568, Japan

ABSTRACT

CIGS solar cells were fabricated on a hybrid back contact comprised of a TCO layer (ZnO:Ga (GZO)) and Mo layers. It was discovered that an additional Mo layer introduced underneath the TCO layer promotes sodium diffusion through the TCO back contact into the upper CIGS absorber layer. Improvement in V_{OC} and J_{SC} values relative to those of sodium-free solar cells was achieved with the Mo/GZO/Mo hybrid back contact as a result of the enhanced sodium diffusion.

INTRODUCTION

Chalcopyrite $Cu(In,Ga)Se_2$ (CIGS) and related multinary compounds are attracting much attention due to their applicability in highly efficient and cost-effective solar cell modules[1-3]. In these systems, alkali metals, especially sodium (Na), are widely known to play an important role in improving photovoltaic performance[4, 5]. Alkali doping *via* diffusion from a soda-lime glass (SLG) substrate is one of the easiest ways to incorporate alkali metals into the CIGS absorber, and high efficiencies (>20%) have been achieved using the doping methods[1,6].Transparent conductive oxides (TCOs), such as In_2O_3, SnO_2, and ZnO, enable formation of bifacial or tandem transparent structures. In addition, they enable light trapping with textured structures that can be readily formed by optimization of the deposition conditions or through chemical etching. However, when a TCO is introduced as the back contact on an SLG substrate, the preferable diffusion of alkali metals is blocked by the oxide layer[7, 8].

In this study, it was discovered that sodium diffusion through the TCO back contact into the upper CIGS absorber layer was remarkably promoted when an additional Mo layer was introduced underneath the TCO back contact (ZnO:Ga (GZO)). The concentration of diffused sodium into CIGS absorbers deposited on variously structured back contacts (GZO, Mo, Mo/GZO, GZO/Mo, Mo/GZO/Mo) on SLG substrates were studied using secondary ion mass spectrometry (SIMS) and electron probe microanalyzer (EPMA). The photovoltaic (PV) properties of CIGS solar cells fabricated on variously structured back contacts (Mo/GZO, Mo/GZO/Mo, and Mo) on SLG substrates were also evaluated.

EXPERIMENT

In this study, back contacts comprised of Mo, GZO, Mo/GZO, GZO/Mo, and Mo/GZO/Mo structures were used to fabricate CIGS solar cells. Mo and GZO layers were fabricated using radio-frequency (13.56 MHz) magnetron sputtering and reactive plasma deposition, respectively. Each electrode was fabricated on an SLG substrate. The structures of the typical CIGS solar cells fabricated in this study are shown in Fig. 1. The CIGS films were deposited on the back contact/SLG structures using a three-stage process[4, 9, 10]. The highest deposition temperature was limited to 520 °C, and the thickness of the CIGS absorbers was ~1.8 μm. Ga/III and Cu/III ratios were ~0.4 and ~0.93, respectively. CdS buffer layers and i-ZnO and n-ZnO layers were deposited via chemical bath deposition and radio frequency magnetron sputtering, respectively. Al grids, which served as the front electrodes, were fabricated on top of the n-ZnO layer via vacuum evaporation. The active area of each cell was 0.48 cm^2. No antireflection coating was used. Sodium composition and its depth profiles were evaluated using EPMA and SIMS with Cs$^+$ primary ions, respectively.

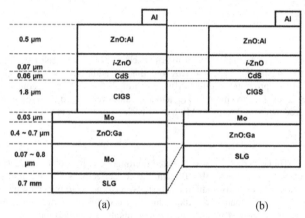

(a)   (b)

Fig. 1 Schematic structure of CIGS solar cells grown on
(a) Mo/GZO/Mo/SLG and (b) Mo/GZO/SLG.

DISCUSSION

Sodium compositions for CIGS thin films fabricated on variously structured back contact/SLG substrates, evaluated by EPMA with an accelerating voltage of 5 kV are shown in Fig. 2. No sodium was observed for the sample with the GZO layer introduced directly on SLG substrate, represented by the layered structures GZO/SLG and Mo/GZO/SLG. On the other hand, introduction of an additional Mo layer underneath the GZO layer, between the GZO layer and SLG substrates, enhanced sodium diffusion. About 1 mole% of sodium was observed in the CIGS layer grown on the GZO/Mo/SLG or Mo/GZO/Mo/SLG substrate. The Mo layer between the CIGS layer and the GZO layer was introduced for some samples for the purpose of obtaining preferable ohmic contacts between the back contact layer and the upper CIGS absorber and did not affect sodium diffusion. No remarkable change in sodium diffusion, $i.e.$, change in sodium concentration   in   the   CIGS   absorber,   was   observed   with   an

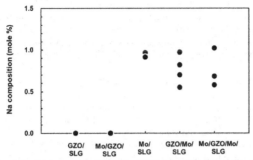

Fig. 2. Sodium compositions for CIGS thin films fabricated on variously structured back contact/SLG substrates, evaluated by EPMA with an accelerating voltage of 5 kV.

introduction of the Mo layer between the GZO and the CIGS absorber. This clearly indicates that the Mo layer introduced between the GZO layer and the SLG substrate plays an important role for enhancing sodium diffusion through the GZO layer into the upper CIGS absorber.

Sodium SIMS depth profiles for the CIGS solar cells fabricated on variously structured back contact/SLG substrates are shown in Fig. 3. Reduced sodium diffusion was observed in the GZO back contact and the upper CIGS absorber, in which the sodium concentration remarkably decreased. On the other hand, a noteworthy level of sodium diffusion was observed when an additional Mo layer was inserted underneath the GZO layer in agreement with the EPMA results

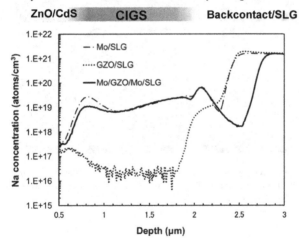

Fig. 3. Sodium SIMS depth profiles for CIGS solar cells fabricated on different back contact/SLG structures.

discussed above. As a result of this enhanced sodium diffusion into the CIGS absorber, a sodium concentration of $\sim10^{19}$ atoms/cm^3, which is comparable with that in a CIGS absorber grown on a conventional Mo/SLG structure, was achieved. SEM images of the CIGS layers grown on the variously structured back contact/SLG structures are shown in Fig. 4. A remarkable enlargement of the CIGS grains was observed in the sample with a GZO back contact, which is typical in alkali-free samples[11].

Current density-bias (JV) curves and photovoltaic (PV) parameters for the CIGS solar cells grown on the Mo, Mo/GZO, and Mo/GZO/Mo back contacts fabricated on SLG substrates are shown in Fig. 5 and Table I, respectively. An obvious reduction in the V_{OC} values was observed for the Na-free solar cell grown on the Mo/GZO/SLG substrate, which is typical in alkali-free samples. On the other hand, the solar cell grown on the Mo/GZO/Mo/SLG substrate, with the Mo layer inserted underneath the GZO layer, showed relatively higher V_{OC} and J_{SC} values. PV properties comparable to those of conventional solar cells (CIGS/Mo/SLG) were also obtained, even with the TCO back contact.

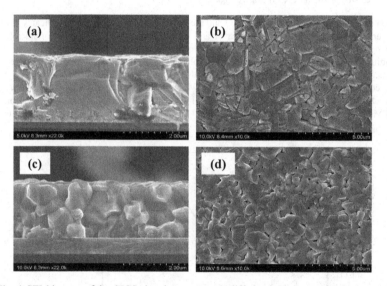

Fig. 4. SEM images of the CIGS absorbers grown on different back contact/SLG structures: (a) cross section and (b) surface images for CIGS/GZO/SLG; (c) cross section and (d) surface images for CIGS/Mo/GZO/Mo/SLG.

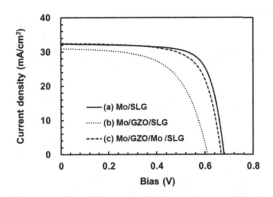

Fig. 5. JV curves for the CIGS solar cells grown on (a) Mo, (b) Mo/GZO, and (c) Mo/GZO/Mo back contacts fabricated on SLG substrates.

Table I. PV parameters for the CIGS solar cells grown on Mo, Mo/GZO, and Mo/GZO/Mo back contacts fabricated on SLG substrates.

Back contacts	Eff. (%)	V_{OC} (V)	J_{SC} (mA/cm^2)	FF
Mo	16.1	0.68	32.3	0.74
Mo/GZO/Mo	15.1	0.67	32.5	0.70
Mo/GZO	11.3	0.61	31.0	0.60

CONCLUSION

Sodium diffusion through a TCO layer into a CIGS absorber was found to be remarkably enhanced when an additional Mo layer was introduced beneath the TCO layer. Na concentration of ~10^{19} atoms/cm^3, which is comparable with that in standard CIGS solar cells, was observed in the CIGS absorber fabricated on a Mo/GZO/Mo hybrid back contact, whereas a Na content three orders of magnitude smaller was found in the CIGS absorber grown on the Mo/GZO back contact. Although the mechanism of sodium diffusion promotion in the combined TCO/Mo back contact has not yet been fully clarified and further studies are needed, two possibilities can be proposed: (1) the formation of Na and Mo compounds and (2) a change in the crystalline quality of the TCO due to the presence of the Mo layer, which may enhance the diffusion of alkali metals. An improvement in the V_{OC} and J_{SC} values relative to those of the Na-free solar cells was achieved with the Mo/GZO/Mo hybrid back contact. PV properties comparable to those of conventional solar cells (CIGS/Mo/SLG) were also obtained, even with the TCO back contact. This result opens new possibilities for the fabrication of more functionalized structures using a TCO layer as the back contact.

ACKNOWLEDGEMENT

This work was supported by the New Energy and Industrial Technology Development Organization (NEDO) under the Ministry of Economy, Trade and Industry (METI).

REFERENCES

[1] M. Powalla, P. Jackson, W. Witte, D. Hariskos, S. Paetel, C. Tschamber, W. Wischmann, Solar Energy Mat. Solar Cells, in press.
[2] S. Niki, M. Contreras, I. Repins, M. Powalla, K. Kushiya, S. Ishizuka, and K. Matsubara, Prog. Photovolt: Res. Appl., 18 (2010) 453.
[3] EMPA press release (18 January 2013) <http://www.empa.ch/plugin/template/empa/*/131441>.
[4] M. A. Contreras, B. Egaas, P. Dippo, J. Webb, J. Granata, K. Ramanathan, S. Asher, A. Swartzlander, and R. Noufi, Proceedings of the 26th IEEE Photovoltaic Specialists Conference, Anaheim (IEEE, New York, 1997), p. 359.
[5] L. Kronik, D. Cahen, and H. W. Schock, Adv. Mater. 10, 31 (1998).
[6] P. Jackson, D. Hariskos, E. Lotter, S. Paetel, R. Wuerz, R. Menner, W. Wischmann and M. Powalla, Prog. Photovolt: Res. Appl., 19, 894 (2011).
[7] F. Guanghui, D. Jiafeng, P. Donghui, H. Ouli, Journal of Non-Crystalline Solids, 112, 454 (1989).
[8] F.-J. Haug, D. Rudmann, G. Bilger, H. Zogg, A. N. Tiwari, Thin Solid Films, 403-404, 293 (2002).
[9] A. M. Gabor, J. R. Tuttle, D. S. Albin, M. A. Contreras, R. Noufi, and A. M. Hermann, Appl. Phys. Lett., 65, 198 (1994).
[10] R. Hunger, K. Sakurai, A. Yamada, P.J. Fons, K. Iwata, K. Matsubara, S. Niki, Thin Solid Films, 431-432, 16-21 (2003).
[11] S. Ishizuka, A. Yamada, M. Monirul Islam, H. Shibata, P. Fons, T. Sakurai, K. Akimoto, and S. Niki, J. Appl. Phys, 106, 034908 (2009).

66

Mater. Res. Soc. Symp. Proc. Vol. 1538 © 2013 Materials Research Society
DOI: 10.1557/opl.2013.1007

Formation of Ga_2O_3 barrier layer in $Cu(InGa)Se_2$ superstrate devices with ZnO buffer layer

Jes K. Larsen, Peipei Xin, William N. Shafarman

Institute of Energy Conversion, University of Delaware, Newark, DE, 19716, USA

ABSTRACT

The junction formation when $Cu(InGa)Se_2$ is deposited onto ZnO in a superstrate configuration (glass/window/buffer/$Cu(InGa)Se_2$/contact) is investigated by x-ray photoelectron spectroscopy and analysis of device behavior. When $Cu(InGa)Se_2$ is deposited on ZnO, a Ga_2O_3 layer is formed at the interface. Approaches to avoid the formation of this unfavorable interlayer are investigated. This includes modifications of the process to reduce the thermal load during deposition and improvement of the thermal stability of the ZnO buffer layer. It was demonstrated that both lowering of the substrate deposition temperature and deposition of the ZnO buffer layer at elevated temperature limits the Ga_2O_3 formation. The presence of Ga_2O_3 at the junction does affect the device behavior, resulting in a kink in JV curves measured under illumination. This behavior is absent in devices with limited Ga_2O_3 formation.

INTRODUCTION

$Cu(InGa)Se_2$ devices with superstrate configuration are interesting for several reasons. The superstrate configuration opens the possibility to deposit the window layer at elevated temperature to improve its optoelectronic properties. This device structure, furthermore, makes it easier to texture the window layer for light trapping and engineer the reflectance of the back contact as needed for cells with thin, < 1 µm, absorber layers. At the same time this approach eliminates the need for a transparent back cover, which potentially leads to cost reduction.

A critical issue for the superstrate device structure is to control the junction formation when $Cu(InGa)Se_2$ is deposited onto the buffer layer. Since $Cu(InGa)Se_2$ is typically deposited at elevated temperature, inter-diffusion and reactions at the interface play an important role. In the present study, $Cu(InGa)Se_2$ is deposited by co-evaporation onto a ZnO buffer layer under various conditions. The chemical interactions are investigated by x-ray photoelectron spectroscopy (XPS) depth profiling and the relationship to device behavior is investigated.

$Cu(InGa)Se_2$ devices with a superstrate structure have been studied previously. Early studies using CdS buffer layers, typically used in substrate $Cu(InGa)Se_2$ devices, demonstrated CdS/$Cu(InGa)Se_2$ interdiffusion at elevated deposition temperatures[1,2]. To eliminate the problem of interdiffusion, CdS was replaced by a ZnO buffer layer, which did improve device performance[2]. The best reported superstrate device, reaching 12.8 % efficiency, utilized ZnO as the buffer layer[3]. Two groups have reported on interactions at the interface between $Cu(InGa)Se_2$ and ZnO in superstrate devices[4,5]. In both studies it was demonstrated that ZnO reacts with $Cu(InGa)Se_2$ during absorber deposition, leading to the formation of a Ga_2O_3 layer at the interface. Considering the difference in electron affinity between Ga_2O_3[6] and $Cu(InGa)Se_2$[7] a large spike of 0.9 eV is expected in the conduction band alignment at the Ga_2O_3/$Cu(InGa)Se_2$ interface. The Ga_2O_3 layer, therefore, acts as a barrier for photocurrent, reducing the performance of superstrate devices with a ZnO buffer layer. The present study investigates approaches to avoid formation of Ga_2O_3 at the interface.

EXPERIMENT

A set of samples are prepared to investigate the process conditions to prevent Ga_2O_3 formation. A summary of the sample preparation conditions is collected in table 1. As a reference baseline, a sample using room temperature deposited ZnO and $Cu(InGa)Se_2$ deposited at 550 °C is included (sample A). Two different approaches to prevent Ga_2O_3 formation are tested. One approach aims to improve the stability of the ZnO buffer layer by increasing the substrate temperature to 500 °C during sputtering (sample B). The other approach aims to limit the $ZnO/Cu(InGa)Se_2$ interaction by reducing the thermal load during the absorber deposition (samples C and D). For sample C, the thermal load is reduced by lowering the substrate temperature to 450 °C during $Cu(InGa)Se_2$ deposition. For sample D, the absorber is deposited at 350 °C followed by a 1 minute annealing step at 550 °C[8].

Table 1: Conditions for preparation of samples. The thickness of all samples prepared for devices is 2 μm.

Sample	ZnO T_{ss}	Cu(InGa)Se$_2$ T_{ss}	Description
A	RT	550 °C	Baseline
B	500 °C	550 °C	High temperature ZnO
C	RT	450 °C	Low temperature absorber
D	RT	350 °C (1min 550 °C anneal)	1 minute absorber annealing

Substrates for all samples are sodalime glass (SLG) sputter-coated with indium tin oxide (ITO). 100nm ZnO buffer layers are deposited by rf sputtering onto SLG/ITO substrates. $Cu(InGa)Se_2$ absorbers are deposited for 60 min onto the SLG/ITO/ZnO substrates by a single stage co-evaporation process. All samples are grown under Cu-poor conditions with $[Cu]/([In]+[Ga]) = 0.7 - 0.9$ and with $[Ga]/([In]+[Ga]) = 0.3$, measured by x-ray fluorescence. For all devices the absorber thickness is 2 μm. To complete devices, 200 nm gold back contacts (0.4 cm^2) are e-beam evaporated onto the $Cu(InGa)Se_2$. For XPS depth profile studies, another run with similar processing conditions is used, but only 100 nm $Cu(In,Ga)Se_2$ is deposited (samples AX, BX, CX, DX). This is done to more easily reach the interface by argon sputter etching.

Devices made from the samples in table 1 are investigated by current-voltage (JV) measurements and temperature dependent current voltage (JVT) measurements. The thin counterpart of each sample is studied by XPS depth profiling.

RESULTS and DISCUSSION

The presence of Ga_2O_3 at the $ZnO/Cu(InGa)Se_2$ junction can be verified by XPS depth profile measurements. Figure 1 shows four representative spectra of the Ga 3d peak recorded in the bulk of $Cu(InGa)Se_2$ layer and near the $Cu(InGa)Se_2$/ZnO interface.

The Ga 3d peak is sensitive to changes in the chemical environment and contains several components that must be deconvoluted. In order to analyze the data, the background is removed with a Shirley algorithm and the peak is fitted with four Lorenzian-Gaussian components. Based on literature values,[9] the four contributions to the peak are ascribed to In 4d (17 eV), Ga 3d in $Cu(InGa)Se_2$ (18 eV), Ga 3d in Ga_2O_3 (20.5 eV), and O 2s (23 eV).

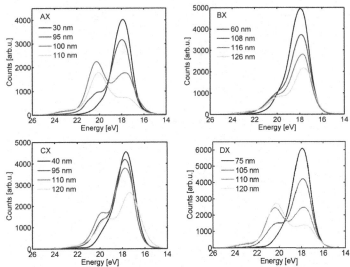

Figure 1: XPS spectra measured in the bulk of Cu(InGa)Se₂ (first spectrum) and near the interface region (last three spectra). The legend indicates the approximate sputter depth.

Figure 2 shows depth profiles of the contributions to the Ga 3d signal from the components ascribed to Ga 3d in Cu(InGa)Se₂ and Ga 3d in Ga₂O₃, respectively. The baseline sample (AX) shows a peak in the Ga₂O₃ signal at the interface between the Cu(InGa)Se₂ and ZnO layers, where the signal of Cu(InGa)Se₂ drops off. This is a clear demonstration of Ga₂O₃ formation under baseline deposition conditions. For the sample where ZnO is deposited at 500 °C (BX) a significantly different behavior is observed. As the signal from Cu(InGa)Se₂ drops off, the component ascribed to Ga₂O₃ also disappears. This shows that Ga₂O₃ formation is limited for this sample, presumably because of increased stability of the ZnO buffer layer. Since the Ga₂O₃ signal does not show a peak, it is assumed that a continuous Ga₂O₃ layer is not formed. Sample CX, for which the absorber is deposited at 450 °C, shows a weak increase in the Ga₂O₃ signal at the interface region, indicating that some amount of Ga₂O₃ is formed. However, the absence of a strong well-defined peak suggests that the reaction is limited but not fully prevented in this case. The final sample (DX) shows a strong increase in the Ga₂O₃ component at the interface. This behavior is a clear indication of Ga₂O₃ formation and demonstrates that the short 1 minute annealing step is sufficient to form Ga₂O₃ comparable to the baseline sample.

Devices are prepared from the samples described in table 1 in order to investigate the relationship between Ga₂O₃ formation and device properties. JV curves of these samples are shown in figure 3. It is noted that the overall performance of all devices is poor. The interest of the JV curves for this study is, however, the shape of the plots rather than the device performance. It is seen that sample A and sample D both have a kink in the light curve. A kink in the JV curve can be attributed to a barrier for the photocurrent[7,10].

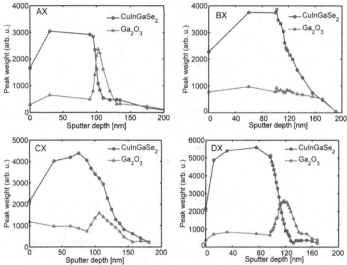

Figure 2: XPS depth profiles of samples AX,BX,CX,DX mentioned in table 1. The curves show the contributions of the Ga 3d peak ascribed to Cu(InGa)Se$_2$ and Ga$_2$O$_3$.

Samples AX and DX, that are comparable to these two samples, clearly show the presence of a Ga$_2$O$_3$ layer. Since a barrier in the conduction band is expected between Ga$_2$O$_3$ and Cu(InGa)Se$_2$, it is proposed that the kink is due to a photocurrent barrier at this interface. Neither sample B or C display a kink in their JV curves. A limited amount of Ga$_2$O$_3$ was detected in the comparable samples BX and CX, further indicating that the kink is related to the presence of Ga$_2$O$_3$ at the junction. Samples B and C, where formation of Ga$_2$O$_3$ is expected to be limited according to the XPS measurements, do, however, not exhibit better device performance than the devices containing a Ga$_2$O$_3$ interlayer. These devices are dominated by a very low V$_{OC}$, which could be caused by several factors such as Na deficiency[11] or a cliff in the band alignment at the Cu(InGa)Se$_2$/ZnO interface[7].

JVT measurements were performed to study the dominant recombination pathway in the superstrate devices. By extrapolating a linear fit of the V$_{OC}$ vs. T graph to T = 0, the activation energy E$_a$ for the diode saturation current J$_0$ can be determined[12]. If the activation energy determined with this approach is lower than the band gap of the absorber, it is an indication that the device is limited by interface recombination[12]. For samples investigated in this study with [Ga]/([In]+[Ga]) = 0.3 eV the band gap is 1.2 eV.

Parameters obtained from JVT measurements are shown in table 2. When JVT measurements were performed, the samples were first cooled to -60 °C and then heated to 120 °C. It was observed that the device properties changed as the sample temperature reached around 90 °C. When cooling samples after the heating step it was observed that the

70

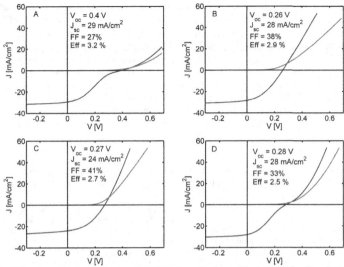

Figure 3: JV measurements of device A,B,C,D described in table 1. The blue line is the measured under 1 sun illumination and the red is measured in the dark.

Table 2. Activation energies of the saturation current extracted from JVT measurements and device parameters obtained before and after JVT measurement.

Sample	E_a [eV]		Efficiency [%]		V_{OC} [V]		J_{sc} [mA/cm^2]		FF [%]	
	Initial	Final	Initial	Final	Initial	Final	Initial	Final	Initial	Final
A	0.45	0.82	3.2	4.3	0.4	0.4	29	29	27	36
B	0.40	0.80	2.9	3.5	0.26	0.31	28	29	38	38
C	0.48	0.88	2.7	5.1	0.27	0.41	24	24	41	52
D	0.40	0.88	2.5	4.4	0.28	0.39	28	28	33	40

activation energies had changed appreciably compared to the initial value from measurements performed before the sample was heated above 90 °C. Two different activation energies could therefore be determined (see table 2). The change in activation energy is accompanied by an improvement of the device performance. Especially notable is the increase in V_{OC}, while J_{SC} remains unchanged. It can be speculated that the improvement is related to a change in interface defects, but the exact nature of the origin remains unclear. From the fact that all activation energies determined by JVT measurements are well below the bandgap of the absorber, it can be concluded that all devices are dominated by interface recombination. This is independent on the amount of Ga_2O_3 formed at the interface and is still the case after heating the samples.

CONCLUSIONS

Several approaches to prevent formation of Ga_2O_3 during co-evaporation of Cu(InGa)Se$_2$ on ZnO at elevated temperature have been investigated. It was demonstrated that formation of Ga_2O_3 was not prevented by reduction of annealing time at 550 °C to one minute. Two approaches did, however, limit the extent of Ga_2O_3 formation. Lowering substrate temperature to 450 °C during $Cu(InGa)Se_2$ deposition did reduce the amount of Ga_2O_3, while deposition of ZnO at 500 °C further limits Ga_2O_3 formation. The presence of Ga_2O_3 at the junction does affect the device behavior. In devices where Ga_2O_3 is present, a kink is observed in room temperature JV measurements. This is interpreted as a barrier for the photocurrent caused by a spike in the conduction band alignment between $Cu(InGa)Se_2$ and Ga_2O_3. The approaches successfully preventing Ga_2O_3 formation do, however, not lead to improvement of device performance. In the presence or absence of Ga_2O_3, the superstrate devices using ZnO buffer layers presented here are dominated by interface recombination. Interestingly it was discovered that the device performance increased during JVT measurements due to a change in the activation energy of the saturation current.

ACKNOWLEDGMENTS

The authors thank Kevin Hart for ITO and gold contact deposition. This research is supported by the DOE F-PACE program. This material is based, in part, upon work supported by the Department of Energy.

Disclaimer. "This report was prepared as an account of work sponsored by an agency of the United States Government. Neither the United States Government nor any agency thereof, nor any of their employees, makes any warranty, express or implied, or assumes any legal liability or responsibility for the accuracy, completeness, or usefulness of any information, apparatus, product, or process disclosed, or represents that its use would not infringe privately owned rights. Reference herein to any specific commercial product, process, or service by trade name, trademark, manufacturer, or otherwise does not necessarily constitute or imply its endorsement, recommendation, or favoring by the United States Government or any agency thereof. The view and opinions of authors expressed herein do not necessarily state or reflect those of the United States Government or any agency thereof."

REFERENCES

1. T. Yoshida and R.W. Birkmire, Proc. 11th European Communities Photovoltaic Solar Energy Conf. 811, 811 (1992).
2. T. Nakada, T. Mise, T. Kume, and A. Kunioka, 2nd World Conference and Exhibition on Photovoltaic Solar Energy Conversion 413 (1998).
3. T. Nakada and T. Mise, Proceedings of the 17th E.C. Photovoltaic Solar Energy Conference 1027 (2001).
4. T. Nakada, Thin Solid Films 480-481, 419 (2005).
5. M. Terheggen, H. Heinrich, G. Kostorz, F.-J. Haug, H. Zogg, and A.. N. Tiwari, Thin Solid Films 403-404, 212 (2002).
6. J. Robertson, Journal of Vacuum Science & Technology B: Microelectronics and Nanometer Structures 18, 1785 (2000).
7. R. Scheer and H.W. Schock, Chalcogenide Photovoltaics: Physics, Technologies, and Thin Film Devices (Wiley-VCH, 2011).
8. J.D. Wilson, R.W. Birkmire, and W.N. Shafarman, in 2008 33rd IEEE Photovolatic Specialists Conference (IEEE, 2008), pp. 1–5.
9. C.W. Wagner, Handbook of X-ray Photoelectron Spectroscopy (Physical Electronics Division, Perkin-Elmer Corp., 1979).
10. T. Minemoto, T. Matsui, H. Takakura, Y. Hamakawa, T. Negami, Y. Hashimoto, and T. Uenoyama, Solar Energy Materials & Solar Cells 67, 83 (2001).
11. D. Rudmann, A.F. da Cunha, M. Kaelin, F. Kurdesau, H. Zogg, A.N. Tiwari, and G. Bilger, Applied Physics Letters 84, 1129 (2004).
12. S.S. Hegedus and W.N. Shafarman, Progress in Photovoltaics: Research and Applications 12, 155 (2004).

Kesterite

Mater. Res. Soc. Symp. Proc. Vol. 1538 © 2013 Materials Research Society
DOI: 10.1557/opl.2013.1005

Effects of Growth Conditions on Secondary Phases in CZTSe Thin Films Deposited by Co-evaporation

Douglas M. Bishop, Brian E. McCandless, Thomas C. Mangan, Kevin Dobson, and Robert Birkmire

Institute of Energy Conversion, University of Delaware, 451 Wyoming Rd, Newark, DE 19711, U.S.A.

ABSTRACT

High temperature multi-source co-evaporation has been the most successful approach to fabricate record efficiency Cu(InGa)Se$_2$ devices, yet many groups have been unable to replicate this success when transferring these methods to the Cu$_2$ZnSnSe$_4$ system. The difficulties stem from the dramatic differences in the thermochemical properties which result in decomposition and loss of volatile species, such as Zn and SnSe, at temperatures needed for growth. In co-evaporation, decomposition and element loss must be managed throughout the entire growth process, from the back contact interface to the final terminating surface of the film. The beginning and ending phases of deposition encompass different kinetic regimes suggesting a phased approach to growth may be helpful. A series of depositions with different effusion profiles were used to demonstrate the effects of decomposition during different stages of growth. Secondary phase detection can be challenging in CZTSe, but a combination of SEM imaging and thin cross-section depth profile by EDS were found to best identify and locate the secondary phases that occur during different phases of growth for co-evaporated Cu$_2$ZnSnSe$_4$ films.

Deposition with a uniform incident flux followed by shuttered vacuum cool-down yielded films with a ZnSe phase at the absorber/Mo interface and Cu-rich composition at the surface of the exposed film. Devices from these absorber layers never exceeded conversion efficiencies of 1%. Decomposition at the surface could be prevented by continuing effusion of Se and Sn during the cool-down of the substrate. Resulting films demonstrated more faceted grains as well as significantly improved device performance. Secondary phases that traditionally form at the back contact during the beginning of growth were minimized by decreasing the substrate temperature to 300°C during the initial stages of deposition which reduced the ZnSe formed at the Mo interface. The thermochemical origin of the secondary phases will be discussed and the performance of representative devices will be presented.

INTRODUCTION

Cu$_2$ZnSnS$_4$ and Cu$_2$ZnSnSe$_4$ (CZTS and CZTSe, respectively) have generated significant interest as earth abundant absorber materials for photovoltaic devices in the last half decade. The kesterite crystal structure can be derived from chalcopyrite structure, however, the thermochemical properties are quite different from Cu(InGa)Se$_2$, leading to difficulties when growth and annealing strategies are translated directly to CZTSe [1-2]. In stark contrast to CIGS, record device efficiencies have often been achieved by solution-growth methods [3]. Research teams attempting high-temperature multi-source co-evaporation have reported mixed results with only one team (NREL) reporting >5% device efficiency without a separate controlled annealing process [4]. Co-evaporated films have often exhibited secondary phases at the front and rear of the film, which is a result of phase decomposition and element loss during different stages of growth [2, 5–7]. A key advantage of co-evaporation at film formation temperatures is the potential to drive reaction chemistry towards desired compositions and properties by controlling the elemental incident fluxes throughout the deposition. Thus, despite the challenges in co-evaporation, with the proper understanding of the thermodynamics and kinetics, a procedure can be developed to avoid decomposition and manage element loss to obtain high efficiency CZTSe devices.

This paper highlights the thermodynamic challenges of the CZTSe system with a simple constant effusion deposition. Different time/temperature deposition profiles were used to control the location and presence of secondary phases at the front and back of the films.

EXPERIMENTAL

Films were deposited by multi-source co-evaporation from the elements on 1"x 1" SLG/Mo substrates held at 300-500°C depending on deposition. Base pressure was 1×10^{-6} torr before growth. After deposition, devices were fabricated with the typical CIGS stack, SLG/Mo/CZTSe/CdS/ZnO/ITO and a Ni/Al grid on the front surface. Raman spectroscopy and x-ray diffraction (XRD) analysis showed only slight variations between films and little definitive evidence of secondary phases, but imaging the film surface and cross-section by scanning electron microscope (SEM) coupled with energy dispersive spectroscopy (EDS), were able to clearly illustrate decomposition and the presence of secondary phases. An Auriga 60 Crossbeam SEM was used. An in-lens secondary electron detector was used with 3kV beam voltage which was able to clearly distinguish the presence and location of ZnSe inclusions from scratch and focused ion beam (FIB) cross-sections. Film compositions were obtained using x-ray florescence (XRF).

RESULTS AND DISCUSSION

As has been demonstrated in the literature, although Reaction 1 is spontaneous (Gibbs free energy < 0); the reverse reaction, decomposition of CZTSe, can still occur at an appreciable rate at temperatures needed for film processing. This is due to the distribution of energies and the relatively small difference in Gibbs free energy between CZTSe and binary reactants [8], [9], [10].

$$Cu_2Se \, (s) + ZnSe \, (s) + SnSe \, (g) + \tfrac{1}{2} \, Se_2 \, (g) \leftrightarrow Cu_2ZnSnSe_4 \, (s) \qquad \text{[Reaction 1]}$$

At solid state equilibrium, the products of decomposition would be quickly diminished by the more favorable forward reaction to create CZTSe, but in thin films with significant exposed surfaces, SnSe and Se can be lost due to high volatility, rendering the forward reaction to create CZTSe impossible. In this way, entire CZTSe films can decompose into Cu_2Se and ZnSe at temperatures above 500°C [9, 10]. This thermodynamic process occurs regardless of atmosphere, however a higher background pressure can slow the rate of SnSe loss by allowing some lost SnSe to re-adsorb. A SnSe-containing atmosphere, or effusion of Sn and Se in the case of co-evaporation, can also keep the surface at equilibrium by replacing species lost to volatility.

The vapor pressure of Zn is four orders of magnitude higher than SnSe at 500°C, and consequently Zn loss is also observed in many CZTSe deposition processes from elemental sources. However, Zn loss has not proven to be problematic. The reaction to form ZnSe, which occurs >330°C, stabilizes any Zn from vaporization. In this case, the reverse reaction back to elemental Zn is extremely unfavorable due to the very low Gibbs free energy of ZnSe. Formation of Cu-Zn alloys can theoretically, also act to stabilize Zn, though the presence of Cu-Zn has not been detected in co-evaporated films.

From analysis of the final film, under our experimental conditions, 50% of the deposited Zn and 85% of the deposited Sn was lost during deposition at 500°C. A higher loss of Sn is observed despite the much lower vapor pressure of Sn/SnSe. This can be attributed to a combination of the rates of reaction to stabilize SnSe (by forming $Cu_2ZnSnSe_4$ or Cu_2SnSe_3) and also the aforementioned reverse reactions that can occur throughout the deposition.

The rate of SnSe loss changes during three growth regimes at the beginning, middle, and end of the growth process. During the first stage Sn and Se deposit on the Mo substrate with little CZTSe being formed. In this stage SnSe is formed quickly and lost very rapidly as the SnSe has not been stabilized by formation of CZTSe or Cu_2SnSe_3 [2]. The sticking coefficient also depends on lattice mismatch and bonding to the deposition surface, so it may be expected that the Mo substrate may also increase SnSe loss compared to later stages of film growth. In the second stage, during the majority of growth, Sn and Se are depositing on CZTSe and other binary intermediates. During this stage, SnSe is lost through either vaporization of the binary or decomposition of CZTSe, although SnSe is also somewhat stabilized through the formation of CZTSe. In the third stage, when growth has finished and the absorber layer is cooling, SnSe can only be lost through CZTSe decomposition. Since decomposition and the creation of secondary phases are directly related to SnSe loss, these regimes have a significant effect on the final absorber layer composition and the location of secondary phases.

The two main control variables during deposition include effusion rate and substrate temperature. Initial films were grown with a constant effusion onto a 500°C substrate with enough excess Sn and Se to achieve the desired Cu-poor, Zn rich CZTSe composition. In the control method (Method 1, see Figure 1), films were shuttered immediately after growth, while in a second method, to replace lost SnSe and prevent CZTSe decomposition, effusion of volatile Sn and Se was continued but ramped down until films cooled below 350°C (Method 2, see Figure 2).

Since SnSe loss is particularly rapid in the first stage of growth, a lower substrate temperature was used to reduce volatility. For these depositions, elements were initially deposited at 300°C, before being ramped to 500°C in order to drive the reaction and promote grain growth (Method 3, see Figure 4).

Decomposition during Film Cool-down

The most apparent result of decomposition occurs on the surface of the film. During growth, excess effusion of volatile species compensates for SnSe loss, but after film growth has stopped, as the films cool down, decomposition and element loss continues unless further of Sn and Se are supplied.

In Method 1 (Figure 1A), a constant elemental effusion was used and substrates were shuttered after deposition and the substrate temperature cooled to room temperature before venting the chamber. Although this simple procedure can produce good quality films for CIGS, CZTSe films were found to be shunted, with a Cu_2Se secondary phase detected by XRD. The shunting and overall device performance were not fully restored with a KCN etch. Zn-rich regions were also found on the film surface through SEM and EDS mapping at low beam energies. The surface morphology from method 1 appears to be more rounded with facets, steps and islands on the surface which would be consistent with decomposition, as volatile species are removed more rapidly from corners and step edges (Figure 2A) [11].

In Method 2 (Figure 1B) the effusion of Sn and Se were ramped down during the cool-down of the film until the surface was below 350°C. The film morphology exhibits a more faceted grain structure (Figure 2B) and produced devices that exhibited much less shunting (shunt conductance typically decreased by 2x-10x), and higher efficiencies (>4% compared with <1%), suggesting that CZTSe surface decomposition has not occurred. Zn was effused for 0-5min during cool-down based on literature results observing improved cell results for slightly Zn rich termination[4].

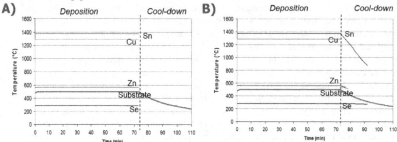

Figure 1: Source and substrate temperature during deposition. A) Method 1 with no effusion during substrate cool-down. B) Method 2 with effusion of Sn and Se during cool-down

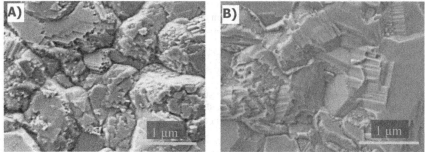

Figure 2: SEM images of CZTSe surface with different deposition methods. A) Method 1 with no effusion during cool-down showing. B) Method 2 with effusion during cool-down showing more faceted grains.

Decomposition at the Back Contact

Similar thermodynamic issues also present problems at the back side of the film during the beginning stages of film growth. For films deposited at 500°C, a layer of ZnSe is formed at the back contact as has been observed in the literature [5,6,12]. The ZnSe layer was confirmed by an EDS depth profile of a 100 nm thick cross-section created by a FIB (Figure 3).

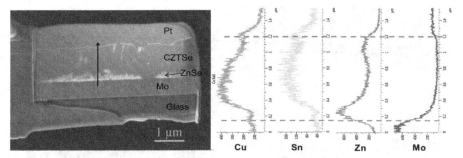

| | Cu | Sn | Zn | Mo |

Figure 3: 100nm cross-section of CZTSe absorber layer created by FIB and EDS depth profile.

ZnSe can be clearly illustrated by bright charging regions in secondary electron SEM images when in-lens secondary electron detection is used. The in-lens detector has better detector efficiency of low energy electrons and has higher signal-to-noise ratio allowing higher contrast for impurities and in this case, charging [13]. When the traditional secondary electron detector was used, the contrast elucidating ZnSe is not observed.

It is important to note that typical planar EDS measurements will often miss secondary phases at the back contact due to a sampling volume that highly weights the top micron of the film. For this reason all film compositions were measured by XRF which was calibrated by ICP.

The origin of the ZnSe layer is due to the higher volatility and loss of SnSe in the initial stages of deposition [2, 6]. A new deposition regime was developed to prevent SnSe loss particularly during the early stages of growth by beginning the deposition at a lower temperature (Method 3, Figure 4). The deposition was started with the substrate at 300°C and later ramped to 500°C for 20 minutes to drive the reaction and promote grain growth. Elemental source temperatures were correspondingly increased to accommodate the lower sticking at the higher substrate temperature. At 300°C, SnSe vapor pressure is three orders of magnitude lower than at 500°C, which allows Sn to incorporate into the growing film from the beginning of the deposition and decrease formation of the ZnSe layer.

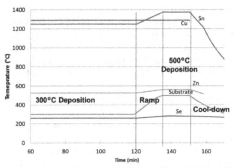

Figure 4: Source and substrate temperature during deposition for Method 3. Deposition began with lower substrate temperature to minimize Sn loss.

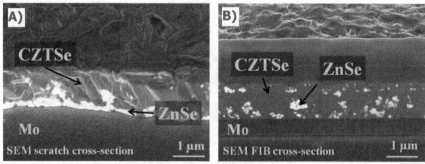

Figure 5: SEM images of CZTSe deposited with different conditions. A) SEM scratch cross section of CZTSe deposited at 500°C (Method 2) B) SEM FIB cross-section deposited beginning at 300°C (Method 3).

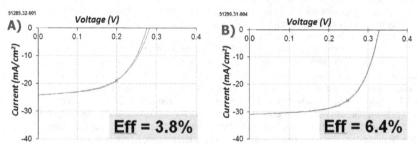

Figure 6: Corresponding cell results for CZTSe films deposited at A) constant 500°C substrate (Method 2), and B) deposited beginning at 300°C (Method 3).

Figure 5A and B shows cross-section SEM images of CZTSe films grown beginning at 500°C and beginning at 300°C respectively. Both films clearly contain ZnSe inclusions, however, in the deposition started at 300°C the precipitation occurred more evenly throughout the depth of the film and does not form a contiguous layer at the back contact. Previous reports in the literature have not observed ZnSe inclusions in the middle of the film, however this may be due to difficulty of detection [14]. The presence of ZnSe in the bulk is likely due to high Zn levels, beyond the tolerance of the narrow single phase region for CZTSe [12]. In depositions at 500°C, it is likely favorable for ZnSe to continue to grow from where it nucleated at the back contact, rather than forming precipitates in the middle of the film, in order to minimize interfacial energy.

Despite the similar compositions, Zn/Sn ratio of ~1.3, and Cu/(Zn+Sn) ~ 0.85, substantial improvement in J_{SC} and V_{OC} is seen (Figure 6) with films deposited at 300°C. Previous literature reports have suggested ZnSe at the back contact has little effect on cell performance though the nearly contiguous ZnSe layer seen above appear to decrease Voc and Jsc. [12] . Given the low carrier lifetime in CZTSe, increasing the carrier distance to the back contact could lead to significant carrier recombination [4]. The ZnSe inclusions in the middle of the cell appear to have less effect on cell performance and may act mainly as "dark space" but do not block current as would be expected from a semi-contiguous ZnSe layer.

It is worth noting the "improved" devices still have not been fully optimized, as noted by the clear presence of 2nd phases. The record co-evaporated device (9.15%) used a Cu-rich growth (although similar high-performing devices were created from without Cu-rich growth) and also a number of additional steps including excess Na, air post-oxidation annealing, and an anti-reflective coating which may help explain part of the gaps in Voc and Jsc[4].

CONCLUSIONS

Improved materials and devices have been demonstrated by modified deposition regimes that address the thermodynamic challenges of CZTSe multi-source evaporation. The use of in-lens SEM imaging has proven a useful tool for identifying the presence and location of ZnSe secondary phases.

Decomposition can be easily observed by the morphology on the surface of films cooled in a vacuum without continued supply of Sn and Se. The loss of SnSe must be addressed by effusion of volatile elements during all stages of growth and cool-down.

Different kinetics in the beginning phases of deposition can result in a ZnSe layer at the back contact which can block current. This layer, also related to loss of Sn, can be addressed by reducing SnSe volatility by lowering the substrate temperature at the beginning of growth. With an improved understanding of material decomposition and chemistry throughout the film growth, deposition can be further optimized.

ACKNOWLEDGMENTS

The authors would like to acknowledge E. I. du Pont de Nemours and Company for partial funding of this work.

REFERENCES

[1]   B.-A. Schubert, B. Marsen, S. Cinque, T. Unold, R. Klenk, S. Schorr, and H.-W. Schock, "Cu2ZnSnS4 thin film solar cells by fast coevaporation," *Progress in Photovoltaics: Research and Applications*, vol. 19, no. May 2010, pp. 93–96, 2011.

[2]   A. Redinger and S. Siebentritt, "Coevaporation of Cu2ZnSnSe4 thin films," *Applied Physics Letters*, vol. 97, no. 9, p. 092111, 2010.

[3]   T. K. Todorov, J. Tang, S. Bag, O. Gunawan, T. Gokmen, Y. Zhu, and D. B. Mitzi, "Beyond 11% Efficiency: Characteristics of State-of-the-Art Cu2ZnSn(S,Se)4 Solar Cells," *Advanced Energy Materials*, p. n/a–n/a, Aug. 2012.

[4]   I. L. Repins, C. Beall, N. Vora, C. DeHart, D. Kuciauskas, P. Dippo, B. To, J. Mann, W.-C. Hsu, A. Goodrich, and R. Noufi, "Co-evaporated Cu2ZnSnSe4 films and devices," *Solar Energy Materials and Solar Cells*, pp. 1–6, Feb. 2012.

[5]   S. Ahn, S. Jung, J. Gwak, A. Cho, K. Shin, K. Yoon, D. Park, H. H. Cheong, and J. H. Yun, "Determination of band gap energy Eg of Cu2ZnSnSe4 thin films: On the

discrepancies of reported band gap values," *Applied Physics Letters*, vol. 97, no. 2, p. 021905, 2010.

[6] A. Redinger, K. Hönes, X. Fontané, V. Izquierdo-Roca, E. Saucedo, N. Valle, A. Pérez-Rodriguez, and S. Siebentritt, "Detection of a ZnSe secondary phase in coevaporated Cu2ZnSnSe4 thin films," *Applied Physics Letters*, vol. 98, no. 10, p. 101907, 2011.

[7] N. Vora, J. Blackburn, I. L. Repins, C. Beall, B. To, J. Pankow, G. Teeter, M. Young, and R. Noufi, "Phase identification and control of thin films deposited by co-evaporation of elemental Cu, Zn, Sn, and Se," *Journal of Vacuum Science & Technology A: Vacuum, Surfaces, and Films*, vol. 30, no. 5, p. 051201, 2012.

[8] J. J. Scragg, P. J. Dale, D. Colombara, and L. M. Peter, "Thermodynamic Aspects of the Synthesis of Thin-Film Materials for Solar Cells," *Chemphyschem : a European journal of chemical physics and physical chemistry*, pp. 3035–3046, Apr. 2012.

[9] A. Redinger, D. M. Berg, P. J. Dale, and S. Siebentritt, "The consequences of kesterite equilibria for efficient solar cells," *Journal of the American Chemical Society*, vol. 133, no. 10, pp. 3320–3, Mar. 2011.

[10] J. J. Scragg, T. Ericson, T. Kubart, M. Edoff, and C. Platzer-Björkman, "Chemical Insights into the Instability of Cu 2 ZnSnS 4 Films during Annealing," *Chemistry of Materials*, vol. 23, no. 20, pp. 4625–4633, Oct. 2011.

[11] R. C. Snyder and M. F. Doherty, "Faceted crystal shape evolution during dissolution or growth," *AIChE Journal*, vol. 53, no. 5, pp. 1337–1348, May 2007.

[12] W.-C. Hsu, I. L. Repins, C. Beall, C. DeHart, G. Teeter, B. To, Y. Yang, and R. Noufi, "The effect of Zn excess on kesterite solar cells," *Solar Energy Materials and Solar Cells*, vol. 113, pp. 160–164, Jun. 2013.

[13] B. J. Griffin, "A comparison of conventional Everhart-Thornley style and in-lens secondary electron detectors: a further variable in scanning electron microscopy.," *Scanning*, vol. 33, no. 3, pp. 162–73.

[14] D. M. Berg, "KESTERITE EQUILIBRIUM REACTION AND THE DISCRIMINATION OF SECONDARY PHASES FROM CU2ZNSNS4," University of Luxembourg, 2012.

Mater. Res. Soc. Symp. Proc. Vol. 1538 © 2013 Materials Research Society
DOI: 10.1557/opl.2013.1006

Is it Possible to Grow Thin Films of Phase Pure Kesterite Semiconductor?
A ZnSe case study

Phillip J. Dale[1], Monika Arasimowicz[1], Diego Colombara[1], Alexandre Crossay[1], Erika Robert[1], and Aidan A. Taylor[2]

[1]University of Luxembourg, Laboratory for Energy Materials, 41 rue du Brill, L-4422, Belvaux, Luxembourg.
[2]Physics Department, Durham University, South Road, Durham, DH1 3LE, UK.

ABSTRACT

The kesterite semiconductor $Cu_2ZnSnS(e)_4$ is seen as a suitable absorber layer to replace $Cu(In,Ga)Se_2$ in thin film solar cells, if thin film photovoltaics are to be deployed on the terawatt scale. Currently the best devices, and hence the best kesterite absorber layers are grown away from stoichiometry and are zinc rich and copper poor, presumably leading to the formation of $ZnS(e)$. However, it has been shown that secondary phases present in an absorber layer reduce device performance. If growth in Zn rich conditions seems to be mandatory, then any secondary phases formed should be grown on the surface of the absorber layer so that they may be easily removed by etching. Therefore, it is important to know how and why secondary phases form, and if possible, how to segregate them to the surface of the absorber layer.
Here we show that ZnSe is formed at the initial stages of absorber formation from annealing metal stacks in selenium vapor. Further we demonstrate that the way the precursor metals are distributed on the substrate leads to different absorber layer performances in full devices. The importance of selenium vapor pressure is highlighted in respect to the order of selenisation of the metals, Zn before Cu. Additionally, the importance of selenium and tin selenide vapor pressure during annealing is reviewed with regard to avoiding a decomposition of the $Cu_2ZnSnSe_4$ to ZnSe and Cu_2Se phases. Regardless of the atmosphere above the absorber, the reaction of the absorber with molybdenum appears unavoidable without the use of a passivation strategy. Counter-intuitively, it is demonstrated that for our absorber layers grown under Zn-rich conditions, removal of the ZnSe is harmful for device performance.

INTRODUCTION

Kesterite semiconductors are currently investigated for use as the p-type absorber layer material within thin film solar cells. The advantage of kesterite, $Cu_2ZnSn(S,Se)_4$, based material over chalcopyrite, $Cu(In,Ga)Se_2$, is that it replaces the low earth abundance element indium with the abundant zinc and tin. The current disadvantage is the lower photovoltaic power conversion efficiency of the kesterite absorber layer when placed in a device structure, 11.1% [1], as compared to the chalcopyrite layer, 20.4 % [2]. One simple, and maybe naive, reason for the lower efficiency is the fact that the current absorber layers in record devices are apparently grown in conditions where the layer consists of a kesterite phase and further secondary phases.
Secondary phases are known to be detrimental to the performance of the device and are commonly found on the top of the absorber, in between grains of the absorber, or between the absorber and the back contact (figure 1(a)). Commonly, all secondary phases act by reducing the total amount of light absorbed in the absorber layer reducing the number of possible excited charge carriers and thus reducing device efficiency. For kesterite selenides ($Cu_2ZnSnSe_4$,

CZTSe), ZnSe islands on the surface of the absorber block charge collection reducing the photocurrent produced under illumination [3]. Conducting Cu_2Se is shown to grow on and between $Cu_2ZnSnSe_4$ grains acting as a shunting path in the photovoltaic device [4], in exactly the same manner as for $Cu(In,Ga)Se_2$. Etching away a surface Cu and Sn containing phase, suspected to be the low band gap material Cu_2SnSe_3, from the top of an absorber layer significantly increased the open circuit voltage and fill factor of the device [5]. Similar problems with secondary phases are expected for the kesterite sulfide. For example Cu_2SnS_3 has a low band gap compared to Cu_2ZnSnS_4 [6] and would act to lower the open circuit voltage of a device if present on top of the absorber layer. Therefore due to the harmful effect of the secondary phases on device performance it would seem expedient to avoid their formation in the first place. The chemical composition of the absorber layers in the best devices reported so far is never in the single phase region of the phase diagram, but rather in a two or three phase region. Figure 1(b) shows the composition of the best devices marked as symbols in a kesterite selenide phase diagram adapted from reference [7]. The extent of the kesterite single phase region is marked as an oval at the intersection of the 0.33 mole fraction lines of ZnSe, Cu_2Se, and $SnSe_2$. All the best devices are in a region of Cu deficiency and Zn richness, with the record 11.1% device additionally being slightly Sn rich. Therefore, according to the phase diagram, admittedly obtained under equilibrium growth conditions, all the thin film absorbers should contain the detrimental ZnSe phase. Alternatively, it could be considered that the best devices are single phase, and that the single phase existence region of the current phase diagram is not large enough. The group of Lafond have provided some evidence that for the kesterite sulfides the single phase region may be somewhat larger than that presented by Olekseyuk et al. [8, 9].

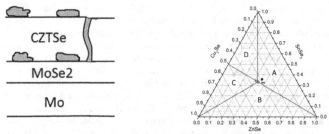

Figure 1(a) Schematic diagram of where secondary phases are found to occur during the synthesis of $Cu_2ZnSn(S,Se)_4$ absorber layers. Grey areas denote secondary phases. **(b)** Phase triangle showing the composition of the absorber layers from the current best devices (squares) [1, 10, 11]. Each solid line represents a narrow region with CZTSe + one secondary phase: clockwise from the top + $SnSe_2$, + ZnSe, + Cu_2Se, + Cu_2SnSe_3. Letter labeled regions limited by solid lines exhibit CZTSe + two secondary phases. A : + $SnSe_2$ + ZnSe, B : + Cu_2Se + ZnSe, C : + Cu_2Se + Cu_2SnSe_3, D : + $SnS(e)_2$ + Cu_2SnSe_3.

The question then arises why are the best devices grown in the Zn rich region? One of the simplest arguments is that growing Zn rich avoids the formation of the ternary Cu_2SnSe_3 which appears to have a lower band gap than CZTSe and thus gives a lower open circuit voltage [12].

The challenge appears to be to grow single phase kesterite and to avoid the formation of detrimental secondary phases. For example if a growth under Zn rich conditions is a strict requirement for high quality absorber layers, then, at the end of the synthesis, any excess ZnSe should be present on the top surface so that it may be removed by etching.

To achieve the requirement of putting secondary phases on the top surface we must investigate, understand, and eventually control the growth of the absorber layer. For any synthesis, secondary phase formation and removal can occur (1) during the formation of CZTSe (2) at elevated temperatures after CZTSe formation and (3) as post treatment once the absorber layer is cooled. The identification of secondary phases (4) depends critically on using the right characterization techniques. Step (1) depends on the particular synthesis route, but (2) and (3) are universal.

This manuscript will examine steps 1 – 3 for the case of ZnSe in the CZTSe absorber using exemplary data from our laboratory and some literature data for comparison. The aim of this manuscript is to highlight key areas during the synthesis process where secondary phase formation is likely and suggest how to reduce or control the phases formed. The manuscript is not exhaustive due to the numerous synthesis methods used to make absorber layers. Here, a two stage synthesis of CZTSe absorber layers is presented, namely the electrodeposition of metals followed by annealing in a selenium and tin selenide atmosphere. For step (1) we demonstrate that Zn is the first metal to selenize. We also show that the way the Zn is distributed affects device performance. For step (2) we review previous work from our laboratory and others demonstrating how ZnSe and other secondary phases can be formed both at the front and back of the device, even if starting from stoichiometric $Cu_2ZnSnSe_4$. For step (3) we demonstrate that contrarily to the literature case, removing ZnSe by HCl etching is actually bad for our particular samples.

Experimental

Electrodeposition of Cu1/Sn/Cu2/Zn layers onto Mo coated glass used the same procedure as that given in reference [13], except for CuSnZn precursors which were co-electrodeposited from an agitated basic electrolyte using galvanostatic control. Baseline precursors do not contain the Cu2 layer. Metal alloying was performed on some samples using a rapid thermal annealing oven (RTP AS-One 100) at 350°C in a N_2 atmosphere for 30 minutes [14]. Absorbers were formed by selenization in a selenium and tin selenide atmosphere as described in reference [15]. Kinetic experiments to examine the rate of selenium uptake were carried out in a rapid thermal annealing oven using individual metals situated approximately 6 cm from the Se source. Energy dispersive x-ray spectroscopy (EDX) (oxford instruments) was used to quantify the selenium uptake. HCl, KCN and Br_2-methanol etching were performed in a similar manner to [5].

Standard scanning electron microscopy (SEM) cross sectional images were obtained simply by fracturing the substrate and directly imaging with an electron beam (Hitachi SU-70). Improved images were obtained by using a focused ion beam microscope (FIB, FEI Helios Nanolab 600). In order to protect the surface of the absorber during cross-sectioning with the Ga^+ ion beam, Pt was deposited with the electron beam onto the site-of-interest. Once a satisfactory thickness of Pt had been achieved, a 30 kV ion beam was used to cut a 4 µm deep trench into the absorber with the deepest region of the trench coincident with the Pt-protected area. Following the trench cut, 16, 8 and 5 kV polishing steps were performed on the face of the absorber; this reduces the extent of the ion damage and improves the quality of the cross-section image. Imaging of the cross-sections was carried out with the ion beam at 30 kV and 9 pA. The ion beam was used in preference to the electron beam as it provides superior resolution and contrast between the phases. Chemical composition depth profiles were obtained using glow discharge optical emission spectroscopy (courtesy of NEXCIS).

Absorber layers were completed to give the final structure of Mo/CZTSe/CdS/i-ZnO/Al:ZnO/Ni:Al. Current voltage measurements were obtained in the dark and under standard AM 1.5 illumination (100 mWcm^{-2}, 25°C).

Results

1.Phase segregation during CZTSe formation

In this section results are shown on the early phase segregation of ZnSe during the formation of CZTSe from electrodeposited metals. At the end of the section ZnSe phase segregation will be briefly discussed for co-evaporation synthesis routes.

For the two stage CZTSe absorber synthesis consisting of metal deposition followed by annealing in selenium vapor, does the arrangement of the metals in the precursor matter for the formation of the absorber layer? From the Cu:Sn:Zn phase diagram it is known that at 250 °C CuSn and CuZn alloys form and that no CuSnZn ternary alloy exists [16]. In the following we show that the geometrical arrangement of the metals during annealing affects the growth of the absorber, and that this has some consequences on the device efficiency.

Two types of metal precursors were deposited, type A consisting of four layers Cu/Sn/Cu/Zn and type B consisting of a single layer with Cu, Sn, and Zn all co-electrodeposited simultaneously. The cross sectional microstructure of the two types is shown in figure 2. Type A precursors exhibit a bilayer structure with slightly smaller grains on the upper layer. Type B precursors appear to consist of a single layer with grains extending from the substrate to the surface. Chemical depth profiles of the layers are overlaid on the SEM micrographs. Type A precursors show two chemically distinct regions with the upper layer containing copper and zinc and the lower layer containing copper and tin, consistent with previous results[13, 17]. On the other hand, type B precursors appear chemically more uniform through the depth. X-ray diffraction measurements (not shown) confirm the presence of Cu$_5$Zn$_8$ and Cu$_6$Sn$_5$ phases in the type A precursor. These phases actually form at room temperature given enough time [13], or they can be formed more rapidly by annealing for short times at moderate temperatures [14].

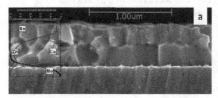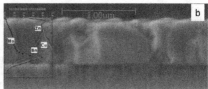

Figure 2. SEM images of (a) Type A: Cu/Sn/Cu/Zn stack pre-alloyed (adapted from ref [17]) and (b) Type B: co-electrodeposited CuSnZn layer. Overlaid on both images are elemental compositions normalized to the highest concentration of each element.

86

The alloying of the metals is unavoidable, but type A have the minimum interfacial contact between the Cu_5Zn_8 and Cu_6Sn_5 alloys, whilst type B presumably have significantly more contact. Thus type A and type B represent two possibilities in either segregating zinc and tin from one another as much as possible, or mixing them intimately.

The next part of the synthesis involves the reaction of the precursors with selenium vapour. For the reaction of Cu:In:Ga precursors to form $Cu(In,Ga)Se_2$ it is well documented that In reacts with Se before Ga, see for example [18]. A kesterite precursor type A sample was annealed for 60 s at 500 °C and allowed to rapidly cool. A chemical composition depth profile of this sample is shown in figure 3(a) [17]. The depth profile shows a distinctly layered structure with mainly Cu and Sn at the back with Zn and Se at the front. XRD measurements (not shown) confirm the presence of ZnSe and the absence of CZTSe as the unique kesterite reflexes are not observed in the diffractogram. It appears that at short times the Zn metal preferentially reacts with the selenium rather than the Cu. To investigate whether this is a kinetic effect i.e. Zn reacts faster than Cu with Se, or a thermodynamic effect i.e. there is a greater gain in Gibbs free energy (ΔG_r) by the reaction of Zn with Se than Cu, the kinetics of selenization of the individual metals was studied. Figure 3(b) shows the atomic ratio of Se/metal as a function of annealing time at 500 °C for just Cu layers and just Zn layers [17]. The data actually show that elemental Cu reacts slightly faster than elemental Zn with selenium, suggesting that reaction kinetics cannot explain the early formation of ZnSe seen in figure 3(a). However, thermodynamic calculations on the other hand based on the reaction of selenium with Cu_5Zn_8 show that the formation of ZnSe over Cu_2Se is more favourable as indicated by the larger negative Gibbs free energy of formation for ZnSe than Cu_2Se per mol of reacting Se_2 (-304 kJ ZnSe, -200 kJ Cu_2Se at 550°C) [17].

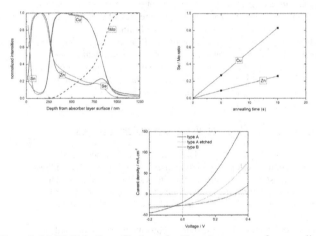

Figure 3. (a) SIMS depth profile through a Type A precursor after annealing for 60 s at 500°C in the presence of a selenium vapor source 6 cm away from the substrate, adapted from [17]. (b) elemental Selenium/metal ratio as measured by EDX for the individual metals Cu and Zn as a function of annealing time at 500 °C, adapted from [17]. (c) Current Voltage characteristics for Type A and Type B precursors annealed under the same conditions and completed into devices. Dashed line indicates a Br_2-Methanol etched absorber made from a type A precursor.

Therefore, the most likely reason for the preferential reaction of selenium with Zn rather than Cu, when both elements are present is related to thermodynamics. The reaction of most/all the zinc to form ZnSe leads it to be phase separated with a minimum interfacial area to the other metals. This effect is enhanced when the metal selenide formation is limited by the supply of selenium.

Stoichiometric type A and copper poor zinc rich (Cu/(Zn+Sn) = 0.8, Zn/Sn = 1.2) type B precursors were annealed using exactly the same conditions in a tube furnace with selenium powder for half an hour. The resulting absorber layers were etched in KCN and completed into full devices. Figure 3(c) shows the resulting current voltage characteristics of the devices under illumination. Type A precursors show very poor device characteristics compared to type B, even though they contained a relatively lower amount of zinc. One hypothesis is that the difference between type A and type B is probably that in type A ZnSe forms a continuous layer on the surface and in type B the ZnSe is more dispersed and has more opportunity to react. Etching of the top surface of a type A absorber with Br_2-methanol improved the short circuit current and the open circuit voltage perhaps suggesting the removal of a blocking phase. However, the etched absorber still did not reach the level of performance of the type B.

ZnSe formation is not restricted to those absorber layers synthesised from metal precursors. For example precursors co-evaporated at low temperature and then annealed at higher temperatures exhibit nanometre scale ZnSe precipitates which appear to form networks within the bulk of the absorber layer [19]. The co-evaporated precursor consisted of a zinc rich mixture of all the kesterite selenide elements. During the second heating step the precursor crystallizes into CZTSe. The excess ZnSe forms a secondary phase, in accordance with the phase diagram (figure 1(b),[7]). The precipitated ZnSe appears not to be able to segregate to the front surface, and thus is not easily removable. One probable reason for the lack of driving force to segregate is the similar lattice constants of the unit cell of CZTSe and ZnSe.

The above experiments show that the ordering of the metal alloy phases within the metal precursor appears to be important, as well as the reactivity of the metals with selenium, as exemplified for Cu and Zn. Two precursors with minimum and maximum interface between the tin and zinc alloys produce significantly different device efficiencies. In the next section the reasons for secondary phase formation during the high temperature annealing of even exactly stoichiometric CZTSe absorber layers is discussed.

2.Secondary phase development after CZTSe formation at elevated temperatures

The following section reviews why just after CZTSe formation, even when it starts exactly stoichiometric, it is still likely to develop secondary phases. All absorber layer synthesis methods contain a thermal processing step where formed kesterite is still in an environment of ca. 450 / 580°C. At these elevated temperatures thermodynamic chemical equilibrium must be considered. The question to be answered here is whether the CZTSe is thermally stable with respect to the gas phase and additionally to the substrate.

All chemical compounds exist in equilibrium with their constituents and semiconductors such as CZTSe or $Cu(In,Ga)Se_2$ are no different. The chemical equation for this equilibrium is expressed as follows [20]

$$Cu_2ZnSnSe_{4(s)} = Cu_2Se_{(s)} + ZnSe_{(s)} + SnSe_{2(s)} \qquad (1)$$

What is important to know is how far to one side or the other is the equilibrium. If the equilibrium is far to the left then CZTSe is stable but if the equilibrium is not so far to the left then the binary species on the right do exist even if in miniscule amounts. This would not be a significant problem for the absorber layer if the volume fraction of these binary species were extremely low. The only reason for there to be a problem is if the binary species themselves were unstable or volatile. It turns out that $SnSe_2$ is in its own equilibrium as shown in reaction 2

$$SnSe_{2(s)} = SnSe_{(s)} + Se_{(g)} \tag{2}$$

The equilibrium of this second reaction is such that it does not lie extremely to the left, and thus a proportion of $SnSe_2$ exists as SnSe and Se. The consequence of the second reaction for equation (1) is that the equilibrium is pulled to the right, meaning that more CZTSe is converted to the binary selenides in order to produce the $SnSe_2$ which is being lost in reaction 2. Ordinarily, the loss of $SnSe_2$ could be stopped by only supplying excess selenium, if SnSe were non-volatile. It turns out that SnSe is volatile too and is lost to the gas phase if not deliberately supplied [15, 20]. Therefore to keep reaction 2 and consequently reaction 1 to the left, maximizing the kesterite phase, both selenium and tin selenide are required in the gas phase. If selenium and tin selenide are not supplied in the gas phase, tin selenide will be lost leaving behind the secondary phases Cu_2Se and ZnSe.

To avoid the loss of SnSe and the formation of the secondary phases it becomes clear that the annealing environment is crucial. The annealing apparatus must either supply SnSe and Se or stop the escape of the volatile species. Cold annealing chamber surfaces or vacuums are therefore particularly unhelpful for minimizing the loss of volatile species from the semiconductor. Interestingly, the kesterite absorber can be formed by co-evaporation giving high efficiencies of 9.1 % [10]. However, the vapor pressure of selenium required to keep the kesterite stable is some five to ten times higher than for a comparable $Cu(In,Ga)Se_2$ process [21, 22]. Annealing apparatuses which allow heating under equilibrium conditions, i.e. fixed volumes, will by their very design not allow the loss of volatile species, and may thus be more suitable for growing the CZTSe semiconductor.

Theoretical thermodynamic considerations also showed that secondary phase formation may occur via a reaction between the CZTSe absorber layer and its common substrate, molybdenum [23]. The key step to understanding why this might be is to consider the Gibbs free energies of reaction of the materials involved, in particular the energy to convert CZTSe into its constituent binary selenides, the energy to convert $SnSe_2$ to SnSe, and the energy of formation of $MoSe_2$. The basic reaction is given in reaction 3 and the important energetic quantities are shown in the frost diagram of figure 4.

$$2Cu_2ZnSnSe_{4(s)} + Mo_{(s)} = 2Cu_2Se_{(s)} + 2ZnSe_{(s)} + 2SnSe_{(s)} + MoSe_{2(s)} \tag{3}$$

The frost diagram shows that relatively little Gibbs free energy is required to convert two moles of $SnSe_2$ to SnSe compared to the energy gained by the formation of one mole of $MoSe_2$. Thus the reaction 3 is likely if the energy required to convert CZTSe to binary selenides is small [23].

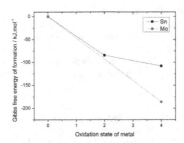

Figure 4 Frost diagram showing the Gibbs free energy of formation for molybdenum and tin selenides at 550 °C.

Experimental evidence for the reaction of kesterite sulfide with the molybdenum substrate was shown in reference [24] where they observed the formation of ZnS, Cu_2S, and SnS. The same reaction is predicted to occur for the selenide [23]. To avoid or minimize the reaction of kesterite with Mo, a barrier layer placed between them, such as a thin TiN layer, has been shown to reduce the amount of $MoSe_2$ formed and thus presumably the relative fraction of secondary phases [25].

In summary, $Cu_2ZnSnSe_4$ is not stable at typical annealing temperatures with respect to its gas phase interface unless in equilibrium with selenium and tin selenide. Further, and perhaps more problematic is the fact that $Cu_2ZnSnSe_4$ is not stable with respect to the common molybdenum substrate and secondary phases can develop between the absorber layer and the back contact which are not removable by etching.

3. Post treatment to remove secondary phases

Once the absorber layer has cooled down, secondary phases on the surface of the layer may be removed by etching. Those secondary phases at the back of the absorber or between the grains are harder to remove. KCN has been shown to remove Cu_2Se phases [4], HCl removes ZnSe [5] and bromine-methanol appears to etch ZnSe [5], Cu_2SnSe_3 [5], and even $Cu_2ZnSnSe_4$ itself [26]. For the case of HCl etching to remove a ZnSe phase on top of annealed co-evaporated precursors, the photovoltaic parameters of the device made with an etched absorber were found to be superior as compared to an unetched. Therefore, as the baseline absorber layers in the LEM laboratory are formed from zinc rich precursors, it was decided to etch them in HCl before KCN to remove any surface ZnSe.

Figure 5(a) shows the current voltage behavior of a device prepared using the baseline precursor process with a deliberately copper poor, zinc rich absorber (precursor composition Cu/(Zn+Sn) = 0.6, Zn/Sn = 1.3) which has only been etched in KCN and an absorber etched in HCl and subsequently KCN. The two devices exhibit similar short circuit currents but different parallel resistances, fill factors, and open circuit voltages. Essentially, etching in HCl appears to make the device performance worse. Inspection of a similarly etched absorber layer in the SEM shows that some ZnSe still remains on the surface (image not shown) demonstrating that HCl is not effective in removing large amounts of ZnSe. One hypothesis as to why not all the ZnSe is removed, is that the etching process stops due to a passivating elemental selenium layer forming over the remaining ZnSe. To try to understand the difference in the device's performance, electrochemical analyses were made on absorber layers after different etch times in HCl. The electrochemical measurements measured the electrical current passed from the absorber layer to a redox electrolyte at a fixed potential where on a metallic electrode the redox species is reduced and thus considerable current flows. However, if the electrode consists of a p-type semiconductor absorber layer without pinholes and without conductive phases, the electrical current passing should be low due to the lack of electrons. On the other hand, if there are pinholes and/or conductive phases then the electrical current will be intermediate to that of a semiconductor and that of a metallic electrode. Figure 5(b) shows the current flow due to the reduction of the redox electrolyte as a function of etch time. It can be seen that the lowest current flow is achieved with no etching. Any HCl etching leads to an increase in reduction of the redox electrolyte indicating that conducting phases are being exposed or pinholes are being created, so that MoSe$_2$ (a degenerate semiconductor [27]) is exposed. KCN etching on the other hand only caused minor variations on the measured reduction current (not shown).

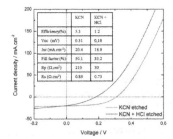 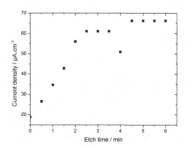

Voltage / V   Etch time / min

Figure 5. (a) JV characteristics of electrodeposited and annealed absorbers etched in (i) KCN and (ii) HCl and KCN. **(b)** Electrochemical current of the absorber layer as a function of HCl etch time.

Cross sectional microscopy of the absorber layers is key to understanding these results. Simply breaking the absorber layer in two and viewing the cross section was insufficient. Only by preparing a cross section as for transmission electron microscopy with protective Pt coatings did the delicate microstructure become apparent. Figure 6 shows a cross section of a baseline absorber layer with protective Pt coating. All the phases present were made by considering the element compositions as measured by point EDX. As expected CZTSe is the dominant phase extending from the MoSe$_2$ to the Pt protective layer. However, a significant amount of ZnSe can be observed, with much brighter contrast, on the surface of the absorber layer. Critically the ZnSe also appears as a skin above some cavities. Therefore if this ZnSe is etched away, the cavities are opened and expose the underlying MoSe$_2$ phase. This result combined with the electrochemical experiment explains why the device performance is worse for the absorber layer etched in HCl. The device has a low parallel resistance, and hence poor fill factor, because of the possibility of electrical shorting through to the underlying MoSe$_2$ layer.

91

Figure 6. FIB cross-section through the absorber created by electodeposition and annealing. Imaging was performed with a 30 kV ion beam.

In summary, although ZnSe has been found to be a phase which blocks photocurrent its removal is not always beneficial. In the case presented here, removal of the secondary phase leads to exposure of a large number of pinholes. The identification of the reason for the presence of pinholes is on-going, but before ZnSe phase removal can be considered as a way of improving the absorber layer, the absorber growth process must be improved.

CONCLUSIONS

It is proposed that the current kesterite-based thin film devices are not reaching their full potential in terms of power conversion efficiency. One reason for this is that even the best devices are grown in such a manner that secondary phases are present, and secondary phases are known to be detrimental to device performance. Therefore understanding how to grow kesterite absorber layers where any secondary phases formed should appear on the top surface of the absorber layer is important. Secondary phases on the top surface of the absorber layer are accessible for chemical etching, providing that the secondary phases have been suitably identified.

Three different stages of secondary phase formation and removal have been identified. In stage one it was demonstrated for metal precursors annealed in selenium vapor that secondary phase formation can be affected by simply the ordering of the metals in the precursor layer and that the partial pressure of selenium in the annealing atmosphere is important. In stage two, universal for all preparation routes the thermodynamic stability of the kesterite with respect to the annealing atmosphere and the substrate were reviewed. Importantly, chalcogen and tin chalcogenide must be supplied in the gas phase to avoid secondary phase formation at the front surface of the absorber layer. Strategies for avoiding secondary phases at the back contact due to reaction of the kesterite with the molybdenum must be developed. In stage three, common wet etchants were reviewed and it was shown, contrarily, that the removal of ZnSe from the surface of our absorber layers is in fact detrimental to device performance due to the opening up of cavities in the absorber layer exposing the $MoSe_2$ at the back contact.

ACKNOWLEDGMENTS

The authors thank the Helmholtz Zentrum in Berlin for device finishing, and the use of the SEM apparatus through the CRP Lippmann (Luxembourg). The authors acknowledge the financial support of the Luxembourgish Fonds National de la Recherche and the EU seventh framework programme FP7/2007-2013 under grant no. 284486.

REFERENCES

1. Todorov, T.K., et al., *Beyond 11% Efficiency: Characteristics of State-of-the-Art Cu2ZnSn(S,Se)(4) Solar Cells.* Advanced Energy Materials, 2013. **3**(1): p. 34-38.
2. http://www.empa.ch/plugin/template/empa/1351/131438/---/l=2. Downloaded March 2013.
3. Watjen, J.T., et al., *Direct evidence of current blocking by ZnSe in Cu2ZnSnSe4 solar cells.* Applied Physics Letters, 2012. **100**(17).
4. Tanaka, T., et al., *Existence and removal of Cu2Se second phase in coevaporated Cu2ZnSnSe4 thin films.* Journal of Applied Physics, 2012. **111**(5).
5. Mousel, M., et al., *HCl and Br2-MeOH etching of Cu2ZnSnSe4 polycrystalline absorbers.* Thin Solid Films, (0).
6. Berg, D.M., et al., *Thin film solar cells based on the ternary compound Cu2SnS3.* Thin Solid Films, 2012. **520**(19): p. 6291-6294.
7. Dudchak, I.V. and L.V. Piskach, *Phase equilibria in the Cu2SnSe3-SnSe2-ZnSe system.* Journal of Alloys and Compounds, 2003. **351**(1-2): p. 145-150.
8. Lafond, A., et al., *Crystal Structures of Photovoltaic Chalcogenides, an Intricate Puzzle to Solve: the Cases of CIGSe and CZTS Materials.* Zeitschrift Fur Anorganische Und Allgemeine Chemie, 2012. **638**(15): p. 2571-2577.
9. Olekseyuk, I.D., I.V. Dudchak, and L.V. Piskach, *Phase equilibria in the CU2S-ZnS-SnS2 system.* Journal of Alloys and Compounds, 2004. **368**(1-2): p. 135-143.
10. Repins, I., et al., *Co-evaporated Cu2ZnSnSe4 films and devices.* Solar Energy Materials and Solar Cells, 2012. **101**: p. 154-159.
11. Barkhouse, D.A.R., et al., *Device characteristics of a 10.1% hydrazine-processed Cu2ZnSn(Se,S)4 solar cell.* Progress in Photovoltaics, 2012. **20**(1): p. 6-11.
12. Siebentritt, S., *Why are kesterite solar cells not 20% efficient?* Thin Solid Films, (0).
13. Scragg, J.J., D.M. Berg, and P.J. Dale, *A 3.2% efficient Kesterite device from electrodeposited stacked elemental layers.* Journal of Electroanalytical Chemistry, 2010. **646**(1-2): p. 52-59.
14. Arasimowicz, M., M. Thevenin, and P.J. Dale. *The effect of soft pre-annealing of differently stacked Cu-Sn-Zn precursors on the quality of Cu2ZnSnSe4 absorbers.* in *Materials Research Society Spring Meeting.* 2013. San Francisco: MRS.
15. Redinger, A., et al., *The Consequences of Kesterite Equilibria for Efficient Solar Cells.* Journal of the American Chemical Society, 2011. **133**(10): p. 3320-3323.
16. Chou, C.Y. and S.W. Chen, *Phase equilibria of the Sn-Zn-Cu ternary system.* Acta Materialia, 2006. **54**(9): p. 2393-2400.
17. Arasimowicz, M., et al., *Elucidation of the Kesterite Semiconductor Formation Mechanism from Metal Stacks and the Effect of Annealing Parameters.* submitted, 2013.
18. Hergert, F., et al., *A crystallographic description of experimentally identified formation reactions of Cu(In,Ga)Se-2.* Journal of Solid State Chemistry, 2006. **179**(8): p. 2394-2415.
19. Schwarz, T., et al., *Atom probe study of Cu2ZnSnSe4 thin-films prepared by co-evaporation and post-deposition annealing.* Applied Physics Letters, 2013. **102**(4).

20. Redinger, A., et al. *Route towards high efficiency single phase Cu<inf>2</inf>ZnSn(S, Se)<inf>4</inf> thin film solar cells: Model experiments and literature review.* in *Photovoltaic Specialists Conference (PVSC), 2011 37th IEEE.* 2011.
21. Islam, M.M., et al., *Impact of Se flux on the defect formation in polycrystalline Cu(In,Ga)Se-2 thin films grown by three stage evaporation process.* Journal of Applied Physics, 2013. **113**(6).
22. Hsu, W.-C., et al., *Reaction pathways for the formation of Cu2ZnSn(Se,S)(4) absorber materials from liquid-phase hydrazine-based precursor inks.* Energy & Environmental Science, 2012. **5**(9): p. 8564-8571.
23. Scragg, J.J., et al., *Thermodynamic Aspects of the Synthesis of Thin-Film Materials for Solar Cells.* Chemphyschem, 2012. **13**(12): p. 3035-3046.
24. Scragg, J.J., et al., *A Detrimental Reaction at the Molybdenum Back Contact in Cu2ZnSn(S,Se)(4) Thin-Film Solar Cells.* Journal of the American Chemical Society, 2012. **134**(47): p. 19330-19333.
25. Shin, B., et al., *Control of an interfacial MoSe2 layer in Cu2ZnSnSe4 thin film solar cells: 8.9% power conversion efficiency with a TiN diffusion barrier.* Applied Physics Letters, 2012. **101**(5).
26. Guetay, L., et al., *Lone conduction band in Cu2ZnSnSe4.* Applied Physics Letters, 2012. **100**(10).
27. Cummings, C.Y., et al., *CuInSe2 precursor films electro-deposited directly onto MoSe2.* Journal of Electroanalytical Chemistry, 2010. **645**(1): p. 16-21.
28. Schorr, S., et al., *The complex material properties of chalcopyrite and kesterite thin-film solar cell absorbers tackled by synchrotron-based analytics.* Progress in Photovoltaics, 2012. **20**(5): p. 557-567.
29. Just, J., et al., *Determination of secondary phases in kesterite Cu2ZnSnS4 thin films by x-ray absorption near edge structure analysis.* Applied Physics Letters, 2011. **99**(26).
30. Fontane, X., et al., *In-depth resolved Raman scattering analysis for the identification of secondary phases: Characterization of Cu2ZnSnS4 layers for solar cell applications.* Applied Physics Letters, 2011. **98**(18).
31. Djemour, R., et al., *Detecting ZnSe secondary phase in Cu2ZnSnSe4 by room temperature photoluminescence.* Applied Physics Letters, 2013. **accepted**.
32. Watjen, J.T., et al., *Cu out-diffusion in kesterites-A transmission electron microscopy specimen preparation artifact.* Applied Physics Letters, 2013. **102**(5).
33. Vora, N., et al., *Phase identification and control of thin films deposited by co-evaporation of elemental Cu, Zn, Sn, and Se.* Journal of Vacuum Science & Technology A, 2012. **30**(5).
34. Redinger, A., et al., *Detection of a ZnSe secondary phase in coevaporated Cu2ZnSnSe4 thin films.* Applied Physics Letters, 2011. **98**(10).
35. Boscher, N.D., et al., *Atmospheric pressure chemical vapour deposition of SnSe and SnSe2 thin films on glass.* Thin Solid Films, 2008. **516**(15): p. 4750-4757.

Mater. Res. Soc. Symp. Proc. Vol. 1538 © 2013 Materials Research Society
DOI: 10.1557/opl.2013.1015

Polarization dependent Raman spectroscopy characterization of kesterite Cu2ZnSnS4 single crystals

D. O. Dumcenco[1], Y. P. Wang[1], S. Levcenco[2], K. K. Tiong[3] and Y. S. Huang[1]

[1]Department of Electronic Engineering, National Taiwan University of Science and Technology, Taipei 106, Taiwan

[2]Helmholtz Zentrum Berlin f'ur Materialien and Energie GmbH, D-14109 Berlin, Germany

[3]Department of Electrical Engineering, National Taiwan Ocean University, Keelung 202, Taiwan

ABSTRACT

The vibrational properties of kesterite Cu2ZnSnS4 (CZTS) single crystals were studied by polarization-dependent Raman scattering measurements. The CZTS crystals grown by chemical vapor transport technique using iodine trichloride as a transport agent consist of several mirror-like planes. The detailed analysis of the experimental spectra obtained from different planes allows determining the symmetry assignment of the observed Raman-active modes. The wavenumber values of Raman-active modes are compared with the results of recent theoretical calculations. The presented data are useful for examination of CZTS absorber films applied for solar cells to clarify the existence of structural or phase inhomogeneities.

INTRODUCTION

The quaternary compound Cu2ZnSnS4 (CZTS) semiconductor has attracted great interest due to its potential applications for sustainable thin-film solar cell devices [1-3]. CZTS has large direct band gap (E_g ~ 1.5 eV), high absorption coefficients ($>10^4$ cm^{-1}), intrinsic p-type conductivity and low thermal conductivity [4-6]. Solar cells based on selenium-containing Cu2ZnSn(S,Se)4 compound have demonstrated efficiencies above 10% [7]. Comparatively to the other important solar cell materials (i.e. In, Ga, etc.) [8], the composition of naturally abundant and inexpensive elements such as Zn and Sn, makes CZTS particularly attractive candidate for large-scale commercial application [9].

From theoretical point of view, recent detailed theoretical investigations provided some vibrational properties of CZTS [10,11]. Experimentally, some infrared (IR) [12] and Raman [11-14] spectroscopy experiments on CZTS thin films were also performed for the analysis of the crystal structure. However, there are still little detailed experimental data on vibrational properties of CZTS single crystals [15].

The detailed experimental results of polarization-dependent backscattering Raman spectroscopy on CZTS single crystals are presented. The CZTS crystals grown by chemical vapor transport (CVT) technique using iodine trichloride (ICl3) as a transport agent are consisted of several mirror-like planes. The observed modes detected from (100), (001), (110) and (112) planes were classified by applying the selection rules to the Raman active modes. Also, experimental results have been compared with the data of recent theoretical calculations [10,11].

EXPERIMENT

Using CVT technique with ICl_3 as a transport agent, single crystals of CZTS with well-defined faces were grown [6]. From energy dispersive X-ray analysis, the average atomic ratio of Cu:Zn:Sn:S was found to be closed to 2:1:1:4. The crystals were X-ray pre-oriented with respect to the selected directions and polarizations of the incident and scattered light.

Room-temperature Raman measurements were performed at the back-scattering configuration on a Renishaw inVia micro-Raman system with 1800 grooves/mm grating and optical microscope with 50x objective. A linearly polarized Ar^+ laser beam of the 514.5 nm excitation line with a power of ~1.5 mW was focused into a spot size ~5 µm in diameter. For the polarization measurement of the scattering light, a polarizer and a half-wave plate were used. Prior to the measurement, the system was calibrated by means of the 520 cm^{-1} Raman peak of a polycrystalline Si.

THEORY

CZTS can be crystallized in kesterite (KS/space group $I\bar{4}$) or stannite (ST/space group $I\bar{4}2m$) structure. The KS and ST crystallorgraphic forms are very close with the only difference in the distribution of the cations in the tetrahedral sites [16]. As a result, within experimental broadening of the diffraction peaks, it is a challenge to differentiate KS and ST structures by using X-ray diffraction. Phonon dispersion in solids is sensitive to the coupling between atoms within the lattice. For dispersion measurements, either neutron diffraction [17] or Raman spectroscopy [10,11] can be utilized. Specifically, Raman scattering is observed at wavenumbers corresponding to the phonon modes at the Γ point. Moreover, Raman spectroscopy is a convenient and widely available technique.

There are total of 24 vibrational modes, since CZTS has eight atoms per primitive cell. Amongst these modes, there are Raman active modes, i.e. 15 for KS structure and 14 for ST [10, 11]. A and A_1 modes for KS and ST, respectively, result from symmetric vibrations of only anion lattice and they are responsible for the strongest lines observed in the experimental Raman spectra of CZTS [11, 13]. In B_1 modes of ST, half of the Cu atoms is displaced toward the positive Z axis and the other half is displaced toward the negative Z axis, while the Zn and Sn atoms remain stationary [10]. In the meantime, anions move only in XY plane. In B modes of KS and B_2 modes of ST, the cations only move along Z direction, whereas these cations move only within XY plane in E modes for the both structures. Thus the main difference between KS and ST structures is related to the nonexistence of B_1 mode structure caused by absence of the individual layers of Cu in KS.

Polarization-dependent backscattering measurements have been carried out for accurate determination of the positions and origin of Raman active modes. The widely used Porto notation [18] has been utilized in this study for the designation of the crystal and polarization directions. To determine the origin of the observed Raman modes, the polarization selection rules for backscattering configuration along [100], [001], [110] and [112] crystallographic directions of the crystals with KS and ST structure have been used [19]. Thus for (100) and (001) planes the orthogonal system O with $X = [100]$, $Y = [010]$, $Z = [001]$ axes have been used; for (001) and (110) planes – system O' with $X' = [110]$, $Y' = [\bar{1}10]$, $Z' = [001]$ axes; for (112) plane

– system O'' with $X'' = [112]$, $Y'' = [\bar{1}\,10]$, $Z'' = [\bar{1}\,\bar{1}\,1]$ axes. Raman tensors for KS and ST structures in the orthogonal system O are well known [20]. Transformation of the Raman tensors from the principal axes (system O), with the help of a general rotational tensor, gives a possibility to find the polarization selection rules for O' and O'' systems [19].

Results

Polarization-dependent Raman spectra from (110) plane (inset of Fig. 1) of the CZTS are shown in Fig. 1. For both $\bar{X}'(Y',Y')X'$ and $\bar{X}'(Z',Z')X'$ configurations, the features at 334 and 285 cm^{-1} have been clearly distinguished. In the case of KS/ST, the difference between configurations is related to $B(TO)/B_2(TO)$ modes [15]. Subtracting the spectrum of $\bar{X}'(Z',Z')X'$ configuration from $\bar{X}'(Y',Y')X'$, where both are normalized to A mode, the resultant spectrum clearly shows several features. The difference between the intensities of A modes at $\bar{X}'(Y',Y')X'$ and $\bar{X}'(Z',Z')X'$ is related to the different values of a and b used in Raman tensors [15]. The selection rules confirm disparity of feature X to B_1 mode of ST. Moreover, unidentified feature X at 280 cm^{-1} is lower than frequency of B_1 mode expected at 291.1 (324.1) cm^{-1} from theoretical calculations [10,11]. In contrast, comparison of the modes observed at 352, 245 and 160 cm^{-1} with the literature data on theoretical calculations allow us to assign them as B (TO) modes of CZTS with a KS structure. According to the polarization selection rules, two prominent features at 334 and 285 cm^{-1} are related to A mode. There is an additional small feature at 306 cm^{-1} that is also probably associated with A mode.

Figure 2 shows the polarization-dependent results from (001) plane (inset of Fig. 2). In accordance with the selection rules for KS [15], the spectra for both $\bar{Z}'(X',X')Z'$ and $\bar{Z}(X,X)Z$ configurations contain A and B(LO) modes. This two configurations are just characterized by the intensities of B(LO) modes. Subtracting the spectrum of $\bar{Z}(X,X)Z$ configuration from $\bar{Z}'(X',X')Z'$, the resultant spectrum clearly shows three B(LO) modes at

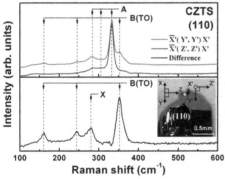

Figure 1. Polarization-dependent Raman spectra at backscattering configuration from (110) plane using orthogonal system O'. The subtracted spectrum shows clear unknown X and B(TO) modes.

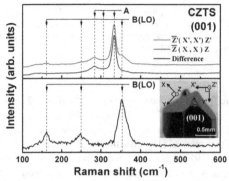

Figure 2. Polarization-dependent Raman spectra at backscattering configuration from (001) plane using orthogonal system O. The subtracted spectrum shows clear B(LO) modes.

353, 250 and 162 cm^{-1}. Furthermore, there is no any trace of unidentified feature X has been observed in subtracted spectrum. For ST at $\overline{Z}(X,X)Z$ configuration B_1 mode is allowed and as a result it should be detected in subtracted spectrum. This indicates that unidentified feature X doesn't correspond to B_1 mode and the investigated sample has a KS structure.

Polarization-dependent data for (100) and (110) plane (inset of Fig. 3), respectively, at $\overline{X}(Y,Z)X$ and $\overline{X}'(Y',Z')X'$ configurations are shown in Fig. 3. In accordance with selection rules for KS [15] for both configurations, E(TO) and E(LO) modes are allowed. The main difference is due to to the relative intensity of between E(TO) and E(LO) modes for $\overline{X}(Y,Z)X$ and $\overline{X}'(Y',Z')X'$. In despite of the selection rules, the trace of the most intensive A mode is still observed (Fig. 3). The reason of the forbidden A mode observation will be

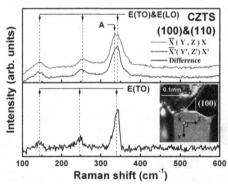

Figure 3. Polarization-dependent Raman spectra at backscattering configuration from (100) and (110) planes using orthogonal system O and O', respectively. The subtracted spectrum shows clear E(TO) modes.

discussed later. To determine the position of E(TO) and E(LO) modes, the subtracted spectrum from (100) plane at $\overline{X}(Y,Z)X$ and (110) plane at $\overline{X'}(Y',Z')X'$ has been obtained. As a result, the features at 341, 246 and 143 cm^{-1} correspond to E(TO) modes. Taking into account the values for E(TO), the position of E(LO) modes are found to be 346, 255 and 145 cm^{-1}.

DISCUSSION

The values of the Raman active modes determined by polarization-dependent measurements from different crystallographic planes correlate well to the data obtained from theoretical calculations [10,11]. As one can see, the experimental results correspond for KS structure much better than it is for ST. That is consistent with polarization-dependent selection rules of Raman-active modes. Should be noted also that some of the theoretically predicted B(TO) and B(LO) modes as well as E(TO) and E(LO) modes are undetected experimentally. In present work, the modes below 130 cm^{-1} can not be determined because of the properties of edge filter used in Raman system. The small deviation from the polarization selection rules related to the presence of a trace of the most intensive A mode forbidden theoretically, at some configurations can be explained by the extinction ratio of the polarizer. Moreover, the origin of feature X as well as of the features higher than 360 cm^{-1} is unclear at present.

From previous Raman measurements of CZTS compounds obtained by different synthesized methods [9,11-14,21-22], the most intense A mode has been observed between 331 and 338 cm^{-1}. The value of 338 cm^{-1} determined for CZTS films is the most common one [9,11,13,14,22]. However, the most recent publication on KS CZTS single crystal shows the presence of the intensive A mode centered at 336 cm^{-1} [23]. From polarization-dependent measurements of CZTS powder samples [24], with changing the location of the measuring spot the Raman spectrum is dominated by 331 cm^{-1} or 337 cm^{-1} related to the existence of local structural inhomogeneities. While for the presented CZTS single crystal, A mode shows the same position measured even from different planes that confirms the homogeneity of the sample.

The difference observed between the position of A mode for bulk and thin film materials is still unclear. But comparing different CZTS samples one can observe a trend. Once A mode shifts to the lower value, the unidentified mode at 371 cm^{-1} shifts to the higher value and its intensity increases [25]. In this work A mode is centered at 334 cm^{-1} and the mode at 371 cm^{-1} is weaker. Recently, the observation of the additional Raman feature from CZTS polycrystals has been reported, and it was attributed to the coexistence of kesterite and disordered kesterite phases [26]. Hence more work needs to be done to determine the shift phenomena of A mode and the origins of the additional features.

CONCLUSIONS

Polarization-dependent Raman spectra from (100), (001), (110) and (112) planes of CZTS single crystals are measured. From a comprehensive analysis of the experimental spectra and comparison with the results of theoretical calculations, the positions and symmetry assignment of the observed Raman features is determined and identified. The results reveal that observed CZTS single crystals have been crystallized in the kesterite structure. The obtained data is helpful to clarify the existence of structural or phase inhomogeneities in CZTS absorber films utilized for solar cells.

ACKNOWLEDGMENTS

Research supported by the National Science Council of Taiwan under projects NSC 100-2112-M-011-001-MY3 and NSC 101-2811-M-011-002.

REFERENCES

1. A. Weber, H. Krauth, S. Perlt, B. Schubert, I. Kötschau, S. Schorr, H. W. Schock, *Thin Solid Films* **517**, 2524-2526 (2009).
2. H. Katagiri, K. Jimbo, W.S. Maw, K. Oishi, M. Yamazaki, H. Araki, A. Takeuchi, *Thin Solid Films* **517**, 2455-2560 (2009).
3. G. Suresh Babu, Y.B. Kishore Kumar, P. Uday Bhaskar, V. Sundara Raja, *Sol. Energy Mater. Sol. Cells* **94**, 221-226 (2010).
4. K. Ito and T. Nakazawa, *Jpn. J. Appl. Phys.* **27**, 2094-2097 (1988).
5. N. Kamoun, H. Bouzouita, B. Rezig, *Thin Solid Films* **515**, 5949-5952 (2007).
6. S. Levcenco, D. Dumcenco, Y.P. Wang, Y.S. Huang, C.H. Ho, E. Arushanov, V. Tezlevan, K.K. Tiong, *Opt. Mater.* **34**, 1362-1365 (2012).
7. D. A. R. Barkhouse, O. Gunawan, T. Gokmen, T. K. Todorov, and D. B. Mitzi, *Prog. Photovolt: Res. Appl.* **20**, 6 (2012).
8. D. Dumcenco and Y.S. Huang, *Opt. Mater.* **35**, 419-425 (2013).
9. K. Wang, O. Gunawan, T.K. Todorov, B. Shin, S.J. Chey, N. A. Bojarczuk, D. Mitzi, S. Guha, *Appl. Phys. Lett.* **97**, 143508 (2010).
10. T. Gürel, C. Sevik, T. Çağin, *Phys. Rev. B* **84**, 205201 (2011).
11. A. Khare, B. Himmetoglu, M, Johnson, D.J. Norris, M. Cococcioni, E.S. Aydil, *J. Appl. Phys.* **111**, 083707 (2012).
12. M. Himmrich and H. Haeuseler, *Spectrochim. Acta A* **47**, 933-942 (1991).
13. M. Altosaar, J. Raudoja, K. Timmo, M. Danilson, M. Grossberg, J. Krustok, E. Mellikov, *Phys. Status Solidi A* **205**, 167-170 (2008).
14. P.A. Fernandes, P.M.P. Salomé, A.F. da Cunha, *J. Alloys Compd.* **509**, 7600-7606 (2011).
15. S. Siebentritt and S. Schorr, *Prog. Photovolt: Res. Appl.* **20**, 512 (2012).
16. S. Chen, X.G. Gong, A. Walsh, S. Wei, *Phys. Rev. B* **79**, 165211 (2009).
17. S. Schorr, *Sol. Energy Mater. Sol. Cells* **95**, 1482-1488 (2011).
18. T.C. Damen, S.P.S. Porto, B. Tell, *Phys. Rev.* **142**, 570-574 (1966).
19. Polarization Selection Rules provided by Bilbao Crystallographic Server http://www.cryst.ehu.es/cryst/polarizationselrules.html
20. H. Kuzmany, *Solid-State Spectroscopy. An Introduction*, (Springer, Berlin, 1998).
21. Y. Wang and H. Gong, *J. Electrochem. Soc.* **158**, H800-H803 (2011).
22. A.J. Cheng, M. Manno, A. Khare, C. Leighton, S.A. Campbell, E.S. Aydil, *J. Vac. Sci. Technol. A* **29**, 051203 (2011).
23. S. Levcenko, V.E. Tezlevan, E. Arushanov, S. Schorr, T. Unold, *Phys. Rev. B* **86**, 045206 (2012).

24. X. Fontané, V. Izquierdo-Roca, E. Saucedo, S. Schorr, V.O. Yukhymchuk, M.Ya. Valakh, A. Pérez-Rodríguez, J.R. Morante, *J. Alloys Compd.* **539**, 190-194 (2012).
25. H. Yoo and J.H. Kim, *Thin Solid Films* **518**, 6567-6572 (2010).
26. M. Grossberg, J. Krustok, J. Raudoja, T. Raadik, *Appl. Phys. Lett.* **101**, 102102 (2012).

Mater. Res. Soc. Symp. Proc. Vol. 1538 © 2013 Materials Research Society
DOI: 10.1557/opl.2013.1000

Influence of sodium-containing substrates on Kesterite CZTSSe thin films based solar cells

Giovanni Altamura[1,3,*], Charles Roger[1], Louis Grenet[1], Joël Bleuse[1,2], Hélène Fournier[1], Simon Perraud[1], Henri Mariette[1,2,3]

[1] CEA Grenoble, 17 rue des Martyrs, 38054 Grenoble Cedex 9, France
[2] Institut Néel - CNRS, 25 rue des Martyrs, 38054 Grenoble Cedex 9, France
[3] Joseph Fourier University, 38041 Grenoble Cedex 9, France

*Corresponding author. Tel.: +33(0)4-38-78-14-68. E-mail: giovanni.altamura@cea.fr

Abstract — This work deals with the influence of sodium on the properties of CZTSSe material and solar cells. For that purpose, two types of substrates are compared, one with low sodium content (borosilicate glass), the other one with higher sodium content (soda-lime glass). In each case the Na-content in the CZTSSe is quantified through the substrate by secondary ion mass spectroscopy analysis. Photoluminescence spectroscopy indicates that better quality material is achievable when increasing the Na-content in the CZTSSe. The material characterization results are compared to the photovoltaic properties.

Index Terms — $Cu_2ZnSn(S_{1-x}Se_x)_4$, CZTSSe, CZTS, CZTSe, Sodium, Kesterite, thin film, solar cell.

I. INTRODUCTION

The scientific community has tried to tackle the question of the sustainability of the necessary terawatt-scale power generation capacities. Photovoltaics is a cost-effective solar energy technology which could meet the increase in electricity demand, while avoiding the worst effects of climate change.

Since the $Cu_2ZnSn(S_{1-x}Se_x)_4$ (CZTSSe) semiconductor compound consists in the relatively inexpensive, earth abundant, and nontoxic elements Cu, Zn, Sn, S, and Se, a considerable effort is focused on improving its efficiency.

Some of the CZTSSe characteristics are:
– a bandgap predicted [1] to be between 1.0 eV ($Cu_2ZnSnSe_4$) and 1.5 eV (Cu_2ZnSnS_4);
– an absorption coefficient larger than 10^4 cm^{-1} [1];
– a good tolerance of its crystallographic structure to shifts from the stoichiometric composition [2].

In addition, the knowledge gathered on $Cu(In_{1-x}Ga_x)Se_2$ (CIGS) can be used and adapted to CZTSSe solar cells yielding efficiencies up to 11% [3]. In the case of CIGS solar cells, it is well known that the presence of sodium in the absorber layer is beneficial [4] and necessary to obtain high conversion efficiencies [5]. It mainly improves device performance through the fill factor (FF) and open circuit voltage (V_{oc}) [6]. Although the exact action of Na on the structural and electronic properties of the absorber material is not clearly understood, defect passivation at the grain boundaries and the increase of doping are nowadays the most favored explanations [7].

This paper reports on the characterization methods of CZTSSe layer grown on borosilicate (BS) and soda-lime glass (SLG) substrates, and on electrical performances of solar cells built utilizing these same layers.

II. EXPERIMENTAL DETAILS

CZTSSe has been synthesized on two different Mo-coated SLG (12 wt% Na_2O) and BS (0.01 wt% Na_2O) substrates from a ZnS/Cu/Sn stack of precursors. Hereafter S1 is the CZTSSe grown on SLG, and S2 the

CZTSSe grown on BS. A 340-nm thick layer of ZnS is deposited via a RF-sputtering system in a Plassys MP400 at 1×10^{-3} mbar of Ar at room temperature, while the metallic layers of Cu (120 nm) and Sn (160 nm) are grown by high vacuum electron-beam evaporation in a Plassys MEB550S deposition chamber at 5×10^{-7} mbar and at room temperature.

The selenization of precursors to synthesize the CZTSSe layers occurs in a tubular furnace under a N_2 atmosphere. Selenium is provided in the form of pellets disposed beside the precursor stack [8]. The elements ratios $[Zn]/[Sn] = 1.3$, $[Cu]/([Zn]+[Sn]) = 0.77$ and $[Se]/([S]+[Se]) = 0.8$ have been measured by energy dispersive spectroscopy (EDS) for both samples.

Solar cells have been fabricated as follows: after the CZTSSe absorber processing, a CdS buffer layer is deposited via chemical bath deposition, followed by magnetron sputtering of i-ZnO and Al-doped ZnO which acts as the n-type window layer.

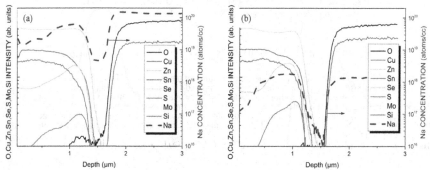

Figure 1: Na concentration measured by SIMS for CZTSSe on different Mo-coated SLG (a) and BS (b). O, Cu, Zn, Sn, Se, S, Mo, Si concentrations are in arbitrary units.

III. RESULTS AND DISCUSSION

Secondary ion mass spectroscopy (SIMS) depth profiles reveal varying Na concentrations inside the absorbers (dashed line in Fig. 1). The lower Na concentrations were found in S2 in a range between 10^{17}-10^{18} atoms/cm³, whereas for S1 the Na concentration is larger than 10^{19} atoms/cm³. It is noticeable as well that in both samples the concentration of Na into CZTSSe is higher close to the Mo interface where the Zn content is higher. This increase in Zn concentration is related to a Zn(S,Se) phase, as demonstrated in ref. [8], which itself could be related to the higher Na concentration in this region, as suggested by Schwarz et al. [9].

The Raman spectra of S1 and S2 depicted in Fig. 2a show that the A1 CZTSe peak (198 cm^{-1}) and A1 CZTS peak (326 cm^{-1}) of S1 are more pronounced compared to the ones of S2 (respectively at 194 cm^{-1} and 323 cm^{-1}). Since the CZTSSe Raman intensities are usually related to the concentration of the secondary phases [10], it is possible to identify CZTSSe with higher Na concentration (S1) as a material with better quality than CZTSSe with a lower Na content. Secondary CZTSe peaks (237 cm^{-1} in S1 and 230 cm^{-1} in S2) together with a Cu$_2$Se peak (259 cm^{-1} in S2) are also detected.

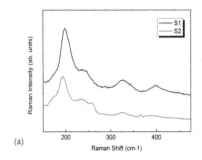 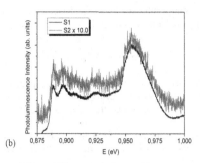

(a)   (b)

Figure 2: Raman (a) and PL (b) spectra for CZTSSe on different Mo-coated glasses. Raman spectra are acquired in backscattering configuration with a 10 mW/cm^2 532 nm green laser excitation. The sample temperature and excitation power density were 7 K and 818 mW/cm^2 for PL.

The photoluminescence (PL) spectra of the S1 and S2 samples measured at T=7 °K are presented in Fig. 2b. For both S1 and S2, the spectra are relatively complex with emission occurring at several distinct peak wavelengths. Although the two spectra show the same features (the main luminescence peak occurs at 0.955 eV), the measurements performed in the present study indicate that the PL spectrum of S1 is much more intense than the S2 one, where the Na concentration is almost two orders of magnitude lower. This reflects the influence of Na, which improves the radiative over non-radiative recombination rates.

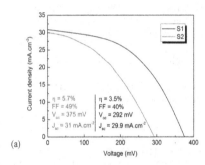 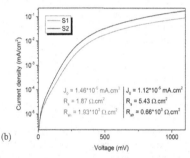

(a)   (b)

Figure 3: Light (a) and dark (b) J-V curves of Mo/CZTSSe/CdS/ZnO/ZnO:Al solar cell on different Mo-coated glass substrates.

The J-V characteristics of the best CZTSSe cells on both Mo-coated SLG and BS are depicted on the Fig. 3, showing that photovoltaic performances improve when the Na content in CZTSSe increases.

The light J-V curve (Fig. 3a) shows an efficiency boost for S1 (5.7%) compared to S2 (3.5%); a gain in both FF (9%) and J_{sc} (1.1 mA.cm^2) is also achieved. The main benefit from the Na content increase in CZTSSe lies in the increase of V_{oc} which is promoted from 292 mV (S2) to 375 mV (S1) with an absolute gain of 83 mV although almost the same shunt resistance is detected in both samples. The losses in V_{oc} for S2 could arise from a rather high J_0 of the devices. Dark J-V curves (Fig. 3b) reveal a lower series resistance (1.87 $\Omega.cm^2$) for S1 than for S2 (5.43 $\Omega.cm^2$), which could be one reason for the gain in short circuit current.

IV. CONCLUSIONS

The effect of Na on the electro-optical properties was addressed through characterization of the finished devices using Raman spectroscopy, photoluminescence and J-V characteristics. The SIMS analysis shows that relatively high concentration of Na is found in CZTSSe grown on SLG compared to CZTSSe grown on BS.
Results indicate the beneficial effect of Na, evidenced by increases in the photovoltaic performances (above all open-circuit voltage and efficiency) and PL intensity.

V. ACKNOWLEDGEMENTS

The authors thank Jeff Mayer from EAG Labs for the SIMS analysis.

REFERENCES

[1] C.P. Chan, H. Lam, C. Surya, Solar Energy Mater. Solar Cells 94, 207 (2010)
[2] L. Choubrac, A. Lafond, C. Guillot-Deudon, Y. Moëlo, S. Jobic, Inorg. Chem. 51, 3346 (2012)
[3] Teodor K. Todorov, Jiang Tang, Santanu Bag, Oki Gunawan, Tayfun Gokmen, Yu Zhu, David B. Mitzi, Advanced Energy Materials, Volume 3, Issue 1, pages 34–38, January, 2013
[4] A. Chirila et al., "Highly efficient Cu(In, Ga)Se2 solar cells grown on flexible polymer films," Nature Materials, September, pp. 1-5, 2011.
[5] K. Granath, M. Bodega, and L. Stolt, "The effect of NaF on Cu(In, Ga)Se2 thin film solar cells," Solar Energy Materials, vol. 60, pp. 279-293, 2000.
[6] R. J. Matson, J. E. Granata, S. E. Asher, and M. R. Young, "Effects of substrate and Na concentration on device properties, junction formation, and film microstructure in CuInSe[sub 2] PV devices," AIP Conference Proceedings, no. October, pp. 542-552, 1999.
[7] R. Caballero et al., "The influence of Na on low temperature growth of CIGS thin film solar cells on polyimide substrates", Thin Solid Films, vol. 517, no. 7, pp. 2187-2190, Feb. 2009.
[8] L. Grenet et al., Solar Energy Materials and Solar Cells, vol. 101, pp. 11-14, Jun. 2012.
[9] T. Schwarz, O. Cojocaru-Miredin, P. Choi, M. Mousel, A. Redinger, S. Siebentritt, D. Raabe, APPLIED PHYSICS LETTERS 102, 042101 (2013)
[10] Alex Redinger, Katja Hönes, Xavier Fontané, Victor Izquierdo-Roca, Edgardo Saucedo, Nathalie Valle, Alejandro Pérez-Rodríguez, and Susanne Siebentritt, "Detection of a ZnSe secondary phase in coevaporated Cu2ZnSnSe4 thin films", Appl. Phys. Lett. 98, 101907 (2011)

Mater. Res. Soc. Symp. Proc. Vol. 1538 © 2013 Materials Research Society
DOI: 10.1557/opl.2013.1024

Air-stable solution processed $Cu_2ZnSn(S_x,Se_{(1-x)})_4$ thin film solar cells: influence of ink precursors and preparation process

Xianzhong Lin[1]*, Jaison Kavalakkatt[1], Martha Ch. Lux-Steiner[1,2] and Ahmed Ennaoui[1]*
[1]Helmholtz-Zentrum Berlin für Materialien und Energie,
Hahn-Meitner-Platz 1, 14109 Berlin, Germany
lin.xianzhong@helmholtz-berlin.de, ennaoui@helmholtz-berlin.de
[2]Freie Universität Berlin, Germany

ABSTRACT

Quaternary semiconductors, Cu_2ZnSnS_4 and $Cu_2ZnSnSe_4$ which contain only earth-abundant elements, have been considered as the alternative absorber layers to $Cu(In,Ga)Se_2$ (CIGS) for thin film solar cells although CIGS-based solar cells have achieved efficiencies over 20 %. In this work we report an air-stable route for preparation of $Cu_2ZnSn(S_x,Se_{(1-x)})_4$ (CZTSSe) thin film absorbers by a solution process based on the binary and ternary chalcogenide nanoparticle precursors dispersed in organic solvents. The CZTSSe absorber layers were achieved by spin coating of the ink precursors followed by annealing under Ar/Se atmosphere at temperature up to 580°C. We have investigated the influence of the annealing temperature on the reduction or elimination of detrimental secondary phases. X-ray diffraction combined with Raman spectroscopy was utilized to better identify the secondary phases existing in the absorber layers. Solar cells were completed by chemical bath deposited CdS buffer layer followed by sputtered i-ZnO/ZnO: Al bi-layers and evaporated Ni/Al grids.

INTRODUCTION

Although $Cu(In,Ga)Se_2$ (CIGS)-based thin film solar cells have achieved efficiencies over 20 % [1], the use of indium which is expensive somehow restrict the large scale development of CIGS-based thin film solar cells. It is crucial to develop new absorbing materials, which contain only earth-abundant constituents, for the further utilisation of the solar energy. $Cu_2ZnSn(S_xSe_{(1-x)})_4$ (CZTSSe), containing only earth-abundant constituents and exhibiting optimal optoelectronic properties for solar energy conversion such as tuneable direct optical band gap from 1.0-1.5 eV and high absorption coefficient in the visible range, has been considered as the alternative to CIGS as absorber layers for thin film solar cells. The solution-processed CZTSSe-based solar cells have achieved great success[2-6]. For example, hydrazine-based solution processed CZTSSe-based solar cells have reached 11 %[5] which is the record efficiency for CZTSSe-based solar cells so far. It is well known that the optoelectronic properties of materials are closely related to qualities of the materials such as crystalline quality, chemical compositions and phase purity. However, according to the phase diagram reported by Olekseyuk et al. [7], the existence region of the phase pure CZTS is quite small, which implies that if the preparation conditions for materials is not well optimized, the occurrence of secondary phases like ZnS and Cu_2SnS_3 will be unavoidable. Moreover, these secondary phases may be detrimental to the device because they act as either recombination centre (when the band gap is smaller than that of CZTSSe absorber) or increase the series resistance of the absorber layer[8]. Therefore, it is of great importance to optimise the preparation conditions for growing CZTSSe absorber layers with high quality before their further application in devices. We have demonstrated that CZTS thin film absorber layer can be prepared from Cu_3SnS_4 and ZnS nanoparticle ink precursors followed by annealing under Ar-H_2S (5 %) atmosphere[9]. To further optimise the preparation conditions, we investigated the influence of the preparation conditions on the structural properties of CZTSSe thin film absorbers by X-ray diffraction (XRD) combined with Raman spectroscopy.

EXPERIMENT DETAILS

The preparation process of CZTSSe thin films includes four main steps: ink formulation, deposition of precursor thin films, heat treatments of the precursors' films and annealing. The ink formulation was processed by mixing Cu_3SnS_4 (CTS), ZnS and SnS nanoparticles with a ratio of 1:1.75:0.7. The synthetic procedures of the nanoparticle precursors could be found in Ref [10, 11]. The composition of the thin films can be controlled by changing the ratio of the nanoparticle precursors. The second step was to deposit the Cu-Zn-Sn-S precursor films by spin coating of the nanoparticle inks at a certain rotating speed. The third step is heat treatment at 170 °C and 350 °C for 2 min, respectively. The aim of the heat treatment process is to remove the organic solvent as well as part of the surfactants surrounded the nanoparticles from the precursor layers. In addition, the heat treatment process also helps to dense the film on the substrates otherwise the deposited layers may be dissolved back into the solvent again when the next spin coating processes. The substrates used were glass substrates or 500 nm thick molybdenum coated soda lime glass substrates. It should be noted that prior the deposition of precursor layers, the glass substrates were cleaned sequentially in an ultrasonic bath in acetone, ethanol and distilled water for 15 min. By repeating the procedures of spin coating and heat treatment processes, the designed thickness can be achieved. Finally, the resulting precursor films were subjected to annealing process at 400 to 580 °C under different selenium-containing atmosphere, which allows the formation of CZTSSe absorbers by reaction of the nanoparticle precursors. The annealing was done in a Gero split-tube furnace with one heating zone. The samples were placed at the centre of the heating zone inside a quartz tube.

X-ray diffraction (XRD) were used to characterize the structural properties, which was operated in the 2θ range from 10 to 90° on a Bruker D8-Advance X-ray diffractometer with CuKα radiation using a step size of 0.02° and a step time of 4 second. For the Raman measurement a Ti: Sapphire-ring-laser was used as an excitation. The wavelength of the laser is fully tenable from 690 nm to 1050 nm. To avoid laser heating the beam power was kept below 3.5 mW. Raman spectra were recorded with a Horiba T64000 triple monochromator system in back scattering configuration with a microscope and a motorized XY stage. The micro-Raman spectroscopy with a 100× objective was performed at room temperature using a wavelength of 747 nm. The electrical characterization has been performed using an in-house class A sun simulator under standard test conditions (AM 1.5G, 100mW/cm² and 23 °C). Quantum efficiency analysis has been performed using an illumination system including two sources (halogen and xenon lamps) and a Bentham TM300 monochromator (Bentham Instruments, Berkshire, UK). Reference measurements were performed on calibrated Si and Ge detectors.

DISCUSSION

Figure1 (d) shows the XRD pattern of a precursor film consisted of four deposited layers on glass substrate with heat treatments mentioned in the experimental section. Nearly all of the peaks can be attributed to the CTS; ZnS, and SnS nanoparticles precursors except the peak marked with circle arising from CuS phase. The appearance of this peak could be due to the reaction between ZnS and CTS resulting in the formation of CZTS and generation of CuS secondary phase (see equation (1)).

$$Cu_3SnS_4 + ZnS \rightarrow Cu_2ZnSnS_4 + CuS \qquad (1)$$

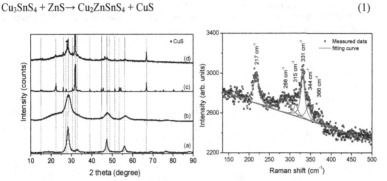

Figure 1 Left: XRD patterns of (a) CTS, (b) ZnS and (c) SnS nanoparticle precursors and precursor films on glass substrate. Right: Raman spectrum of the precursors' film together with the fitting spectrum.

According to Schorr et al.[12] who have investigated the formation of CZTS from the solid state reaction of CuS, SnS and ZnS powder precursors using in-situ high temperature synchrotron X-ray diffraction, the formation of CZTS started just below 300 °C. In addition, recently Hsu et al.[13] have studied the reaction pathway of CZTSSe absorbers prepared from the binary chalcogenides by hydrazine-based solution process, they found that the CZTSSe was formed at 350 °C by the reaction of $Cu_2SnS(Se)_3$ with ZnS. Therefore, it is reasonable that, in our system, the formation of CZTS have already started at 350 °C. However, it is not clear whether the reaction of the SnS precursor with CTS and ZnS has taken place following the equation (2) during the heat treatment stage.

$$2Cu_3SnS_4 + 3\ ZnS + SnS \rightarrow 3Cu_2ZnSnS_4 \qquad (2)$$

To further confirm the formation of CZTS during the heat treatment stage, we performed Raman spectroscopy measurement on the precursors' film. The appearance of 288, 331 and 366 cm^{-1} peaks in the Raman spectrum of the precursor's films (Figure 1 right) confirmed the formation of CZTS in the heat treatment stage. In addition, a strong peak located at 217 cm^{-1} was detected, which can be assigned to the SnS precursor. Other peaks at 315 and 344 cm^{-1} can be attributed to CTS precursors. The peak corresponding to ZnS precursors generally situated at 353 cm^{-1} could not be detected in the precursors' film. The reason for this could be due to the excitation wavelength of the light (747 nm) used for Raman measurements is far below the photon energies, which are necessary for a quasi-resonant excitation of ZnS.

Figure 2 shows the XRD patterns of samples annealed under Ar-Se vapour atmosphere at different tempertures. Since the formation of CZTS has been found to start at the heat treatment stage (350 °C), the formation of CZTSSe should occur when the sample was annealed above 400 °C under Ar-Se atmosphere. Therefore, we indexed the peaks shown in figure 2 to CZTSSe. Nevertheless, the existence of Zn(S,Se) or CTSSe arising from the selenization of ZnS or CTS nanoparticle precursors cannot be excluded. The intensity of the peaks corresponding to the CZTSSe increased gradually with increasing annealing temperature before 540 °C and a significant increase can be oberserved for the sample annealed at 580 °C. Besides the increse of intensity with temperature, the shift of the XRD patterns to lower angle with increasing annealing temperatures was also oberserved, which is due to the expansion of the lattice parameters resulting from the replacement of sulfur with selenium. For instance, the (112) peak position shifted from $2\theta = 27.63°$ to $2\theta = 27.35°$ when the annealing temperature increases from 400 to 580 °C.

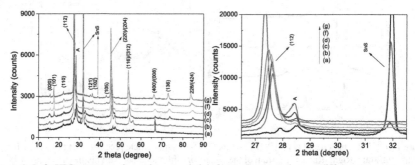

Figure 2. Left: XRD patterns of samples annealed under Ar-Se atmosphere at different temperatures at: (b) 400 °C, (c) 450 °C, (d) 500 °C, (e) 540 °C, and (f) 580 °C. Right: the enlarged part of 2θ between 26.5 and 32.5° of the corresponding samples. For comparion, XRD patten of as-deposited sample (a) was shown below.

Meanwhile, the peak at 31.9° from SnS drcreased dramaticeally when the sample was annealed at 450 °C. Further increase of the annealing temperatures led to the disappearance of this peak, suggesting most of the SnS precursor reacted with the other precursors to form CZTSSe following the reaction route given in equation (3). However, a weak peak at 39.1° correspnding to SnS is still visible even at high annealing temperature, indicating that the SnS phase was left in the resulting films. Since the lost of tin is known during the annealing process, the dcrease of the SnS signal could also resulting from the loss of SnS at higher annealing temperatures [14].

$$CTS + ZnS + SnS \xrightarrow{\Delta Se} CZTSSe \tag{3}$$

Except for those peaks assigned to CZTSSe and SnS, there is another peak located in between 28° and 29° for all the samples, which is named as peak A. This peak is located at 28.44°-28.45° for the samples annealed at 400, 450, 500 and 540 °C, which is slightly smaller than the main diffracion peaks of ZnS and CTS precursors. We attributed this peak to Zn(S,Se) and CTSSe which formed by the reaction of the ZnS and CTS with selenium during the annealing. When the annealing temperature was raised to 580 °C, peak A shifted to 28.52° for this sample. The interplanar distance of peak A in this sample was calculated to be 3.127 Å, which is close to that of (103) planes at 3.129 Å for the CZTSSe sample of 580 (see Table 1). Therefore, this peak can be indexed to (103) of CZTSSe in sample annealed at 580 °C.

The intensity of the peak A decreased with increasing annealed temperature before 500 °C. The explanation for this behavior is that more and more ZnS and CTS precursors reacted with SnS under the Se-conaining atmosphere resulting in the formation of CZTSSe as the increase of annealing temperature. However, for the annealing temperature above 500 °C, nearly no change in term of the intensity for peak A can be observed, indicating that most of the ZnS and CTS reacted with SnS. This result is in aggreement with the variation of SnS signal in the XRD patterns as mentioned above where the main peak of SnS at 31.9° disappeard above 500 °C.

Table 1 Diffraction angles and interplanar spacing distance of (101) and (112) reflections and lattice constants calculated from these two lattice planes. The interplanar spacing distance d_{103} of (103) plane from CZTSSe was also calculated using equation (4) while the d value of reflection A was calculated from the Bragg's Law.

Sample	$2\theta_{(101)}$	d_{101} (Å)	$2\theta_{(112)}$	d_{112} (Å)	a (Å)	c (Å)	d_{103} (Å)	$2\theta_A$ (°)	d_A (Å)
400	27.63	5.008	27.63	3.230	5.597	11.136	3.093	28.44	3.137
450	27.61	5.011	27.61	3.233	5.602	11.140	3.095	28.44	3.137
500	27.57	5.018	27.57	3.238	5.613	11.147	3.098	28.45	3.136
540	27.48	5.036	27.48	3.248	5.622	11.216	3.113	28.45	3.136
580	27.35	5.056	27.35	3.262	5.648	11.265	3.127	28.52	3.129

The lattice constants for all the samples were calculated according to equation (4) using reflection peak of (101) and (112), as shown in Table 1. The lattice constant of a and c increased with increasing annealing temperature which is due to the increasing replacement of S with Se in CZTSSe.

$$\frac{1}{d^2} = \frac{h^2+k^2}{a^2} + \frac{l^2}{c^2} \qquad (4)$$

where d is the is the interplanar distance, a, b and c represent the lattice and h, k, and l the Miller indices.

According to Momose et al. [15], the relationship between [S]/([S] + [Se]) and the diffraction angle follows the Vegard's law [16]:

$$2\theta_{CZTS} = x \cdot 2\theta_{CZTS} + (1-x) \cdot 2\theta_{CZTSe} \qquad (5)$$

where 2θ represents the diffraction angle of the Bragg peak. Regarding to equation (5) and by using the Bragg peak of (112), the value of x in $Cu_2ZnSn(S_x,Se_{(1-x)})_4$ was determined to be 0.37, 0.36, 0.31, 0.25 and 0.16 for samples annealed at 400, 450, 500, 540 and 580 °C respectively.

Figure 3 shows a representative Raman spectrum of the sample annealed at 580 °C. The spectrum shows bimodal behaviour which has also been observed by other groups for the CZTSSe samples [17, 18]. By fitting the spectrum with the software Peak-o-mat using Lorentzian function, seven peaks can be observed. The peak located at 196 cm^{-1} is the characteristic A mode for CZTSe while the peak situated at 327 cm^{-1} is the A mode of CZTS shifted by around 10 cm^{-1} due to the partial replacement of S with Se. Peaks at 155 cm^{-1}, 173 cm^{-1}, and 242 cm^{-1} can also be associated with CZTSSe, which is in agreement with the theoretical calculations from Gürel et al.[19]. The weak peak at 220 cm^{-1} is attributed to SnS while the assignment of the peak at 400 cm^{-1} is assigned to SnSe$_2$.[20, 21] The secondary phases of SnS and SnSe$_2$ were also found in the other samples. However, when the ink composition was adjusted to near stoichiometry, SnS and SnSe$_2$ were not detected by means of XRD and Raman spectroscopy measurements.

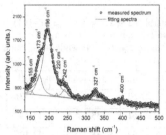

Figure 3. Representative Raman spectrum of sample annealed at 580 °C. The fitting spectra are also shown.

We have fabricated solar cells using the sample prepared from an ink composition of Cu/(Zn+Sn) = 0.72 and Zn/Sn = 1.1. The sample was annealed under Se-containing atmosphere at 580 °C for 25 min. Figure 4(a) shows the J-V characteristic of the best device from this sample under dark and illumination with simulated 1 sun AM 1.5G. The device with an area of 0.5 cm^2 exhibits open circuit voltage of 332.2 mV, short circuited current density of 23.8 mA/cm^2 and fill factor of 35.65 %, yielding an efficiency η=2.82 %. The external quantum efficiency (EQE) curve of the corresponding device illustrates the highest EQE is over 80 %, as shown in Figure 4(b). The band gap of the CZTSSe absorber layer was estimated to be around 1.15 eV from the onset of the EQE data in the infrared region. This value is close to 1.13 eV for the best solar cells reported by Todorov et al.[5].

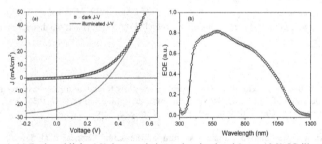

Figure 4. (a) Dark and light *J-V* characteristics under simulated 1 sun AM1.5G illumination, (b) External quantum efficiency curve of the cell.

CONCLUSIONS

CZTSSe thin films have been prepared successfully from the binary and ternary nanoparticle precursors. XRD and Raman spectroscopy analysis confirmed the formation CZTS during the heat treatment stage at 350 °C. Increasing amount of sulfur was replaced by selenium with rising annealing temperature. The crystalline quality of the sample annealed at 580 °C was significantly enhanced compared to the other samples. By adjusting the ink composition and preparation process, solar cells with efficiencies of 2.82 % have been achieved. The optimization of the annealing process as well as the composition of the resulting absorber layers is important for the further improvement of the solar cell performances.

ACKNOWLEDGMENTS

The authors would like to thank J. Klaer for the preparation of Mo substrates, C. Kelch and M. Kirsch for the completion of the devices. One of the authors (X. Z. Lin) gratefully acknowledges the financial support from the China Scholarship Council, HZB and Helmholtz Association. This work was carried out as part of a program supported by the BMBF (Grant 03SF0363B).

REFERENCES

1. P. Jackson, D. Hariskos, E. Lotter, S. Paetel, R. Wuerz, R. Menner, W. Wischmann and M. Powalla, Progress in Photovoltaics: Research and Applications **19** (7), 894-897 (2011).
2. Y. Cao, M. S. Denny, J. V. Caspar, W. E. Farneth, Q. Guo, A. S. Ionkin, L. K. Johnson, M. Lu, I. Malajovich, D. Radu, H. D. Rosenfeld, K. R. Choudhury and W. Wu, J. Am. Chem. Soc. **134**, 15644-15647 (2012).
3. Q. Guo, G. M. Ford, W.-C. Yang, B. C. Walker, E. A. Stach, H. W. Hillhouse and R. Agrawal, J. Am. Chem. Soc. **132**, 17384-17386 (2010).
4. T. K. Todorov, K. B. Reuter and D. B. Mitzi, Advanced Materials **22** (20), E156-E159 (2010).
5. T. K. Todorov, J. Tang, S. Bag, O. Gunawan, T. Gokmen, Y. Zhu and D. B. Mitzi, Advanced Energy Materials **3** (1), 34-38 (2013).
6. W. Yang, H.-S. Duan, B. Bob, H. Zhou, B. Lei, C.-H. Chung, S.-H. Li, W. W. Hou and Y. Yang, Advanced Materials **24** (47), 6323-6329 (2012).
7. I. D. Olekseyuk, I. V. Dudchak and L. V. Piskach, Journal of Alloys and Compounds **368** (1-2), 135-143 (2004).
8. S. Siebentritt, Thin Solid Films, 10.1016/j.tsf.2012.1012.1089 (2013).
9. X. Lin, J. Kavalakkatt, K. Kornhuber, S. Levcenko, M. C. Lux-Steiner and A. Ennaoui, Thin Solid Films **535**, 10-13 (2013).
10. X. Lin, A. Steigert, M. C. Lux-Steiner and A. Ennaoui, RSC Advances **2** (26), 9798 (2012).
11. X. Lin, J. Kavalakkatt, M. C. Lux-Steiner and A. Ennaoui, 27th European Photovoltaic Solar Energy Conference and Exhibition; Frankfurt, Main **3DV**, 2794-2797 (2012).
12. S. Schorr, A. Weber, V. Honkimäki and H.-W. Schock, Thin Solid Films **517**, 2461–2464 (2009).
13. W.-C. Hsu, B. Bob, W. Yang, C.-H. Chung and Y. Yang, Energy & Environmental Science **5** (9), 8564 (2012).
14. A. Weber, R. Mainz and H. W. Schock, Journal of Applied Physics **107** (1), 013516 (2010).

15. N. Momose, M. T. Htay, K. Sakurai, S. Iwano, Y. Hashimoto and K. Ito, Applied Physics Express 5 (8), 081201 (2012).
16. L. Vegard, Zeitschrift für Physik 5 17 (1921).
17. J. He, L. Sun, S. Chen, Y. Chen, P. Yang and J. Chu, Journal of Alloys and Compounds 511 (1), 129-132 (2012).
18. D. B. Mitzi, O. Gunawan, T. K. Todorov, K. Wang and S. Guha, Solar Energy Materials and Solar Cells 95 (6), 1421-1436 (2011).
19. T. Gürel, C. Sevik and T. Çağın, Physical Review B 84 (20) (2011).
20. H. Chandrasekhar, R. Humphreys, U. Zwick and M. Cardona, Physical Review B 15 (4), 2177-2183 (1977).
21. O. P. Agnihotri, A. K. Garg and H. K. Sehgal, Solid State Communications 17, 1537-1540 (1975).

Mater. Res. Soc. Symp. Proc. Vol. 1538 © 2013 Materials Research Society
DOI: 10.1557/opl.2013.1001

Fabrication and Characterization of Low-Cost, Large-Area Spray Deposited Cu₂ZnSnS₄ Thin Films for Heterojunction Solar Cells

Sandip Das, Kelvin J. Zavalla, M. A. Mannan, and Krishna C Mandal*

Department of Electrical Engineering, University of South Carolina, Columbia, SC 29208, USA.

ABSTRACT

Large-area Cu_2ZnSnS_4 (CZTS) thin films were deposited by low-cost spray pyrolysis technique on Mo-coated soda-lime glass (SLG) substrates at varied substrate temperatures of 563-703°K. Deposition conditions were optimized to obtain best quality films and effect of post deposition thermal processing of the as-deposited films under H_2S ambient were investigated. Structural, morphological, and compositional characterization of as-deposited and H_2S treated CZTS absorber layers were carried out by x-ray diffraction (XRD), Raman spectroscopy, scanning electron microscopy (SEM) and energy dispersive x-ray analysis (EDX). Optical and electrical properties were measured by UV-Vis spectroscopy, van der Pauw, and Hall-effect measurements. Films grown at ~360°C substrate temperature showed superior optoelectronic properties, improved stoichiometry and smoother morphology compared to films grown at much higher or lower temperatures. Film properties were significantly improved after the H_2S processing. Our results show that large area high quality CZTS films can be fabricated by low-cost spray pyrolysis technique for high throughput commercial production of CZTS based heterojunction solar cells.

INTRODUCTION

High efficiency, economically sustainable solar cells using earth-abundant and non-toxic materials is necessary to meet TW-scale photovoltaic power generation predicted by 2030 [1]. Current state-of-the-art thin film solar cell technologies based on $CuInGaSe_2$ (CIGS) and CdTe absorbers have reached record efficiencies of 20.3% and 17.3% [2-3]. However, use of toxic Cd in CdTe solar cells threatens the environmental health hindering mass production and supply limitations of scarce and expensive In, Ga in CIGS cells are expected to limit the production capacity of these chalcogen-based solar cells in near future. Therefore, it is necessary to explore new solar absorber materials consisting of earth-abundant, environment-friendly and cheaper constituent elements. Recently, Cu_2ZnSnS_4 (CZTS) has emerged as a potential candidate as an alternative to existing CIGS and CdTe absorbers in thin film solar cells [4-6]. All constituent elements in CZTS are abundant in earth's crust (Cu: 50 ppm, Zn: 75 ppm, Sn: 2.2 ppm, S: 260 ppm) compared to CIGS/CdTe (Cd: 0.11 ppm, In: 0.049 ppm, Ga: 18 ppm; Te: 0.005 ppm), much cheaper and are non-toxic. CZTS is an excellent absorber material with an ideal direct bandgap (E_g ~1.5 eV at 300K), large optical absorption co-efficient ($\alpha > 10^4$ cm^{-1}) and has recently achieved a record efficiency of 11.1% for Se enriched CZTS [6].

Ito and Nakazawa first observed the photovoltaic effect in a CZTS/CTO heterodiode with an open-circuit voltage (V_{OC}) of 165 mV [7]. They prepared the CZTS thin film by atom beam sputtering. Several vacuum and non-vacuum techniques have been investigated to fabricate CZTS absorber including thermal evaporation [4], chemical hot injection method [5], e-beam evaporation [8], sputtering [9], spray pyrolysis [10-13] and electrochemical deposition [14].

Spray pyrolysis is a low-cost technique to fabricate semiconductor thin films offering roll-to-roll fabrication at high throughput for commercial production. However, only a few research groups have studied spray deposition of CZTS thin-films till date [10-13]. Sprayed CZTS films suffer from significant sulfur deficiency which limits their applicability to fabricate solar cells. Therefore, it is important to improve the film properties to realize good quality heterojunction solar cells using spray deposition technique. We investigated the effect of post deposition thermal processing in H_2S environment on the film properties. In this article, we report the fabrication and optimized film properties after H_2S processing of the as-deposited CZTS films.

EXPERIMENTAL

An aqueous solution of 0.01M CuCl, 0.005M $ZnCl_2$, 0.005M $SnCl_4$ and 0.04M thiourea dissolved in de-ionized water was used for spray deposition of CZTS films on soda lime glass (SLG) and Mo-coated SLG substrates. Deposition was carried out at three different substrate temperatures (290°C, 360°C and 430°C) with a constant nozzle to substrate distance of ~18 cm and a spray rate of ~3.0 ml/min using an Iwata CM-C atomizing nozzle. Compressed air (~ 30 PSI) was used as the carrier gas. Substrate temperature was maintained within ± 5°C of the desired deposition temperature. A schematic of the spray deposition system is depicted in fig. 1. Deposition cycles were automated using the in-house built setup by automated temperature control, substrate rotation and a self-programmed actuator for precise control of the spray nozzle.

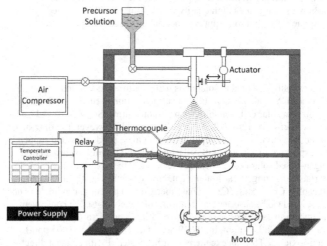

Figure 1. Schematic of the in-house built automated spray deposition system.

As-deposited CZTS films were annealed under H$_2$S flow at 500°C for one hour. Resulting film thicknesses were between 1.2-1.6 μm as measured by a Dektak IIA surface profilometer. X-ray diffraction (XRD) was performed using a Rigaku x-ray diffractometer using Cu-K$_\alpha$ radiation. Surface microstructure and stoichiometry was investigated with a FEI Quanta SEM equipped with EDX. Optical transmittance was recorded using an Agilent Cary 60 UV-Vis spectrophotometer. Electrical properties were measured by van der Pauw technique and Hall-effect measurements using Keithley 7001 digital switching mainframe, Keithley 220 programmable current source, Keithley 485 picoammeter and Keithley 2000 DMM.

RESULTS AND DISCUSSION

XRD measurements showed that the as-deposited and annealed CZTS films exhibited kesterite structure with a preferred [112] orientation. Best crystallinity was observed for films deposited at ~360°C after sulfurization and is shown in fig. 2. The most dominant peaks are indexed in the XRD spectra corresponding to (112), (204/220), (312), (101), (200), (002), (110), (211) planes. It is extremely difficult to distinguish any binary or ternary impurity phases from XRD spectra alone due to close peak proximity. Therefore, Raman spectra were acquired to investigate the existence of possible secondary phases present in the film. The Raman spectra for the H$_2$S treated films produced at 290°C, 360°C and 430°C are shown in fig. 3(a). It is observed that for the film produced at 290°C, CZTS peak is missing and two strong peaks at 295 cm^{-1} and 356 cm^{-1} are dominant in the spectra which can be attributed to tetragonal Cu$_2$SnS$_3$ (CTS) and β-ZnS phases respectively [15]. For the film produced at 430°C, along with CZTS phase (peak at 338 cm^{-1}), a strong peak at 475 cm^{-1} corresponding to Cu$_{2-x}$S phase is identified. Also, a smaller peak at 266 cm^{-1} corresponding to cubic CTS can be observed. The film fabricated at 360°C substrate temperature showed only a major peak around 338 cm^{-1} corresponding to CZTS phase. The broad peak around 338 cm^{-1} was de-convoluted and the best match was found with four individual peaks. Peaks positioned at 287 cm^{-1}, 338 cm^{-1}, 368 cm^{-1} corresponding to CZTS phase and a small peak at 314 cm^{-1} which we attributed to a possible SnS$_2$ phase [15]. However, no other secondary phases were evidenced for the film produced at 360°C. These results clearly indicate that 360°C is the desired range of substrate temperature for single phase CZTS film deposition.

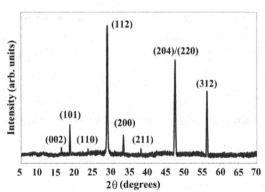

Figure 2. XRD pattern of CZTS thin film deposited at 360°C after H$_2$S treatment

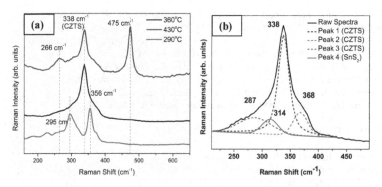

Figure 3. (a) Raman spectra of the films produced at 290°C, 360°C and 430°C after sulfurization; (b) Deconvoluted Raman spectra for the film at 360°C.

Surface microstructures of all films were investigated by scanning electron microscopy (SEM). SEM micrographs of as-deposited and H₂S treated CZTS films prepared at 290°C, 360°C and 430°C are shown in fig. 4 (a-f). As-deposited films were very rough. Film deposited at lower temperature showed small discrete grains which merged together and formed a smoother surface after sulfurization. Films deposited at higher temperature had big particulates on the surface which were significantly reduced after sulfurization. All films exhibited much smoother surface morphology after sulfurization The best film morphology was obtained for the films deposited at 360°C.

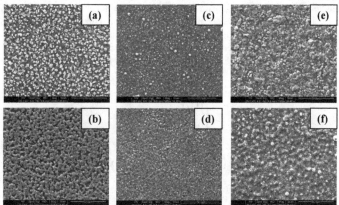

Figure 4. SEM micrographs of as-deposited CZTS films at substrate temperatures of (a, c, e) 290°C, 360°C, 430°C; and (b, d, f) corresponding films after H₂S treatment.

Stoichiometric and compositional analysis of the films were examined by EDX and the results are summarized in Table I. As-deposited films were found to be sulfur deficient and Cu rich. The high contrast white regions in the SEM images correspond to Cu-rich areas, possibly in the form of Cu_xS phases. It is observed that after sulfurization, the concentrations of these particles are reduced significantly resulting improved stoichiometry. The film produced at 360°C showed the best stoichiometry after sulfurization. Hall effect measurement revealed that all films were of p-type conductivity with an average carrier concentration in the order of 10^{19} cm^{-3} and resistivities ranging 0.08-3.8 Ω.cm.

TABLE I
Summary of EDX Results

Sample	Cu (at%)	Zn (at%)	Sn (at%)	S (at%)
290°C (as deposited)	28.73	11.20	14.79	45.28
290°C (sulfurized)	23.86	10.87	11.60	53.67
360°C (as deposited)	29.16	13.12	16.03	41.69
360°C (sulfurized)	24.10	13.26	10.80	51.84
430°C (as deposited)	38.09	14.87	15.60	31.44
430°C (sulfurized)	32.86	13.10	11.55	42.49

Optical absorption coefficient was calculated using eq. (1) from the recorded transmittance and reflectance data. Where, α is the absorption co-efficient, t is the thickness of the film, T_λ and R_λ are the transmittance and reflectance values at any wavelength λ. The absorption co-efficient was in the range of 10^4 - 10^5 cm^{-1}.

$$\alpha = -\frac{1}{t}\ln\left(\sqrt{\frac{(1-R_\lambda)^4 + 4T_\lambda^2 R_\lambda^2 - (1-R_\lambda)^2}{2T_\lambda R_\lambda^2}}\right) \qquad (1)$$

Eq. (2) was used determine the optical bandgaps of the films from the $(\alpha h v)^2$ vs hv plot by taking intercept at the x-axis of linear region extrapolation. Where, E_g is the bandgap of the CZTS films and A is a constant.

$$\alpha.hv = A(hv - E_g)^{1/2} \qquad (2)$$

The $(\alpha h v)^2$ vs hv plots are shown in fig. 5 (a-f). Bandgaps of the films were between 1.42 eV - 1.72 eV at different deposition conditions and sulfurization. It is observed that before sulfurization, the effective bandgaps were found to be in the range of 1.7-1.8 eV and after H_2S treatment, bandgaps approached to 1.42-1.45 eV. This is probably due to the presence of binary/ternary sulfides in the as-deposited films which reduced significantly after sulfurization as evidenced from Raman and EDX data.

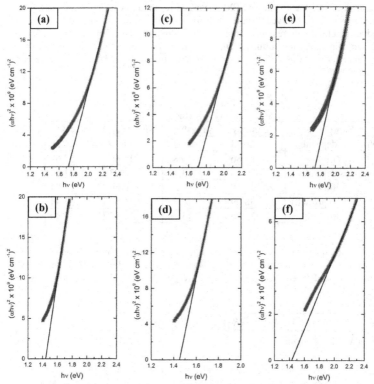

Figure 5. $(\alpha h\nu)^2$ vs. $h\nu$ plot for (a, c, e) as-deposited films at 290°C, 360°C, 430°C and (b, d, f) corresponding films after H_2S treatment.

CONCLUSIONS

We have successfully fabricated CZTS thin-films by a low-cost spray-pyrolysis technique at various substrate temperatures. Best quality films were obtained for films deposited at 360°C and film properties were substantially improved by H_2S treatment of the as-deposited films. Our results indicate that the earth abundant, non-toxic CZTS based heterojunction solar cells can be fabricated by low-cost spray deposition technique for high throughput roll-to-roll commercial production. Presently we are investigating the heterojunction properties of the films and the results will be reported later.

ACKNOWLEDGEMENTS

This work was partially supported by DARPA (grant # N66001-10-1-4031).

REFERENCES

1. C. S. Tao, J. Jiang and M. Tao, *Solar Energy Mater. & Solar Cells.* **95**, 3176-3180, 2011
2. P. Jackson, D. Hariskos, E. Lotter, S. Paetel, R. Wuerz, R. Menner, W. Wischmann and M. Powalla, *Prog. Photovolt: Res. Appl.* **19**, 894-897, 2011.
3. M. A. Green, K. Emery, Y. Hishikawa, W. Warta and E. D. Dunlop, *Prog. Photovolt: Res. Appl.* **20**, 606-614, 2012.
4. K. Wang. O. Gunawan, T. Todorov, B. Shin, S. J. Chey, A. Bojarczuk, D. Mitzi and S. Guha, *Appl. Phys. Lett.* **97**, 143508 (1-3), 2010.
5. Q. Guo, G. M. Ford, W.-C. Yang, B. C. Walker, E. A. Stach, H. W. Hillhouse and R. Agrawal, *J. Am. Chem. Soc.* **132**, 17384-17386, 2010.
6. T. K. Todorov,J. Tang, S. Bag, O. Gunawan, T. Gokmen, Y. Zhu and D. B. Mitzi, *Adv. Energy Mater.* **3**, 34-38, 2013.
7. K. Ito and T. Nakazawa, *Jpn. J. of Appl. Phys.* **27**, 2094-2097, 1988.
8. H. Araki, A. Mikaduki, Y. Kubo, T. Sato, K. Jimbo, W. S. Maw, H. Katagiri, M. Yamazaki, K. Oishi and A. Takeuchi, *Thin Solid Films.* **517**, 1457-1460, 2008.
9. N. Momose, M. T. Htay, T. Yudasaka, S. Igarashi, T. Seki, S. Iwano, Y. Hashimoto and K. Ito, *Jpn. J. of Appl. Phys.* **50**, 01BA02, 2011.
10. N. Nakamaya and K. Ito, *Appl. Surf. Sci.* **92**, 171-175, 1996.
11. N. Kamoun, H. Bouzouita and B. Rezig, *Thin Solid Films.* **515**, 5949-5952, 2007.
12. Y. B. K. Kumar, G. S. Babu, P. U. Bhaskar and V. S. Raja, *Physica Stat. Solidi A.* **207**, 149-156, 2010.
13. V. G. Rajeshmon, C. S. Kartha, K. P. Vijayakumar, C. Sanjeeviraja, T. Abe and Y. Kashiwaba, *Solar Energy.* **85**, 249-255, 2011.
14. J. J. Scragg, P. J. Dale and L. M Peter, *Thin Solid Films.* **517**, 2481-2484, 2009.
15. P. A. Fernandes, P. M. P. Saloméa and A. F. da Cunha, *J. Alloys and Compounds.* **509**, 7600-7606, 2011.

Mater. Res. Soc. Symp. Proc. Vol. 1538 © 2013 Materials Research Society
DOI: 10.1557/opl.2013.1004

The Effect of Soft Pre-Annealing of Differently Stacked Cu-Sn-Zn Precursors on the Quality of Cu$_2$ZnSnSe$_4$ Absorbers

Monika Arasimowicz[1], Maxime Thevenin[2], and Phillip J. Dale[1]

[1]Laboratory for Energy Materials, University of Luxembourg,
41, rue du Brill, L-4422 Belvaux, Luxembourg
[2]Laboratory for Photovoltaics, University of Luxembourg,
41, rue du Brill, L-4422 Belvaux, Luxembourg

ABSTRACT

Cu$_2$ZnSnSe$_4$ p-type semiconductors currently investigated for use in thin film solar cells can be synthesized by firstly depositing a metallic precursor and secondly annealing the precursor in selenium vapor. Differently stacked Cu-Sn-Zn metallic precursors were characterized after a soft annealing at 350°C under nitrogen atmosphere. For the stack where the Sn and Zn were in direct contact with sufficient Cu to form a stable alloy, a bi-layered structure consisting of Cu-Sn on the bottom and Cu-Zn on the top was formed. Contrarily, when Zn was not in direct contact with Cu, the metals diffused to form a stable alloy and the system segregates horizontally, forming a mixed columnar structure. These two types of precursors were selenized under exactly the same conditions to form kesterite absorbers for solar cell devices. Using this approach the improvement from 0.44% power conversion efficiency for the bi-layered precursor to 4.5% for the mixed precursor was achieved.

INTRODUCTION

Vapor phase chalcogenisation of Cu-Sn-Zn metallic precursors is a low cost and scalable method of thin film Cu$_2$ZnSn(S,Se)$_4$ (CZTSSe) fabrication. However different absorber properties were achieved using different stacking orders for the metals [1-3]. None of the published studies explained why some metal stacking orders are successful, whilst others are not. Figure 1 shows the existence of phases in the Cu-Sn-Zn alloy system at 250°C investigated experimentally by Chou et al.[4]. Mainly a liquid phase was observed for the composition relevant to the stoichiometry of kesterite. The existence of liquid Sn, Cu-Sn alloys (bronzes) and Cu-Zn alloys (brasses) was confirmed and no ternary compound could be found in that system.

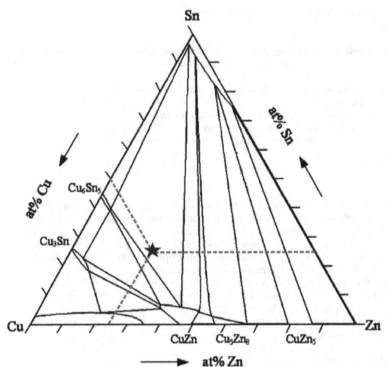

Figure 1. The isotherm of Cu-Sn-Zn phase equilibria at 250°C, (after Chou et al.[4]). The star indicates the stoichiometry of precursors deposited in this study.

Here we consider the distribution of the metals in the precursor. Our hypothesis is that for any metal precursor stack where Sn or Zn are not in direct contact with sufficient Cu to form a stable Cu-Sn alloys and Cu-Zn alloys, they will tend to diffuse to find the Cu that they need. This alloying is slow at room temperature, but accelerated during the heating ramp of the annealing step, before the chalcogen has time to react with the precursors. As a consequence two dimensional or three dimensional metallic structures emerge before the chalcogenisation step and this could lead to different absorber quality. Using mass transport controlled electrodeposition and soft annealing different Cu-Sn-Zn alloys were synthesized. Depending on the way the layers are deposited two dimensional and three dimensional structures can be formed during the soft annealing step. We present chemical and structural experimental evidence on the sub-micron scale for the two dimensional and three dimensional precursor structures, as well as the consequence of this on the quality of the final absorber. For exactly the same annealing conditions it will result in different absorber homogeneity. These new insights can be used to explain why some of the synthesis routines described in literature yield much better efficiencies than others.

EXPERIMENT

Using mass controlled electrodeposition stoichiometric stacked elemental layers in the sequence of Cu1/Sn/Cu2/Zn or Cu/Sn/Zn were deposited on the molybdenum covered soda lime glass substrate as described elsewhere [5,6]. The formation of metallic alloys was accelerated by a soft annealing of as deposited precursors in a rapid thermal annealing system (RTP AS-One, Annealsys). Precursors were annealed under nitrogen gas at 350°C and cooled down to room temperature after 30 minutes. The microstructure of Cu-Sn and Cu-Zn alloys was confirmed by scanning electron microscopy, (SEM Hitachi SU-70) with energy dispersive X-ray spectroscopy detector (EDX Oxford Instruments) by elemental mapping on the cross section of each precursor. Afterwards the precursors were selenized in the presence of Se and SnSe as described by Redinger et al.[7] and incorporated into the complete device of the following structure: SLG/Mo/CZTSe/CdS/ZnO/ZnO:Al/Ni:Al. To determine the properties of solar cells current-voltage (IV) characteristics were measured in the set-up equipped with calibrated halogen lamp.

RESULTS and DISCUSSION

The results are divided into two sections. In the first part we will show the evidence that pre-alloying will stabilize Cu-Sn and Cu-Zn alloys in different arrangements. In the second part we will show the uniformity of the absorbers and the performance of the solar cells built from different precursors.

Precursor characterisation

To enhance the mobility of metals between the layers in the Cu1/Sn/Cu2/Zn stack the ratio between two Cu layers Cu1:Cu2 was differentiated from 1:3 – 1:1 – 3:1 to 1:0. To ensure that each precursor contained the same amount of Cu, Sn and Zn, the deposition of each metal was controlled by the charge passed. Although the stoichiometry of the precursors was the same, the different stacking affected the microstructure of the soft-annealed precursors. SEM micrograph cross sections of the precursor show a bi-layered structure and a columnar structure for pre-alloyed precursors with Cu1:Cu2 of 1:1 and 1:0 respectively, figure 2. The columnar structure is similar to the precursor reported by Ahmed et al. [3]. The distribution of the individual elements was mapped using an EDX detector. For the Cu1:Cu2 ratio of 1:1 a bi-layered structure of Cu_6Sn_5 and Cu_5Zn_8 was observed as the tin and zinc do not need to diffuse far to form their respective Cu alloy stable phases. For any other configuration of metal stack the diffusion of tin and zinc was enhanced and laterally uniform distribution of phases through the films was observed.

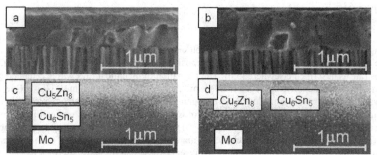

Figure 2. SEM cross section images of pre-alloyed precursor showing bi-layered and columnar structure (a, b) and EDX map of elements, overlaid on the same cross section (c, d). (must be viewed in colour)

The influence of absorber precursor morphology on the final solar cell performance

Figure 3. shows a comparison of the current-voltage JV characteristic of solar cells made from annealed absorber layers starting from the two types of precursors – one with bi-layered structure (black lines) and second with columnar structure (red lines). The parameters of solar cells are listed in the table 1. Surprisingly the bi-layered precursor structure led to a very poor quality device while the columnar precursor structure allowed obtaining a device with 4.5% efficiency, 287mV open circuit voltage, 41% fill factor and 36mA/cm^2 short circuit current.

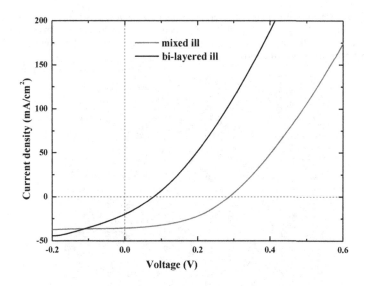

Figure 3. Current-voltage JV curves of solar cells prepared from mixed and bi-layered precursors.

property	bi-layer	Mixed
η (%)	0.4	4.5
V_{OC} (mV)	79	287
FF (%)	28	41
J_{SC} (mAcm^{-2})	-19	-36
A (cm^2)	0.55	0.55

Table I. Parameters of solar cells made from bi-layered and mixed precursors.

To investigate why the bi-layered stack gave significantly worse performance than the mixed precursor, SEM images of both precursors and absorber layers were recorded. Figure 4 shows the top view of the precursors and absorber layers. It was found that the bi-layered structure has round features on the surface with a diameter of 1 to 10 microns on the precursor and 5 to 15 microns on the selenized absorber, as shown on the Fig. 4a and 4c respectively. The cavities and lumps covered around 15% of the surface of the absorber. EDX revealed higher Mo signal on the features in the precursor and no significant changes in the composition of the final absorber. We deduce that the round features are cavities with or without cracks. The cavities reduce charge collection due to the reduced contact between absorber and the back contact. The cracks allow for electrical shorting. These two reasons explain why all devices parameters are worse for the absorber made from the bi-layered precursor. It is not clear whether the cavities are formed due to stress relief or if residual gases from electrodeposition of different layers are released during heating the sample. Absorbers based on the columnar precursors were uniform, without any traces of lumps and cavities.

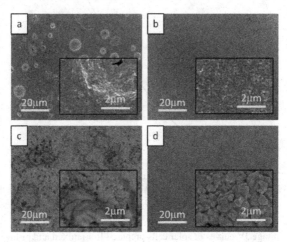

Figure 4. SEM micrograph of the precursors (a and b) and absorbers (c and d) made from bi-layered (left) and mixed (right) precursors respectively. The inset shows 10x higher magnification of the corresponding image.

CONCLUSIONS

We reported on the possibility of manipulating the microstructure of a Cu-Sn-Zn metallic precursors and the further impact on the quality of kesterite thin films and solar cells. Based on the formation of stable alloys between Cu-Sn and Cu-Zn two and three dimensional structures of precursors were grown during soft thermal treatment of the stacks. It was found that the mixed structure of precursor allows the formation of uniform absorbers. Most likely, harmful for the device, cavities and cracks were observed on the bi-layered precursors as well as on the homologous absorbers. Using suitable precursors the efficiency of solar cell was improved from 0.44% for the bi-layered precursor to 4.5% for the mixed precursor.

ACKNOWLEDGMENTS

M.A. and P.D. thank the Fonds National de la Recherché du Luxembourg (FNR) for a funding through the PECOS Attract project. The authors would like to acknowledge the Helmholtz Center in Berlin (HZB) for solar cell finishing.

REFERENCES

1. H. Araki, A. Mikaduki, Y. Kubo, T. Sato, K. Jimbo, W. S. Maw, H. Katagiri, M. Yamazaki, K. Oishi, A. Takeuchi, *Thin Solid Films*, **517**, 1457-1460 (2008).
2. H. Yoo, J. Kim, *Thin Solid Films*, **518**, 6567-6572 (2010).
3. S. Ahmed, K. B. Reuter, O. Gunawan, L. Guo, L. T. Romankiw, H. Deligianni, *Advanced Energy Materials*, **2**, 253-259 (2012).
4. C.-y. Chou, S.-w. Chen, *Acta Materialia*, **54**, 2393-2400 (2006).
5. J. J. Scragg, D. M. Berg, P. J. Dale, *J. Electroanal. Chem.*, **646**, 52-59 (2010).
6. R. Djemour, M. Mousel, A. Redinger, L. Gutay, A. Crossay, D. Colombara, P. Dale, S. Siebentritt, *Applied Physics Letters*, **102**, 222108 (2013)
7. A. Redinger, D. Berg, P. Dale, S. Siebentritt, *Journal of Am. Chem Soc.*, **133**, 3320 (2011).

Thin Film Solar Cells

Mater. Res. Soc. Symp. Proc. Vol. 1538 © 2013 Materials Research Society
DOI: 10.1557/opl.2013.979

Electroluminescence of Cu(In,Ga)Se$_2$ solar cells and modules

U. Rau[1], T. C. M. Müller[1], T. M. H. Tran[1], B. E. Pieters[1], and A. Gerber[1]
[1]IEK5-Photovoltaik, Forschungszentrum Jülich, 52425 Jülich, Germany

ABSTRACT

Fundamental aspects of (electro-)luminescence of Cu(In,Ga)Se$_2$ solar cells and modules are investigated by means of spectrally and spatially resolved measurements. The validity of the reciprocity relation between spectrally resolved electroluminescence emission and photovoltaic quantum efficiency is verified for the case of industrially produced ZnO/CdS/Cu(In,Ga)Se$_2$ heterojunction solar cells. Further we find that photo- and electroluminescent emission in these devices obey a superposition principle only in a limited range of the applied electrical or illumination bias. This range depends on the light soaking history of the sample and extends up to an injected current density of approximately 15 mAcm^{-2} after 3 h of light soaking at a temperature of 400 K. In the state prior to light soaking this range is limited to 4 mAcm^{-2}. At higher bias, a characteristic discrepancy between electroluminescence and electro-modulated photoluminescence appears. We attribute this anomaly to a potential barrier behavior close to the CdS/ Cu(In,Ga)Se$_2$ interface. Metastable defect reactions induced by holes injected into the space charge region partly reduce this barrier. We further find that the luminescence efficiency is enhanced by a factor of 3 by light soaking at 400 K. Spatially resolved electroluminescence measurements conducted during application of voltage or current bias at ambient temperature in the dark are qualitatively compatible with the conclusions drawn from the spectrally resolved measurements.

INTRODUCTION

Electroluminescence (EL) is the complementary physical action to the normal operating mode of a solar cell or module. Therefore, EL imaging [1] is an attractive tool for the characterization of such devices. As a direct semiconductor, Cu(In,Ga)Se$_2$ (CIGS) is especially suitable for this method and EL imaging was used for the analysis of CIGS modules in the past [2-5]. In recent years, EL was widely used to gain quantitative information, e.g., on resistive losses within a solar module or on the fundamental properties of light absorption and emission in a photovoltaic material. Thus, the method is useful in a wide range of materials and length scales.

However, the basic principles and the physical preconditions justifying a simple quantitative interpretation of EL measurements are usually not scrutinized. The present paper starts with a rigorous experimental analysis of basic physical constraints on the luminescence emission that have to be met in order to allow for a straightforward interpretation of spectrally and/or spatially resolved EL measurements. We find that CIGS solar cells only half-way meets these requirements and that the conditions under which a simple quantitative analysis is possible changes with the light soaking history of the sample. Therefore, we use luminescence measurements as a tool for a meaningful analysis of well known but not sufficiently explained metastabilities in CIGS. We find that the present spatially and spectrally resolved measurements are qualitatively compatible with recent theoretical results on the defect physics of CIGS.

THEORETICAL BACKGROUND

The reciprocity relation [6,7] between electroluminescent emission and external photovoltaic quantum efficiency $Q_e(E)$ applies for these cells before and after light soaking. The reciprocity relation reads The reciprocity relation reads

$$\phi_{em}(E) = \phi_{SC}(E,\phi_{exc}) + \phi_{EL}(E,V_j) = \phi_{SC}(E,\phi_{exc}) + Q_e(E)\phi_{bb}(E)\left[\exp\left(\frac{qV_j}{kT}\right) - 1\right], \quad (1)$$

where $\phi_{bb}(E)$ denotes the spectral photon flux density of a black body , V_j the voltage applied to the junction, and kT/q the thermal voltage. In Eq. (1), the emitted photon flux density $\phi_{em}(E)$, as a function of photon energy E, is a superposition of the pure EL emission ϕ_{EL} stimulated by the junction voltage V_j and the short circuit (SC) emission ϕ_{SC} caused by the photoexcitation ϕ_{exc}. Equation (1) describes not only (i) a quantitative relation between ϕ_{em} and $Q_e(E)$, but also states (ii) that that voltage driven (EL) and SC emission superimpose linearly, (iii) that the spectral shape of this emission is unaltered at different bias conditions, and (iv) that the EL emission ϕ_{em} follows a diode law with a diode ideality factor of unity.

The validity of implications (i) to (iv) is especially important for spatially resolved EL measurements where in most cases the camera signal S_{cam} is interpreted using the proportionality [,9].

$$S_{cam} \propto \exp\left(\frac{qV_j}{kT}\right). \quad (2)$$

However, because of the spectrally dependent quantum efficiency $Q_{cam}(E)$ of the camera we have also

$$S_{cam} = \int Q_{cam}(E)\phi_{EL}(E)dE. \quad (3)$$

Therefore, the validity of Eq. (1), with all four implications, is a strict precondition for the usage of Eq. (2). All these implications are experimentally verifiable but, up to present, experimental investigations concentrated on the verification of (i) and (iii) for the cases of wafer based Si solar cells [10,11], Cu(In,Ga)Se$_2$ (CIGS) thin-film cells [2,12], organic solar cells [13], and GaInP/InGaAs/Ge multijunction solar cells [14,15].

At this point it is important to note that the reciprocity relation is derived from the principle of detailed balance, more precisely, from an extrapolation of all equilibrium rate constants towards a non-equilibrium situation. Thus, Eq. (1) connects the result of a small-signal analysis, namely the quantum efficiency, derived relatively close to thermal equilibrium, with the electroluminescent emission of the same device, measured considerably far from equilibrium. Moreover, the collecting/injecting junction enters in the derivation [6,7] of Eq. (1) as a mere boundary condition. This is why for pin type solar cells Eq. (1) is strictly valid only in the limit of high carrier mobilities: The quasi-Fermi levels through the space charge region (SCR) must be flat under any bias voltage as well as a carrier collection efficiency within the SCR must be unity [16]. In addition, non-linear occupation terms dominating radiative recombination via tail states are not necessarily compatible with reciprocity [17] such that the concept of non-unity ideality factor n_{rad} for radiative recombination must be introduced [18]. We then have instead of Eq. (2)

$$S_{cam} \propto \exp\left(\frac{qV_j}{n_{rad}kT}\right). \qquad (4)$$

In a typical CIGS solar cell the SCR is about one third of the absorber thickness [19] putting the electrostatic properties of the device somewhere in between a pure pin-type solar cell and a pn-type cell with a negligible width of the SCR like in wafer based crystalline silicon solar cells. It is also known that tail-like states play a role in non-radiative recombination of these devices [20,21]. For all these reasons, a careful investigation of the validity of Eq. (1) for CIGS based solar cells in all its aspects is necessary to ensure an appropriate evaluation of luminescence measurements of these devices.

METASTABILITIES IN CIGS

CIGS solar cells and modules are subject to metastable changes of their electronic properties upon voltage and/or light bias [22-26]. These metastable changes are an intrinsic property of the CIGS absorber material and are due to the (V_{Cu}, V_{Se}) divacancy complex [27] and the In_{Cu} antisite defect [28]. Since both defects exist in multiple charge states their influence on the electronic behavior of CIGS solar cells is rather complex, e.g., already in equilibrium the divacancy complex has three different charge configurations, namely$(V_{Cu}, V_{Se})^{+}$, $(V_{Cu}, V_{Se})^{-}$, and $(V_{Cu}, V_{Se})^{3-}$, dependent on the position in the band diagram of the ZnO/CdS/CIGS heterostructure [27,29,30] (Fig.1a).

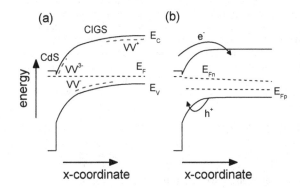

Figure 1. (a) Sketch of the equilibrium band diagram of the CdS/CIGS heterojunction involving the three charge states $(V_{Cu}, V_{Se})^{+}$, $(V_{Cu}, V_{Se})^{-}$, and $(V_{Cu}, V_{Se})^{3-}$ of the divacancy. (b) Under light or current bias excess holes diminish the negative charge close to the CdS/CIGS interface and excess electrons increase the effective doping in the bulk of the material.

Injection of excess minority charge carriers via illumination or electrical bias, i.e. electrons into the neutral bulk, as well as holes into the inverted CdS/CIGS interface region will have a manifold of consequences on the charge distribution and on the electrical properties of the device (Fig.1b). The reaction $(V_{Cu}, V_{Se})^{+} + e^{-} \rightarrow (V_{Cu}, V_{Se})^{-} + h^{+}$ will preferably take place in the neutral bulk and at the edge of the SCR leading to an increase of net doping density thereby reducing the series resistance, as well as the width of the SCR. The latter effect will reduce the overall

135

recombination and lead to what is known as the relaxation (increase) of the open circuit voltage V_{OC} [23-25]. In contrast, the reaction $(V_{Cu}, V_{Se})^{3-} + 2h^{+} \rightarrow (V_{Cu}, V_{Se})^{-}$ taking place closer to the interface region will decrease the net doping and widen the SCR. Furthermore, the capture of holes into the $(V_{Cu}, V_{Se})^{3-}$ state will further diminish the negative charge density in the close to interface region. As a consequence, the collection/injection barrier [31] for electrons at the CdS/CIGS interface will be decreased. Eventually, injection of minority carriers as sketched in Fig. 1b will at least result into two quite different effects: (i) the reduction of the recombination current and (ii) the reduction of the series resistance. Notably, these two effects correspond to the commonly observed features during light soaking of CIGS modules or solar cells: the increase of open circuit voltage and the increase of fill factor. Both effects are also expected to influence the EL emission of a CIGS solar cell or module.

EXPERIMENTAL

The solar modules under investigation (30×30 cm^2, 64 cells of width $w = 4$ mm and length $l = 30$ cm) are produced industrially by an in-line co-evaporation process on a Mo-covered glass substrate, finished by a chemical bath deposition of the CdS layer and by sputtering of the transparent ZnO window layer [32]. The solar cells (4×9 mm^2) used for the spectral investigations are cut from these modules. These samples were mounted into a cryostat and the spectral emission was recorded using a Fourier Transform Infrared Spectrometer equipped with a calibrated liquid nitrogen cooled Germanium detector. The spectral EL measurements were obtained by periodic application of an AC voltage to the solar cell and the use of a lock-in amplifier. For the photoluminescence measurements we used a widened laser illumination and an electro-modulation technique (EM) switching the applied junction voltage between $V_j = 0$ V and the illumination corresponding open circuit voltage $V_j = V_{oc}$. Light soaking was performed at an elevated temperature $T_{LS} = 400$ K for 3 h with illumination intensity of approximately 1 sun to ensure that we arrive at a saturated state. For the spatially resolved EL partly reported [33] the modules were kept in the dark overnight at room temperature prior to the transient measurements. EL pictures were taken during applied current or voltage bias every 5 seconds at ambient temperature using a Si-CCD camera. The recording started immediately after the application of constant voltage or current. Due to the relatively low exposure time no saturated state was obtained by this procedure.

SPECTRALLY RESOLVED LUMINESCENCE ANALYSIS

Figure 2 shows the measured external quantum efficiency $Q_e(E)$ together with the EL emission $\phi_{EL}(E)$ of the same device. Also shown are the data for $Q_e(E)$ and $\phi_{EL}(E)$ recalculated from the respective other measurement with the help of Eq. (1). We see from the data that the predictive power of Eq. (1) is well represented in the relevant overlap region of photon energies 1.15 eV $\leq E \leq$ 1.30 eV, especially by the fact that the $Q_e(E)$ spectrum accurately predicts the maximum of the EL emission. Thus, the reciprocity between Q_e and ϕ_{EL} [implication (i)] is valid not only for high-efficiency CIGS cells from the laboratory [10,12] but also for cells made from industrially produced modules.

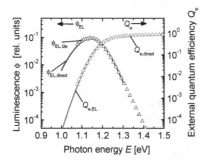

Figure 2. Measured EL spectrum $\phi_{EL,direct}(E)$ obtained from a CIGS solar cell and the external quantum efficiency $Q_{e,EL}(E)$ calculated from $\phi_{EL,direct}(E)$ with the help of Eq. (1) (solid line). Open triangles show the directly measured $Q_{e,direct}(E)$ and the EL spectrum $\phi_{EL,Qe}(E)$ calculated from the experimental $Q_{e,direct}(E)$.

Next we test prediction (ii) and (iii), namely the fact that the spectral shape of the EL emission is unaltered under different bias conditions and corresponds to the spectra obtained by the EM method. Figure 3 compares spectra obtained from EL measurements (Fig. 3a before light soaking, 3b after light soaking) obtained at different applied voltages and current densities with the EM spectra obtained at an illumination level leading to short circuit current densities J_{SC} equaling the current densities from the EL measurements within an error of 4 %. It is seen from Fig. 3 that the spectral shape of the emission is independent from the bias condition and is unaltered between EL and EM. Furthermore, we do not observe changes of the emission spectra by light soaking.

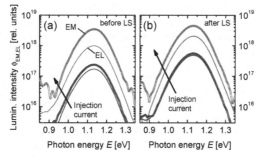

Figure 3. Electroluminescence (EL, solid lines) and electro-modulated photoluminescence (EM, open circles) spectra from a CIGS solar cell under different current injection conditions at temperature $T = 300$ K. (a) Spectra taken before light soaking ($J = 35.85, 7.31$ mA/cm^2, EL, and $J_{SC} = J$, EM), (b) spectra after light soaking for 3 h at an elevated temperature $T_{LS} = 400$ K ($J = 26.01, 6.82$ mA/cm^2, EL, and $J_{SC} = J$, EM).

It is however evident that the intensity of the EM emission is larger than that of the EL emission taken at a current density equaling the short circuit current density of the EM measurement. To investigate this unexpected effect closer we have measured the omnispectral EL and EM for a larger series of injection densities using the Ge detector without monochromator. Because the spectral shape of the emission is unaltered under different bias conditions the detector signal

$$S_{det}(V) = \int Q_{det}(E)\phi_{EL}(E,V)dE \propto f(V) \tag{5}$$

is a function of the external voltage or, likewise of the incident irradiation intensity, regardless of the quantum efficiency Q_{det} of the detector.

Figure 4 shows the integral EL and EM intensities before and after light soaking respective to the applied voltage V (EL) or the measured open circuit voltage V_{OC}. Due to the series resistance the EL data bend towards higher voltages whereas the EM data approximately represent straight lines. A fit to both, the EM data before and after light soaking yields an ideality factor n_{rad} of $n_{rad} \approx 0.96$ which is close but not identical to unity. Moreover, for voltages above 0.6 V both curves increasingly deviate from the straight line.

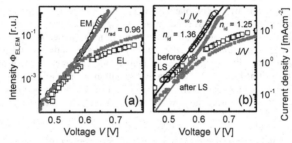

Figure 4: (a) Voltage dependence of the omni-spectral EL $\Phi_{EL}(V)$ before/after light soaking (LS) (open/full squares) and dependence of the EM signal $\Phi_{EM}(V_{oc})$ on the open circuit voltage V_{OC} (open filled circles) obtained from the same CIGS solar cell as in Figs. 2 and 3. The (radiative) ideality factor $n_{rad} = 0.96$ results from the fit (red solid line) to $\Phi_{EM}(V_{oc})$ after and before LS. (b) Short circuit current density vs. open circuit voltage (J_{SC}/V_{OC}) measured simultaneously to the EM measurements and current density vs voltage (J/V) measurement carried out simultaneously to the EL measurements before (open symbols) and after LS (full symbols). The (non-radiative) ideality $n_{id} = 1.36$ and 1.25 (before/after LS) is obtained from the J_{SC}/V_{OC} curve and corresponds to the classical diode ideality factor.

Figure 4b represents the classical dark current voltage J/V curves for both states where the data are taken simultaneously to the EL measurements, as well as short circuit current density vs. open circuit voltage (J_{SC}/V_{OC}) curves measured simultaneously to the EM spectra. Here again the dark J/V curves are entirely dominated by the series resistance. Whereas the J_{SC}/V_{OC} curves allow for an approximate determination of the classical (non-radiative) ideality factor as $n_{id} = 1.36$ and 1.25 (before and after LS, respectively).

Figure 5a plots the luminescent intensities vs. the measured current densities, eliminating in this way the voltage as a parameter and also the influence of the series resistance. We see that the data in a large part of the double logarithmic plot fall on a straight line with a slope Γ that corresponds to the ratio of non-radiative and radiative ideality factor according to

$$\Gamma = \frac{d\Phi_{EM}}{dV_{OC}}\frac{1}{\Phi_{EM}} \bigg/ \left(\frac{dJ}{dV}\frac{1}{J}\right) = \frac{n_{id}}{n_{rad}}.$$ (6)

However, for current densities $J > 4$ mAcm^{-2} and $J > 15$ mAcm^{-2} (before and after light soaking) the experimental data significantly deviate from the straight line and EL/EM data deviate from each other. This significant anomaly is partly healed by light soaking such that after the procedure the device is well behaved in a significant range nearly up to the one sun equivalent of short circuit current density. Figure 5b depicts the external luminescence efficiencies Q_{LED} and Q_{lum}, i.e., the ratios between radiative and non-radiative recombination [6] that are obtained via

$$Q_{LED/lum} \propto \Phi_{EL/EM} / J_{./EM}.$$ (7)

In Fig. 5b it becomes clear that light soaking improves the luminescence efficiencies by about a factor of three. This finding fits into the classical explanation that the effect of persistent conductivity reduces the space charge region and herewith the amount of non-radiative recombination. This effect is already seen by the shift of the J_{SC}/V_{OC} and the J/V curves towards higher voltages in Fig. 4b. At the same time the $\Phi_{EM}(V_{\infty})$ and the $\Phi_{EL}(V)$ curves in Fig. 4a are much less affected by light soaking, except for a reduction in series resistance seen in $\Phi_{EL}(V)$.

The mentioned anomaly is also clearly seen in Fig. 5b where the EL efficiency Q_{LED} appears to saturate at current densities $J > 10$ mAcm^{-2}. Simultaneously, the EM luminescence efficiency Q_{lum} increases. In other words, carrier injection by illumination is increasingly decoupled from carrier collection and non-radiative recombination. This behavior is clearly in conflict with the superposition principle as expressed by Eq. (1).

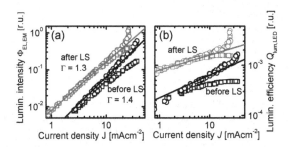

Figure 5. (a) Dependence of the omni-spectral EL Φ_{EL} before/after light soaking (LS) (open/full squares) on the current density J and dependence of the EM signal Φ_{EM} (open/full circles) on the short circuit current density J_{SC}. The slopes Γ of the curves follow from the ratio between non-radiative and radiative ideality factors from Fig. 4 according to Eq. (6). (b) External luminescence quantum efficiencies Q_{LED} calculated from Φ_{EL} and Q_{lum} calculated from Φ_{EM} with the help of Eq. (7). The straight lines are directly calculated from the two slopes Γ in (a) and correspond to $\Gamma-1$, respectively.

SPATIALLY RESOLVED ELECTROLUMINESCENCE ANALYSIS

Figure 6 compares two pictures taken from the central part of a CIGS module at a constant voltage of V_{ext} = 50 V, corresponding to a mean voltage of V_{ce} = 0.78 V per cell, at the beginning (Fig. 6a) and at the end (Fig. 6b) (after 195 s) of the voltage bias soaking experiment. Both pictures show the resistive effect of the ZnO sheet resistance, i.e., the characteristic decay of EL intensity from the right to the left of the cell [2]. However, the comparison also shows that the overall EL intensity has increased during the voltage bias soaking of the sample that has kept in the dark before the experiment for 10 h.

Further, we observe that during the bias soaking experiment at a constant voltage the current driven through the device continuously increases as shown in Fig. 7a as an increase of current density J per single cell area from 23 mAcm^{-2} to 32 mAcm^{-2}. In an analogous experiment the external current was fixed at I = 400 mA, corresponding to a current density J = 34.8 mAcm^{-2}. The second curve in Fig. 7a unveils a decrease of the cell voltage V_{ce} from approximately 850 mV to 790 mV/cell

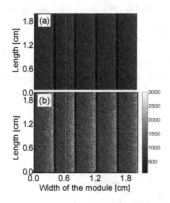

Figure 6. EL image showing a detail of the emission of the mini-module after 5 s (a) and after 195 s (b) of application of a constant voltage bias V_{ext} = 50 V (V_{ce} = 0.78 V). The initial state was prepared by keeping the module in the dark for more than 12 h at ambient temperature.

From the average intensity of EL images taken every 5 s during bias soaking, we derive the time evolution of the average junction voltage V_j with the help of Eq. 2 (Fig. 7b). Note that the extracted values have an unknown offset which is however the same for all images. Note further that the strict applicability of Eq. (2) is strongly called in question by our spectrally resolved measurements discussed above. However, when considering only relatively small voltage differences the error introduced by assuming the strict validity of Eq. (2) remains relatively small. Accepting this uncertainty, we find that in both experiments the voltage V_j increases though much less pronounced (by about 15 mV) in the constant current case. However, the small increase of approximately 3.5 V/cell observed in the junction voltage is interesting because at the same time the overall voltage applied to each cell diminishes by about 60 mV (Fig. 7a). The only way to interpret this finding is that the series resistance of each cell diminishes dramatically during bias soaking. At the same time the voltage drop over the junction increases slightly, i.e., the resistance of the junction increases. Along the same line we interpret the experiment under constant voltage. The overall current increases because the decrease of the series resistance overcompensates the increase of junction resistance. However, because of the constant voltage conditions the junction voltage is allowed to increase more pronouncedly compared to the constant current condition.

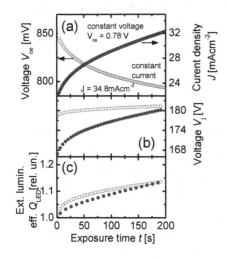

Figure 7. (a) Transients of the average cell voltage V_{ce} (open circles) for an electrical bias soaking experiment under constant current bias ($I = 400$ mA, $J = 34.8$ mAcm^{-2}) and of the current density J (per cell area) for an experiment conducted under constant voltage bias ($V_{ext} = 50$ V, $V_{ce} = 0.78$ V). (b) Junction voltages V_j for the two cases calculated from a series of EL images with the help of Eq. (2). (c) External luminescence efficiencies calculated from the ratio Φ_{EL} / J according to Eq. (7).

We may look at the external luminescence (LED) quantum efficiency Q_{LED} as defined in Eq. (7) to clarify whether or not the observations in both cases are quantitatively similar. Figure 7c shows Q_{LED} normalized to the respective values at the beginning of the experiment to demonstrate that in both cases the increase is about 15 %. Thus, under both conditions we finally detect similar physical changes induced by the electrical bias.

DISCUSSION

Though the spectrally and spatially resolved experiments use quite different methods of injection soaking, the observations fit into a common picture: Long (3 h) light soaking at elevated temperatures $T_{LS} = 400$ K (up to saturation) for the spectrally resolved measurements results in a voltage shift of about 30 mV (Fig. 4b) and an increase of the luminescence efficiencies of up to 300 %. Short time (200 s) voltage or current bias soaking for the spatially resolved measurements leads to a shift of the junction voltage of 13 mV under constant voltage and of 3.5 mV under constant current injection (Fig. 7b). Both experiments result in an increase of the luminescence efficiency by about 15 % (Fig. 7c). Up to this point, all observations are in agreement with the simple picture that injection of electrons by light or electrical bias leads to persistent conductivity. The persistently captured electrons (cf. Fig. 1b) lead to excess negative charges reducing the width of the space charge region and herewith the amount of non-radiative recombination [23-25].

However, the spatially resolved measurements unveiled that the major consequence of light or electrical bias soaking is not the reduction of recombination but the reduction of series resistance. This effect again is expected from persistent conductivity diminishing the series resistance of the CIGS bulk. Yet the detected amount of change is hardly compatible with a mere effect of the CIGS bulk. Rather we assume that an additional series resistance is due to the interfacial barrier that is diminished by injection of holes into the region close to the CIGS/CdS interface [31]. To make things more complicated, the latter effect is likely to be non-linear, i.e. not simply expressed by an additional Ohmic resistance.

The difference of radiative and non-radiative ideality factors found in the spectral luminescence measurements tells us that radiative and non-radiative recombination paths are different. Radiative recombination in CIGS at room temperature is a band-to-band like mechanism involving at most relatively steep band-tails or relatively small spatial band edge fluctuations [12]. This fact is proven by the invariance of the emission spectra under different bias conditions. Also the close-to-unity ideality factor n_{rad} points into that direction. It is also save to assume that the luminescence emission monitors the entire bulk of the CIGS. In contrast, the larger ideality factor n_{id} of the dominating non-radiative recombination path points to the space charge region as the dominant location of recombination involving also deeper states or tail states with large Urbach energy [20]. The difference of the two ideality factors is responsible for the fact that the luminescence efficiencies gradually increase with increasing electrical or light bias. Up to this point, the observations are not surprising: Radiative and non-radiative recombination using different electronic states is a common feature of semiconductor materials.

More surprising is the anomaly detected in the spectral measurements, namely the divergence of EL and EM under high bias conditions (Fig. 5a and b). We assume that under light injection (EM) the split of the quasi-Fermi levels in the bulk of the material is far larger than the open circuit voltage V_{OC} detected at the terminals. This would explain why with increasing illumination intensity V_{OC} does not increase as expected whereas the emitted luminescence does. An electrostatic barrier between the main part of the CIGS bulk and the main region of non-radiative recombination would explain such an effect. In turn, the same barrier would hinder injection of electrons from the junction into the bulk, thereby limiting the EL emission up to the observed saturation of Q_{LED}. We feel that the defect model of Lany and Zunger [27] as shown in Fig.1 eventually could explain our observations. However, more quantitative insight should come from numerical device simulations.

CONCLUSIONS

In summary, spectrally and spatially resolved EL and EM measurements unveil a consistent though complex picture of the metastable behavior of CIGS based devices. The observed divergence of EL and EM under high bias conditions and its dependence on the light soaking history could be a key for the understanding of metastable defects in CIGS and their consequences for the device performance. Thus, luminescence analysis turns out as a powerful tool to analyze the metastable device behavior of CIGS in some detail. Both types of experiments discussed in the present paper, the spatially as well as the spectrally resolved measurements, detect specific key features of the metastable behavior. Metastabilities in CIGS decisively depend on the details of absorber and interface preparation and luminescence methods provide a sensitive way to compare the specific differences.

ACKNOWLEDGMENTS

The present work has been supported by the German Ministry of Environment under contract FKZ0325149B ("PV-IR-EL"). The authors thank R. Schäffler and J. P. Theisen (Manz AG) for providing the CIGS modules as well as C. Grates and M. Schneemann (FZ Jülich) for help with the experiments.

REFERENCES

1. T. Fuyuki, H. Kondo, T. Yamazaki, Y. Takahashi, and Y. Uraoka, *Appl. Phys. Lett.* **86**, 262108 (2005).
2. A. Helbig, T. Kirchartz, R. Schaeffler, J. H. Werner, and U. Rau, *Sol. Ener. Mater. & Sol. Cells* **94**, 979 (2010).
3. G. Brown, A. Pudov, B. Cardozo, V. Faifer, E. Bykov, and M. Contreras, *J. Appl. Phys.* **108**, 074516 (2010).
4. S. Johnston, T. Unold, I. Repins, R. Sundaramoorthy, K. M. Jones, B. To, N. Call, and R. Ahrenkiel, *J. Vac. Sci. & Techn.* A **28**, 665 (2010).
5. M. Paire, L. Lombez, J.-F. Guillemoles, and D. Lincot, *Thin Solid Films* **519**, 7493 (2011).
6. U. Rau, *Phys. Rev. B* **76**, 085303 (2007).
7. U. Rau, *IEEE J. Photov.* **2**, 1697 (2012).
8. K. Ramspeck, K. Bothe, D. Hinken, B. Fischer, J. Schmidt, and R. Brendel, *Appl. Phys. Lett.* **90**, 153502 (2007).
9. P. Würfel, T. Trupke, T. Puzzer, E. Schäffer, W. Warta, and S. W. Glunz, *J. Appl. Phys.* **101**, 123110 (2007).
10. T. Kirchartz, U. Rau, M. Kurth. J. Mattheis, and J. H. Werner, *Thin Solid Films* **515**, 6238 (2007).
11. T. Kirchartz, A. Helbig, W. Reetz, M. Reuter, J. H. Werner, and U. Rau, *Progr. Photov.: Res. Appl.* **17**, 394 (2009).
12. T. Kirchartz and U. Rau, *J. Appl. Phys.* **102**, 104510 (2007).
13. K. Vandeval, K. Tsvingsted, A. Gadisa, O. Inganas, and J. V. Manca, Nature Materials **8**, 904 (2009).
14. T. Kirchartz, U. Rau, M. Hermle, A. W. Bett, A. Helbig, and J. H. Werner, *Appl. Phys. Lett.* **92**, 123502 (2008).
15. S. Roensch, R. Hoheisel, F. Dimroth, and A. W. Bett, *Appl. Phys. Lett.* **98**, 251113 (2011).
16. T. Kirchartz and U. Rau, *Physica Status Solidi* A **205**, 2737 (2008).
17. B. E. Pieters, T. Merdzhanova, T. Kirchartz, and R. Carius, *Sol. Ener. Mater. Sol. Cells* **94**, 1851 (2010).
18. T. C. M. Müller, B. E. Pieters, T. Kirchartz, R. Carius, U. Rau, *Phys. Stat. Sol.* C **9**, 1963 (2012).
19. U. Rau and H. W. Schock, *Appl. Phys.* A **69**, 131 (1999).
20. T. Walter, R. Herberholz, and H. W. Schock, *Solid State Phenomena* **51-52**, 309 (1996).
21. U. Rau, A. Jasenek, H. W. Schock, F. Engelhardt, and T. Meyer, *Thin Solid Films* **361-362**, 298 (2000).

22. M. Igalson and H. W. Schock, *J. Appl. Phys.* **80**, 5765 (1996).
23. U. Rau, M. Schmitt, J. Parisi, W. Riedl, and F. Karg, *Appl. Phys. Lett.* **73**, 223 (1998).
24. P. Zabierowski, U. Rau, and M. Igalson, *Thin Solid Films* **387**, 147 (2001).
25. Th. Meyer, F. Engelhardt, J. Parisi, and U. Rau, *J. Appl. Phys.* **91**, 5093 (2002).
26. J. T. Heath, J. D. Cohen, and W. N. Shafarman, *J. Appl. Phys.* **95**, 1000 (2004).
27. S. Lany and A. Zunger, *J. Appl. Phys.* **100**, 113725 (2006).
28. S. Lany and A. Zunger, *Phys. Rev. Lett.* **100**, 016401 (2008).
29. R. Urbaniak and M. Igalson, *J. Appl. Phys.* **106**, 063720 (2009).
30. S. Siebentritt, M. Igalson, C. Person, and S. Lany, *Progr. Photov.: Res. Appl.* **18**, 390 (2010).
31. R. Kniese, M. Powalla, and U. Rau, *Thin Solid Films* **515**, 6163 (2007).
32. B. Dimmler, M. Powalla, and R. Schaeffler, in: *Proceedings of the 31st IEEE Photovoltaic Specialists Conference* (IEEE, New York, 2005) pp. 189–194.
33. T. M. H. Tran, B. E. Pieters, C. Ulbrich, A. Gerber, T. Kirchartz, and U. Rau, *Thin Solid Films* (2012, published online, http://dx.doi.org/10.1016/j.tsf.2012.10.039).

Mater. Res. Soc. Symp. Proc. Vol. 1538 © 2013 Materials Research Society
DOI: 10.1557/opl.2013.986

Impact of the deposition conditions of window layers on lowering the metastability effects in Cu(In,Ga)Se₂/CBD ZnS-based solar cell

N. Naghavi[1], T. Hildebrandt[1], G. Renou[1], S. Temgoua[1], JF. Guillemoles[1], D. Lincot[1]

[1] Institute of Research and Development on Photovoltaic Energy (IRDEP), 6 quai Watier, 78401 Chatou, France

ABSTRACT

The purpose of the present paper is to focus on the impact of oxygen gas partial pressure during the sputtering of i-ZnO and ZnMgO on the transient behavior of solar cells parameters when a CBD-ZnS buffer layer is used. Based on electrical characterization of cells, we have observed that the effect of light-soaking is different on J-V characteristics depending on the quantity of oxygen present during the first deposition time of the i-ZnO or ZnMgO layers. In fact, we have noticed that, when cells are prepared with standard i-ZnO, the efficiencies are very low and a pronounced transient behavior is observed. However, when the i-ZnO or ZnMgO is first formed by a few nanometers sputtered layer without any additive oxygen, depending on the thickness of this layer, the transient effects strongly decrease. It is then possible to reach efficiencies quite similar to the CdS reference cells, especially with ZnMgO, without any post-treatments.

INTRODUCTION

Currently the highest conversion efficiency CIGSe based thin film solar cells have been obtained using chemical bath deposited CdS buffer layers. However, during the last decade, an important emphasis has being placed on alternative Cd-free materials which could improve the collection of carriers generated by short-wavelength light [1]. Among the alternative buffer materials, zinc-based buffer layers, prepared by Chemical Bath Deposition (CBD), have already demonstrated their potential to lead to high efficiencies solar cells and modules with a 19.4% efficiency for cells [2] and 17.8% efficiency for 30x30 cm² submodules [3].
However, contrary to Cu(In,Ga)Se₂ (CIGSe) based solar cells with a CBD-CdS buffer layer, CBD-ZnS based solar cells require post-air-annealing and light-soaking procedures to achieve optimal conversion efficiencies. For industrial applications, it is better to eliminate these post-treatments [1]. Recent works have identified ZnMgO as a partner for CBD-ZnS buffer layers to highly improve the efficiency of solar cells and strongly reduce their transient behavior[4], [5]. However the role of this layer within the device structure is still not well understood [6]. The purpose of the present contribution is to focus on the impact of oxygen gas partial pressure during the sputtering of i-ZnO and ZnMgO on the transient behavior of solar cells parameters when a CBD-ZnS buffer layer is used. In this paper, based on electrical characterization of cells, we will show that the effect of light-soaking is different on J-V characteristics depending on the quantity of oxygen present during the first deposition time of the i-ZnO or ZnMgO layers, and that it is then possible to reach efficiencies quite similar to the CdS reference cells, especially with ZnMgO, without any post-treatments and just by optimizing the deposition parameters of the window layers.

EXPERIMENT

In order to understand the effects of the impact of oxygen gas partial pressure during the sputtering of i-ZnO and ZnMgO on the transient behavior of solar cells parameters, four sets of experiments have been separately conducted. For all samples, CIGSe layers were deposited by classical three stage co evaporation on glass/Mo substrates. All cells are completed by rf-ZnO:Al. The Zn(S,O,OH) layers (noted ZnS all along this paper for clarity) were deposited by the chemical bath deposition (CBD) technique in a similar way to the CdS buffer layer with an estimated thickness of about 20 to 25 nm [5]. The i-ZnO and $Zn_{1-x}Mg_xO$ layers were deposited by rf magnetron sputtering using a sintered ceramic ZnO or $Zn_{1-x}Mg_xO$ targets with x = 0.27 [4]. The Ar/O_2 concentration during the sputtering process was about 2% for i-ZnO and 0.5% for ZnMgO.

The two first series of experiments have been made using the same Mo/CIGSe/ZnS layers where the first 10 nm of the i-ZnO are deposited under Ar/O_2=2% (w O_2) or without any O_2 under pur Ar (w/o O_2). In the second set of experiments, two series of devices were prepared using the same Mo/CIGSe/ZnS layers where the first 10 nm of the ZnMgO are deposited under Ar/O_2=0.5% (w O_2) or under pur Ar (w/o O_2). For comparison, reference CIGS/CdS/i-ZnO/ZnO:Al devices with the commonly used CBD-CdS buffer were prepared too. The electrical properties of cells are characterized by J(V) curves at 25 °C under illumination (AM 1.5 global spectrum, 1000 W/m^2). Typical post treatments were light soaking of cells during one hour under AM 1.5 global spectrum, 1000 W/m^2.

RESULTS AND DISCUSSION

On figure 1, the effect of 1h light soaking on solar cells parameters of CIGSe cells with different window layers combination are compared to the one without any light soaking.

First the effect of light soaking on classical CdS/i-ZnO with or without 10 nm pure Ar layer was compared. These results are not presented here, but no significant differences are observed for these cells and the light-soaking does not have any real impact on the improvement of cells parameters. Figure 2. a. shows the J(V) curves of a cell with CIGS/ZnS/i-ZnO/ZnO:Al under Ar/O_2 (w O_2). These cells present strongly degraded performances without real improvement after 1h light soaking. On figure 2.b the JV curves of CIGS/ZnS/i-ZnO cells are presented where the first 10 nm of i-ZnO are deposited under pur Ar (w/o O_2) . These cells present efficiencies up to 11% and the exposure to a solar-type spectrum for 60 minutes doesn't lead to any change of the solar cell parameters (**Figure 1**).

When the i-ZnO is replaced by a $Zn_{0.74}Mg_{0.26}O$ layer deposited under a mixture of Ar/O_2(w O_2), as is observed on Figure 1 and 2.c., higher efficiencies than the one with i-ZnO can be obtained after 1h of light soaking. However, for these cells, a 1h light soaking is necessary to improve the FF and V_{oc} of cells and so to reach their high efficiencies. The best efficiencies for these cells remain quite lower than the CdS reference cells.

Finally as is observed on Figure 2.d., the best performances are obtained for cells with the combination of ZnS and $Zn_{0.74}Mg_{0.26}O$ buffer layers where the 10 nm of ZnMgO are deposited under pur Ar (w/o O_2). For these cells no light soaking is necessary to improve the solar cells parameters and high efficiencies are reached without any light soaking.

146

Finally a comparison of all solar cell parameters (figure 1) for different cells shows that the first 10 nm deposition of i-ZnO or ZnMgO under pur Ar (w/o O$_2$).leads to an improvement of all solar cell parameters: J$_{sc}$, V$_{oc}$ & FF. However the best J$_{sc}$, V$_{oc}$ and FF and so the highest efficiencies up to 15.7%, comparable to the one with CdS buffer layer, are reached for cells with ZnMgO. The excellent carrier collection observed is attributable to the wide band gap of the ZnS buffer layer compared to CdS buffer layer.

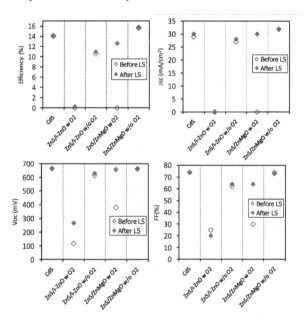

Figure 1. The effect of deposition of the 10 first nm of the i-ZnO or ZnMgO window layer under a mixture of Ar/O$_2$ (w O$_2$) and under pur Ar (w/o O$_2$) on solar cell parameters of CIGS cells before light soacking (LS) :◇ and after 1h of light soaking (LS) :◆.

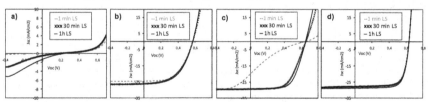

Figure 2. Current–voltage curves of a) CIGS/ZnS/i-ZnO (w O$_2$)/ZnO:Al b) CIGS/ZnS/i-ZnO (w/o O$_2$)/ZnO:Al c)CIGS/ZnS/ZnMgO (w/o O$_2$) /ZnO:Al d)CIGS/ZnS/ZnMgO (w/o O$_2$) /ZnO:Al after 1min, 30 min and 1h of light soacking (LS)

147

DISCUSSION

In order to have a better understanding of the effect of LS on our solar cells the first question to answer is which electrical phenomena at the interface between CIGSe/buffer/window layer can actually affect the device J–V characteristic in these solar cells. While it is certainly too early on the basis of the present evidences to completely prove a model, we can certainly try to point out what are the possibilities, if any, to explain the data. Because of the difference in buffer nature and bandgap (CdS : 2.4 vs ZnS : 3.2 eV), one of the natural cause for the observed difference in electrical behavior is the interface barrier for electrons caused by the low electron affinity of ZnS which can lead to a lowering of the FF in these solar cells. Another natural cause could also be the nature and concentration of interface defects at the CIGSe/buffer/windows interfaces that can affect the Voc of the solar cells. The effect of light soaking in this case can mainly due to a decrease of the defect density at this interface.

We have observed that the deposition of ZnMgO or i-ZnO under a mixture of Ar/O_2 compared to the one under pur Ar seems to be more energetic leading to a higher thickness for the same deposition time. As the thickness of our ZnS buffer layer is relatively thin (25 nm) a more energetic deposition (deposition under Ar/O2) can cause more plasma damages at the CIGSe/ZnS interface than the one deposited under pur Ar. As reported by [6], [7], interface recombination is enhanced by these damages, resulting in decreasing Voc and efficiency. The improvement of cell parameters (essentially the FF and Voc) after light-soaking can be attributed to the recovery of damages at the interface. In fact light soaking leads to the photogeneration of electrons which can either reduce defects at the interfaces and therefore improve the Voc of our solar cells, or lower the interface barriers leading to an improvement of FF.
Another explanation can be that the buffer interface properties can change throughout the i-ZnO or ZnMgO deposition process: defect of O_2 during the first nm windows layer deposition can have a doping effect on the Zn(O,S) buffer layer as illustrated here:
$Zn(O,S) \leftrightarrow Zn(O_{\delta-1},S) + \delta/2\ O_2 \nearrow$

The doping leads to an improvement of the buffer layer conductivity and so to the lowering of the barriers height of window/buffer/CIGS interface. This effect is similar to LS effects and can explain why when the first 10 nm of the ZnO or ZnMgO are deposited under pur Ar, no LS effects are observed.

CONCLUSIONS

In this paper, the importance of the buffer and window layer interface and their deposition process for achieving high efficiency CIGSe thin film solar cells have been shown. Efficiencies higher than CdS based solar cells with CBD ZnS can be obtained without any transient effect just by reducing the O_2 concentration in the deposition of the first few nm of ZnMgO or i-ZnO. We showed that it is possible to strongly reduce the transient effect in Cd-free CIGSe based solar cells just by chemical interface engineering with simply reconsidering the i-ZnO or ZnMgO growth parameters.

148

REFERENCES

[1] N. Naghavi, D. Abou-Ras, N. Allsop, N. Barreau, S. Bücheler, A. Ennaoui, C.-H. Fischer, C. Guillen, D. Hariskos, J. Herrero, R. Klenk, K. Kushiya, D. Lincot, R. Menner, T. Nakada, C. Platzer-Björkman, S. Spiering, A. N. Tiwari, et T. Törndahl, « Buffer layers and transparent conducting oxides for chalcopyrite $Cu(In,Ga)(S,Se)_2$ based thin film photovoltaics: present status and current developments », *Progress in Photovoltaics: Research and Applications*, vol. 18, n° 6, p. 411-433, sept. 2010.

[2] D. Hariskos, R. Menner, P. Jackson, S. Paetel, W. Witte, W. Wischmann, M. Powalla, L. Bürkert, T. Kolb, M. Oertel, B. Dimmler, et B. Fuchs, « New reaction kinetics for a high-rate chemical bath deposition of the $Zn(S,O)$ buffer layer for $Cu(In,Ga)Se2$-based solar cells », *Progress in Photovoltaics: Research and Applications*, vol. 20, n° 5, p. 534-542, août 2012.

[3] M. Nakamura, Y. Chiba, S. Kijima, K. Horiguchi, Y. Yanagisawa, Y. Sawai, K. Ishikawa, et H. Hakuma, « Achievement of 17.5% efficiency with 30x 30cm2-sized $Cu(In,Ga)(Se,S)_2$ submodules », in *2012 38th IEEE Photovoltaic Specialists Conference (PVSC)*, 2012, p. 001807-001810.

[4] D. Hariskos, B. Fuchs, R. Menner, N. Naghavi, C. Hubert, D. Lincot, et M. Powalla, « The $Zn(S,O,OH)/ZnMgO$ buffer in thin-film $Cu(In,Ga)(Se,S)_2$-based solar cells part II: Magnetron sputtering of the ZnMgO buffer layer for in-line co-evaporated $Cu(In,Ga)Se_2$ solar cells », *Progress in Photovoltaics: Research and Applications*, vol. 17, n° 7, p. 479-488, nov. 2009.

[5] C. Hubert, N. Naghavi, O. Roussel, A. Etcheberry, D. Hariskos, R. Menner, M. Powalla, O. Kerrec, et D. Lincot, « The $Zn(S,O,OH)/ZnMgO$ buffer in thin film $Cu(In,Ga)(S,Se)_2$-based solar cells part I: Fast chemical bath deposition of $Zn(S,O,OH)$ buffer layers for industrial application on Co-evaporated $Cu(In,Ga)Se_2$ and electrodeposited $CuIn(S,Se)_2$ solar cells », *Progress in Photovoltaics: Research and Applications*, vol. 17, n° 7, p. 470-478, nov. 2009.

[6] S. Shimakawa, Y. Hashimoto, S. Hayashi, T. Satoh, et T. Negami, « Annealing effects on $Zn1-xMgxO/CIGS$ interfaces characterized by ultraviolet light excited time-resolved photoluminescence », *Solar Energy Materials and Solar Cells*, vol. 92, n° 9, p. 1086-1090, sept. 2008.

[7] D. H. Shin, J. H. Kim, Y. M. Shin, K. H. Yoon, E. A. Al-Ammar, et B. T. Ahn, « Improvement of the cell performance in the $ZnS/Cu(In, Ga)Se_2$ solar cells by the sputter deposition of a bilayer $ZnO: Al$ film », *Progress in Photovoltaics: Research and Applications*, 2012.

149

Mater. Res. Soc. Symp. Proc. Vol. 1538 © 2013 Materials Research Society
DOI: 10.1557/opl.2013.683

The Research and Development of the Third Generation of Photovoltaic Modules

Tingkai Li,
Hunan Gongchuang Photovoltaic Science and Technology Co. Ltd., No.1 Hongyuan Road,
Hengyang, 421005, P.R.China

ABSTRACT

In order to make high efficiency and low cost solar cell modules, the concept of third generation of photovoltaic modules have been provided. The first generation solar cell: Crystal Si solar cell including single crystal and poly-crystal Si solar cell; The second generation solar cell: Thin film solar cell including Si base thin film, CIGS, CdTe and III-V thin films; The third generation solar cell is the future high efficiency and low cost solar cell modules, such as low cost quantum dots solar cell, Si base thin film tandem and triple cell modules, III-V solar cell on Si, HIT solar cell and nanotechnology with no vacuum technique such as printable technologies and etc. This paper reviewed the advantages and disadvantages of each generation of the solar cell modules and technologies and discussed the research and development of the third generation of photovoltaic modules including the detail technology developments.

INTRODUCTION

The scientists have realized that with exhaustion of fossil fuel energy, global warming, population growth, and etc., the world already face to energy challenge & revolution [1][2]. In order to solve this problem, many governments in the world have been developing the very important projects dealing with how to generate renewable energies and how to save and manipulating energies. Table 1 list the world possible demand in next 30 years and potential renewable energies. Based on the table, the potential use of hydroelectric and wind are only 0.9 terawatts (TW) and 2TW respectively; biomass is about $5 - 7$ TW and also need to use 20% of earth land mass. If the biomass used in energy applications may result in world food crisis. On the other hand, the total amount of above energy cannot meet the world energy demand of $30 - 60$ TW. Geothermal energy with high cost and is also not enough. Nuclear energy is dealing with radioactive waste issue, specially, recently the radioactive leaky accident in Japan warning people to consider the nuclear energy safety issues. Therefore, using solar energy may be only

Table 1: Potential renewable energies

Total Energy Demand	30 -60 TW	Hydroelectric	0.9 TW
Wind	2 TW	Biomass	$5 - 7$ TW*
Nuclear	Radioactive waste?	Geothermal	Cost? Not enough
Solar	$50 - 1500$ TW		

*using > 20% of earth land mass.

solution for future energy problem. Based on solar cell research and developments, the solar cell modules can be divided into three generations[3][4]. The first generation solar cell are crystal Si solar cells, include single crystal and poly-crystal Si solar cells. The second generation solar cell are thin film solar cells, include Si base thin film, CIGS, CdTe and III-V thin films solar cells. The high efficiency and low cost Si base thin film solar cell will become main commercial solar cell products in the future. The third generation solar cell is the future solar cell modules with both the higher efficiency and lower cost, as shown in the figure 1. Therefore, how to do both increasing the efficiency and reducing the cost of the solar cell modules are the main research and developments in the future. In this paper we have reviewed the manufacturing progress of various solar cell modules and discussed future researches for the third generation solar modules.

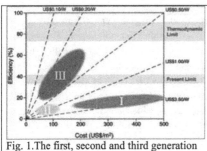

Fig. 1.The first, second and third generation of solar cell modules.

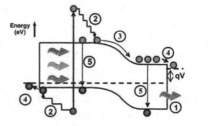

Fig. 2. The loss processes in a standard solar cell.

THE ISSUES AND APPROACHES OF SOLAR CELL EFFICIENCY AND COST

In order to improve the solar cell efficiency, it is should be understand what is the solar energy loss during the photovoltaic processing. Fig. 2 shows the loss processes in a standard solar cell[5]: (1) non absorption of below bandgap photons; (2) lattice thermalization loss; (3) and (4) junction and contact voltage losses; (5) recombination loss (radiative recombination is unavoidable). Among these, the two most important power-loss mechanisms in single bandgap cells area. The inability to absorb photons with energy less than the bandgap (1 in Figure 2) 2. The thermalization of photon energies exceeding the bandgap (2 in Figure 2). These two mechanisms alone amount to the loss of about half of the incident solar energy in solar cell conversion to electricity. Therefore, even though a large number of semiconductor materials show a PV effect, but only a few of them are of sufficient commercial interest because they must satisfy the constraints for minimizing thickness and wide availability. Ideally, the absorber material of an efficient terrestrial solar cell should be a direct bandgap semiconductor with a bandgap of 1.5 eV with a high solar optical absorption (10^5/cm^{-1}), high quantum efficiency of excited carriers, long diffusion length, low recombination velocity, and should be able to form a good electronic junction (homo/hetero/Schottky) with suitably compatible materials. With high optical absorption, the optimum thickness of an absorber in a solar cell is of the order of the inverse of the optical absorption coefficient and thus it must be a thin-film. On other hand, the third-generation options and thermodynamic limits on their efficiency are shown in figure 3.[4]

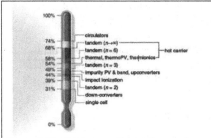
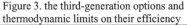

Figure 3. the third-generation options and thermodynamic limits on their efficiency

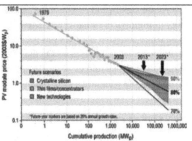

Fig. 4. The tendency of the cost for various solar cell modules.

Upconverters include multi-excitonic approaches are also list in the figure (Note: n is the number of cells in the stack). From the figure, thermodynamic limits of solar cell efficiency can be improved significantly with increasing the numbers of cells in the stack and light concentration. For a single-bandgap device, the thermodynamic limits of solar cell efficiency are only from 31% to 41%, depending on light concentration ratio. But for tandem cells (n→∞) with hot carrier, the cell efficiency could reach at 67% to 86.8% also depending on light concentration. Therefore, making solar cells with tandem, triple junctions is a main approach for third generation of solar cells. Another important issue for third generation of solar cells is cost. Figure 4 shows the tendency of the cost for various solar cell modules. From the figure, the thin film solar cells have a potential lower cost than the first generation of crystal silicon solar cells. Therefore, the third generation of solar cells should be developed based on the second generation of solar cells, which are the thin film solar cell technologies. Dr. Gavin Conibeer [5]provided the following approaches for third-generation photovoltaic. 1. Multiple energy level approaches: The concept of using multiple energy levels to absorb different sections of the solar spectrum can be applied in many different device structures and overcome the thermodynamic limits on their efficiency. 2. Intermediate-level cells: impurity PV and intermediate band solar cells. The approach with these devices is to introduce one or more energy levels within the bandgap such that they absorb photons in parallel with the normal operation of a single-bandgap cell. 3. Multiple carrier excitations: Carriers generated from high-energy photons (at least twice the bandgap energy) absorbed in a semiconductor can undergo impact ionization events resulting in two or more carriers close to the bandgap energy. 4. Modulation of the spectrum: up/down conversion: To create a device that either absorbs a photon of at least twice the bandgap energy or emits two photons incident on the cell (a down-converter), or absorbs at least two below-bandgap photons and emits one above-bandgap photon (an up-converter). 5. Hot carrier cells: To allow absorption of a wide range of photon energies but then to collect the photogenerated carriers before they have a chance to thermalize. 6. Other approaches: To increasing PV efficiency, including quantum antennas, thermophotonics or thermophotovoltaics (TPVs), and circulators. Except of above, recent researchers found that the surface Plasmon and surface nanostructures can improve light absorption significantly. The above suggestions open our mind to think about the any possibility to improve the solar cell efficiency, but no device matches all these goals yet. Combined some above technologies to get higher efficiency is always used in future optimization of solar cell modules. On the other hand, the materials that will used for third generation solar cells should be nontoxic and environmental friendly, not limited in abundance and also compatible with large-scale manufacturing of solar cell modules.

CURRENT STATUS AND DEVELOPMENT OF SOALR CELLS

In order to make the third generation solar cell with high efficiency and lower cost, we should know the current status of solar cells. The currently wide used materials for solar cell applications are mainly single crystal and poly-crystal Si which belong to the first generation of solar cells, and thin film solar cells including amorphous Si, microcrystal Si, CdTe, CuInGaSe, and III-V thin film materials, which are belong to the second generation of solar cells. Figure 5 shows the absorption coefficients of the most solar cell materials. Based on the light absorption coefficients and sunlight spectrum, the best materials for solar cell applications should be Si, CdTe, CuInGaSe, GaAs and other III-V materials.[6] Figure 6 shows the abundance of elements in Earth's crust [7]. From these figure, the rarest elements in the crust are marked in yellow color area. In the above solar cell materials, Si is a richest, but Te, Cd, In, and Se are the rarest elements in the Earth's crust. Additionally, tellurium and selenium have been depleted from the crust due to formation of volatile hydrides.

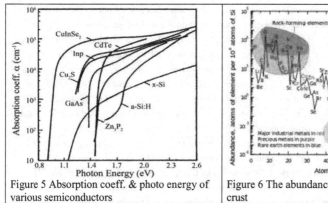

| Figure 5 Absorption coeff. & photo energy of various semiconductors | Figure 6 The abundance of elements in Earth's crust |

For CdTe solar cells, the band gap of CdTe is 1.45 eV, which is best match to the spectrum of sunlight. The absorption coefficient is 10^4/cm. The theoretical efficiency is 28%, and the efficiency of experimental samples reached at 17%, and the commercial sample reached 12%. The cost of commercial CdTe solar cell is the current lowest, as shown in figure 7. However, the

154

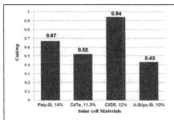

Figure 7. Cost-ownership of various materials for solar cell applications.

Table. 2 The efficiency of CIGS made by various companies

Firm	Efficiency	Production Capacity
Nanosolar	14-14.5%	1.02 Gigawatts
Solyndra	12-14%	110 Megawatts (MW)
HelioVolt	12.2%	20 MW
MiaSole	9-10%	40 MW
Soltec	11%	27.5 MW

Te reserves in the earth's crust is very low, which will results in high price in the future and limitation of CdTe solar cell development. CdTe is a toxic materials and not environmental friendly, Only the first solar produces this solar modules. The low band gap materials matched CdTe have very lower melting point and not stable, which is difficult to make crystal materials for high efficiency tandem cell module applications.

For CuInGaSe (CIGS) solar cells, the $CuInSe_2$ has a bandgap of 1.04 eV. Therefore, Ga substitution for In (Cu(In/Ga)Se2) is probably the best option for increasing the band gap (in between 1.04 eV of $CuInSe_2$ and 1.7 eV of $CuGaSe_2$) to a desired value and to achieve enhanced current conversion efficiency. CIGS has proved to be a leading candidate for photovoltaic applications. It is one of the most absorbing semiconductor materials (absorption coefficient of $3–6 \times 10^5$/cm) and also makes an excellent junction and a solar cell. The efficiency of experimental samples reached at about 20%, and the commercial sample reached 9 -14%, as shown in table 2 [8].

However, based on content of indium in zinc ore stocks, there is a worldwide reserve base of approximately 6,000 tons of economically viable indium[9]. The worldwide indium production is currently 475 tons per year from mining and a further 650 tons per year from recycling. It has been estimated that there is less than 20 years left of Indium supplies, based on current rates of extraction, demonstrating the need for additional recycling. In 2002, the price was US$94 per kilogram. The recent changes in demand and supply have resulted in high and fluctuating prices of indium, which from 2006 to 2009 ranged from US$382/kg to US$918/kg. Additionally Se reserves in the earth's crust is also very low, which results in high price and limitation of CIGS solar cell development. On the other hand, the increasing number of alloy components makes the CIGS layer fabrication an extremely complex process. The film morphology and the resulting conversion efficiency depend on the relative ratio of the component elements in the fabricated alloy layer. The Cu/(In+Ga) and Ga/(In+Ga) ratios are found to be critical in determining the cell efficiency. Therefore, CIGS has challenges in making stoichiometric film: Many groups have employed a vacuum based physical vapor deposition (PVD of In, Ga, and Se at high substrate temperatures, ~500°C) approach to adjust the stoichiometry. To maintain the film stoichiometry,

155

the high temperature post-annealing (at 450-600°C) of CIGS films made by electrodeposition or other methods is usually performed in a Se atmosphere (such as H2Se gas), in a tubular furnace in flowing Ar atmosphere, which results in very high process costs. Currently Electrodeposition is a very cost-effective and simple method for fabrication of large area CIGS films. However, the conversion efficiencies reported for the cells fabricated by this method are by far low. The current understanding on the electron transport and recombination in the absorber layer and the CIGS/CdS interface seems inadequate, demanding more studies to characterize the fundamental properties[10].

For the Si based thin film solar cells [11], the band gap of crystal Si and amorphous Si are 1.1 eV and 1.7 eV respectively, and the band gaps of microcrystal (nano-sized) Si are between 1.1 to 1.7 eV. The Si based compound, such as the band gaps of crystal $Si_{1-x}Ge_x$ ($0 \leq x \leq 1$) are from 0.7 to 1.1 eV, and band gaps of amorphous SiC are about 1.95 eV, which is match to the spectrum of sunlight, as shown in figure 8. The absorption coefficients of amorphous Si:H and amorphous SiGe:H is about 10^5/cm. For single layer (single-junction), unconcentrated silicon cell the theoretical maximum is about 31% (i.e., the Shockley-Queisser limit). By layering multiple cell arrays on top of one another, in order of decreasing band gap energy, the theoretical limit is increased to 41% for a double-junction array and 49% for a triple-junction array. The efficiency of experimental samples reached at 15 - 16%, and the commercial sample reached about 12%, as shown in figure 9. Therefore, the Si base thin films solar cell are the most attractive for current and future photovoltaic applications. Si base thin films have following advantages: 1) Environment friendly and abundant storage capacity, 2) Its strong light absorbers ($\alpha \sim 10^5$/cm), so need less materials only 1% of bulk Si solar cell, 3) The potential efficiency of Si base thin film tandem cell > 15%, and short energy payback time about 1.5 year, 4) Due to the better low light efficiency and the better thermal stability, the actual power generation per unit watt of Si base thin film solar cell is 13 ~ 15% higher than the crystal Si solar cell in the whole year, 5) Good appearance (Color), and specially suitable for Building-Integrated Photovoltaics (BIPV) applications, 6) The Si base thin film a-Si/μc-Si tandem cells show significant improved light introduced degradation and stabilized properties, 7) lower cost than other solar cells, as shown

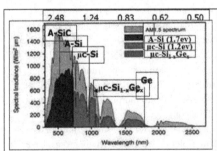

Fig. 8 The bondgap of Si base materials and spectrum of sunlight

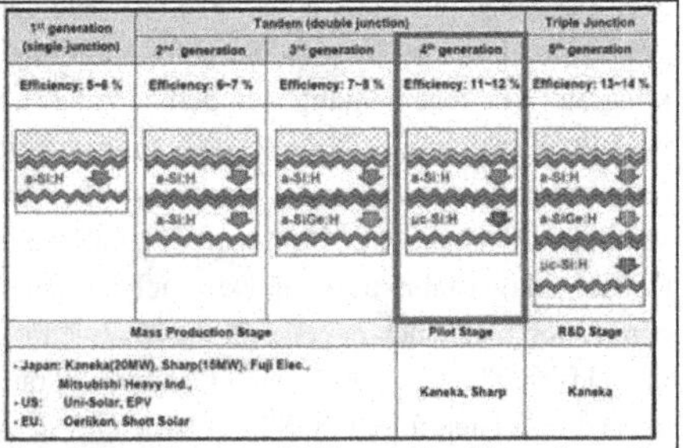

Fig. 9 The history of Si base thin film solar cell development

156

in figure 7 and 8) The manufacturing technologies are mature for large area, high volume thin film solar cell fabrication. Therefore, the Si base thin film solar cell is the most promising candidate materials for the third generation solar cell.

Other most interesting materials for solar cells are III-V materials. In order to reduce the cost ownership, the technologies for III-V grown on Si have been developed. Sharp has developed solar cells that match the concentrator solar cell efficiency world record set by Solar Junction [12]. The technology's staggering 43.5% efficiency from a triple-junction compound solar cell is 1.2% higher than the efficiency of the cells holding the record before March of 2011 (when Solar Junction busted that record). Sharp also reported the world's highest conversion efficiency of non-concentrator solar cell about 36.9%, which used three-junction III-V compound solar cells on silicon [13]. The solar cells have such a triple junction stacked layer structure and the concentrator structures shown in figure 10. The efficiency has been confirmed by AIST

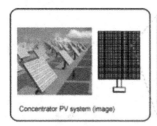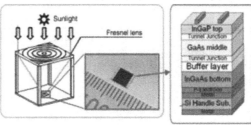

Figure 10 The three-junction III-V compound solar cells on silicon

(National Institute of Advanced Industrial Science and Technology). Due to the lattice and thermal mismatch between III-V materials and Si, as shown in figure 11 and table 3 [14], the buffer layer technologies have been developed to solve these problems. The $Si_{1-x}Ge_x$ (0<x<1)/Ge buffer layer has been widely used for III-V solar cell on Si substrate, such as stack

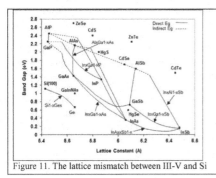

Figure 11. The lattice mismatch between III-V and Si

Table 3 The thermal mismatch between III-V and Si

Materials	TEC (x 10^{-6})	Materials	TEC (x 10^{-6})
Al_2O_3	7.50 (a) 8.50 (c)	InN	4.00
AlN	4.20 (a) 5.30 (c)	InP	4.60
AlAs	5.20	Si	3.59
GaAs	5.40	$Si_{0.65}Ge_{0.3 5}$	3.64 – 5.46
GaN	5.59 (a) 3.17 (c)	SiC(3C)	2.90
GaP	5.30	SiC(6H)	4.20 (a) 4.68 (c)
Ge	5.78	ZnO	2.90 (a)

157

layers of Si/ Si$_{1-x}$Ge$_x$/Ge/InGaAs/ Gaas/ Ingap solar cells. On the other hand, the low temperature GaP buffer layer growth on Si has also been studied recently. Compared with Ge, the GaP has a smaller lattice and thermal mismatch with Si. P rich GaP layer deposition on Si with lower deposition temperature around 500°C can avoid the reaction of Ga with Si, which results in formation of good quality GaP layer. The results are promising for III-V solar cells fabrication on silicon substrates.

The heterojunction with intrinsic thin layer (HIT) solar cell combined both technologies of single crystal and thin film solar cells. The device structure of HIT solar cell is shown in figure 12[15]. Deposition of intrinsic amorphous Si thin layer on both sides of n-type crystal Si to form heterojunction , and then n-type and p-type of amorphous Si continue to deposit on each side to make HIT solar cell device. This device can absorb the direct sunlight and ambient and scattered sunlight from both side of the HIT device, as shown in figure 13, the higher efficiency about 23% has been reached, which is an also very attractive candidate for the third generation of solar cells.

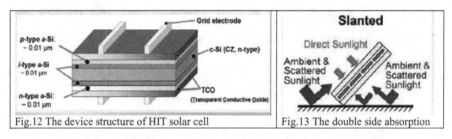

Fig.12 The device structure of HIT solar cell	Fig.13 The double side absorption

Multiple exciton generation (MEG) is a process that can occur in semiconductor nanocrystals, or quantum dots (QDs), whereby absorption of a photon bearing at least twice the bandgap energy produces two or more electron-hole pairs, which can improve the efficiency of solar cells significantly. The photocurrent enhancement arising from MEG in lead selenite (PbSe) QD-based solar cells, as manifested by an external quantum efficiency (the spectrally resolved ratio of collected charge carriers to incident photons) that peaked at 114 ± 1% in the best device measured. The MEG behaviors have also discovered from nanocrystals, or quantum dots of PbSe, PbS, PbTe, CdSe and C$_{60}$ etc.[16, 17]

The most important photovoltaic applications are Building-integrated photovoltaics (BIPV). Building-integrated photovoltaics (BIPV) are photovoltaic materials that are used to replace conventional building materials in parts of the building envelope such as the roof, skylights, or facades. BIPV combines the PV system and building into a whole structure, the building parts not only have the envelope function, but also produce electricity for the building to use. National Renewable Laboratory provide a zero net energy building concept and the project in the future [18], the BIPV is an important part of the project. The advantage of integrated photovoltaics over more common non-integrated systems is that the initial cost can be offset by reducing the amount spent on building materials and labor that would normally be used to construct the part of the building that the BIPV modules replace. These advantages make BIPV one of the fastest growing segments of the photovoltaic industry. In the developed countries, the energy used in the building constructions occupied 30% to 40% of country's total energy consumption, which constrained the economic development. Expected after 50 years, only the power generation from BIPV roofs can provide the world's 1 / 4 power. The BIPV market will be going to very hot, and BIPV construction industry will be the one of the most important emerging industry in the 21st century[ ].

SUMMARY

The third generation of photovoltaic modules should be environment friendly and not limited in abundance and have high efficiency and low cost. In order to reach the target, the following approaches have been provided: 1. The concept of using multiple energy levels to absorb different sections of the solar spectrum can be applied to overcome the thermodynamic limits on their efficiency, 2. Impurity PV and intermediate band solar cells, 3. Multiple carrier excitation, 4. Modulation of the spectrum: up/down conversion, 5. Hot carrier cells, 6. Other approaches: To increasing PV efficiency, including quantum antennas, thermophotonics or thermophotovoltaics (TPVs), and circulators, surface Plasmon, surface nanostructures and quantum dots etc. The recent researches and developments show that the Si base thin film tandem and triple cell modules, III-V solar cell on Si, HIT solar cell and nanotechnology with no vacuum technique such as printable technologies and etc will be very attractive candidates for the third generation of solar cells . The most important photovoltaic applications are Building-integrated photovoltaics (BIPV).

REFERENCE

[1]. R. E. Smalley, IEEE 31st Photovoltaic , Specialist Conference, January 4, (2005).
[2]. John F. Bookout, International Geological Congress, Washington DC; July 10 (1985).
[3]. Green, M. A., Third Generation Photovoltaics: Ultra-High Efficiency at Low Cost, Springer-Verlag, Berlin, (2003).
[4] Nathan S. Lewis, George Crabtree etc. Basic Research Needs For Solar Energy Utilization, Report on the Basic Energy Sciences Workshop on Solar Energy Utilization (2005)
[5] Gavin Conibeer Third-generation photovoltaics, ARC Photovoltaics Centre of Excellence, School of Photovoltaic and Renewable Energy Engineering, University of New South Wales , Sydney, NSW 2052, Australia
[6]. K. L. Chopra1, P. D. Paulson and V. Dutta, Thin-Film Solar Cells: An Overview, Prog. Photovolt: Res. Appl. 12, 69–92 (2004).
[7] http://en.wikipedia.org/wiki/Abundance_of_elements_in_Earth's_crust, From Wikipedia, the free encyclopedia
[8] Beard & Halluin, An Analysis of CIGS Solar Cell Technology, 6 Nanotechnology Law & Business 19 (Spring 2009)
[9] http://en.wikipedia.org/wiki/Indium, From Wikipedia, the free encyclopedia
[10] Viswanathan S. Saji, Sang-Min Lee, and Chi Woo Lee, CIGS Thin Film Solar Cells by Electrodeposition, Journal of the Korean Electrochemical Society, Vol. 14, No. 2, 61-70, (2011)

[11] www.ebookbrowse.com/ch7-thin-film-si-solar-cells-pdf-d14019951

[12] Zachary Shahan, rp report, Sharp Hits Concentrator Solar Cell Efficiency Record, 43.5% Posted on May 31, (2012)

[13]. Sharp report, www.cnbeta.com/articles/161225.htm, (2011).

[14]. Tingkai Li, Chapter 2, III-V Compound Semiconductor: Integration with Silicon-based Microelectronics., CRC publisher, (2010).

[15] Takahiro Mishima, Mikio Taguchi, Hitoshi Sakata, Eiji Maruyama, Development status of high-efficiency HIT solar cells, Solar Energy Materials & SolarCells, 10,1016 (2010)

[16]. Wai-Lun Chan, et al., Science **334**, 1541 (2011).

[17]. Octavi E. Semonin, et al., Science **334**, 1530 (2011).

[18] Ellen Watts AIA, Zero Net Energy Buildings, Next Generation of Bold New Energy Initiatives, June 18, 2010

Mater. Res. Soc. Symp. Proc. Vol. 1538 © 2013 Materials Research Society
DOI: 10.1557/opl.2013.656

Using Dilute Nitrides to Achieve Record Solar Cell Efficiencies

Rebecca Jones-Albertus, Emily Becker, Robert Bergner, Taner Bilir, Daniel Derkacs, Onur Fidaner, David Jory, Ting Liu, Ewelina Lucow, Pranob Misra, Evan Pickett, Ferran Suarez, Arsen Sukiasyan, Ted Sun, Lan Zhang, Vijit Sabnis, Mike Wiemer, and Homan Yuen
Solar Junction, 401 Charcot Avenue, San Jose, CA 95131, U.S.A.

ABSTRACT

High quality dilute nitride subcells for multijunction solar cells are achieved using GaInNAsSb. The effects on device performance of Sb composition, strain and purity of the GaInNAsSb material are discussed. New world records in efficiency have been set with lattice-matched InGaP/GaAs/GaInNAsSb triple junction solar cells and a roadmap to 50% efficiency with lattice-matched multijunction solar cells using GaInNAsSb is shown.

INTRODUCTION

Today's highest efficiency solar cells are III-V multijunction solar cells. For a single junction solar cell, the theoretical efficiency is a function of its band gap [1], and is maximized by optimizing the trade-off between higher current and higher voltage. Lower band gap material converts more photons to electrons but at a lower voltage, with much of the photon energy lost to heat as photo-excited carriers relax to the semiconductor band edges, while higher band gap material produces a higher voltage but converts fewer photons to electrons. Multijunction solar cells allow for higher efficiencies by stacking individual subcells of increasing band gap such that the highest energy light is absorbed by the highest band gap subcell, with lower energy light absorbed by one or more lower subcells. In this way, a larger fraction of the solar spectrum can be absorbed and less of the photon energy is converted into heat. Accordingly, the theoretical efficiency of a multijunction solar cell increases as the number of subcells increases [2]. However, the practical efficiency depends on additional factors, including the band gaps of the subcells, the solar cell design and the material quality.

Historically, the lattice-matched InGaP / (In)GaAs / Ge multijunction solar cell structure produced the highest efficiencies. It was favored because its three subcells are lattice-matched, enabling high material quality and reliability; however, its combination of band gaps is not optimal. This motivated research into new materials and structures with more optimal band gaps and higher potential efficiencies. There was significant interest in the dilute nitrides, specifically GaInNAs alloys, which have a band gap in the range near 1 eV that could not be reached by the more traditional phosphide and arsenide materials when lattice-matched to the standard GaAs and Ge substrates [3,4]. Dilute nitride subcells had the potential to improve multijunction solar cell efficiency by replacing the bottom Ge subcell or becoming a fourth subcell inserted above the Ge subcell [5]. However, previous research found that the minority carrier transport properties in dilute nitride subcells were too poor to improve multijunction solar cell efficiencies [6,7]. As a result, research efforts on monolithic, multijunction solar cells began to focus on lattice-mismatched structures in order to attain more optimal subcell band gaps, including metamorphic and inverted metamorphic solar cells that employ graded buffer layers to change the lattice constant between subcells, and semiconductor bonded solar cells [8-11].

161

Here we demonstrate that high quality, dilute nitride material has been achieved, enabling new world records for solar cell efficiency using lattice-matched structures.

EXPERIMENT

Multijunction solar cells and GaInNAsSb subcells were grown by molecular beam epitaxy (MBE) on GaAs or Ge substrates. Sb, C and H composition were determined by secondary ion mass spectroscopy (SIMS), and strain was measured by x-ray diffraction. Solar cell devices were fabricated and then tested using quantum efficiency and current-voltage (I-V) measurements calibrated to the ASTM G173-03 AM1.5D spectrum. Measurements of GaInNAsSb subcells included an optically thick GaAs filter to approximate the light filtering in a triple junction solar cell. Measurements of solar cell efficiency as a function of concentration ratio from 20 to 1000 suns were performed by the National Renewable Energy Laboratory.

DISCUSSION

Typical multijunction solar cells consist of two or more subcells connected in series by tunnel junctions. Because they are connected in series, the short-circuit current (J_{SC}) produced by the multijunction solar cell is "current-limited" by the subcell that produces the lowest J_{SC}, and the open-circuit voltage (V_{OC}) is roughly the sum of the V_{OC} values of the individual subcells. In addition to J_{SC} and V_{OC}, the solar cell efficiency also depends on the fill factor, which is affected by many factors, including series resistance, shunt resistance and the degree of current-matching between subcells. In order for triple junction solar cells with dilute nitride subcells to improve upon the efficiencies possible from InGaP / (In)GaAs / Ge solar cells, the dilute nitride subcell should produce at least as much J_{SC} as each of the upper InGaP and (In)GaAs subcells so that it does not limit the multijunction J_{SC}, and it should produce a higher V_{OC} than a Ge subcell. Under the AM1.5D spectrum at 1 sun, the subcells should have a J_{SC} of at least 14 mA/cm^2 and a V_{OC} higher than 0.25 V, which is approximately the V_{OC} of a Ge subcell [12]. In this work, the effects of Sb, strain and impurity concentration on the J_{SC} and V_{OC} of subcells of the dilute nitride GaInNAsSb are investigated, and then results and designs of multijunction solar cells using GaInNAsSb subcells are shown.

Ga$_{1-x}$In$_x$N$_y$As$_{1-y-z}$(Sb$_z$) subcells were grown with different Sb compositions (z), and their J_{SC} and V_{OC} were evaluated for use in triple junction solar cells (Fig. 1) [13]. It can be seen that subcells without Sb ($z = 0$), shown as diamonds, have the highest V_{OC} values, but do not produce sufficient J_{SC} to match the J_{SC} of today's InGaP and (In)GaAs subcells, which is depicted by the grey bar. They would significantly current-limit a triple junction solar cell, lowering the efficiency compared to the InGaP / (In)GaAs / Ge structure. In contrast, one of the two subcells with $z = 3 - 6\%$ ("high Sb," shown as squares) does produce sufficient J_{SC}, as has been reported previously [14], as well as four of the five subcells with $z = 0.1 - 3\%$ ("low Sb," shown as triangles). The low Sb subcells have ~100 mV higher V_{OC} values than the high Sb subcells, and thus they would yield the highest efficiencies when incorporated in triple junction solar cells. Subsequent development of low Sb subcells has further improved the J_{SC} and V_{OC} values. These results can be understood in the context of the role of Sb in defect formation. It has been reported elsewhere that Sb inhibits the formation of N-related defects [15], and this data suggests that high levels of Sb introduce additional defects which are deleterious for solar cell performance.

162

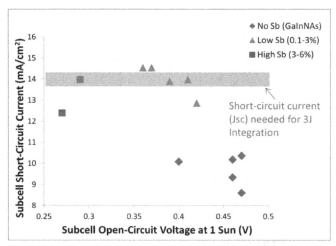

Figure 1: Short-circuit current and open-circuit voltage of GaInNAs(Sb) subcells under the AM1.5D spectrum at 1 sun filtered by an optically thick GaAs subcell.

GaInNAsSb subcells with low Sb were grown with compressive and tensile strain to study the effect of strain on performance. Figure 2 shows that the nature of strain in the pseudomorphic GaInNAsSb material affects the subcell V_{OC} but does not significantly impact the J_{SC}. Among low Sb subcells, those with compressive strain (triangles) have higher V_{OC} than those with tensile strain (diamonds).

SIMS was used to look at the H and C impurity concentrations in the GaInNAsSb subcells in this work. The concentrations of both H and C were below 10^{16} cm^{-3}, which was the baseline of the SIMS measurement. In contrast, dilute nitride materials grown by metalorganic chemical vapor deposition are known to incorporate H and C from the organic precursors and are known to have high background doping levels due to high concentrations of carbon [14, 16, 17].

Low-Sb-GaInNAsSb subcells were incorporated into lattice-matched, triple junction solar cells with an InGaP top subcell and GaAs middle subcell, and efficiencies were measured by the National Renewable Energy Laboratory (NREL) under the AM1.5D spectrum. The solar cells achieved new world records for solar cell efficiency. An efficiency of 43.5% at 415 suns was verified by NREL in 2011 and an efficiency of 44.0% at 947 suns was verified by NREL in 2012 (Fig. 3) [18, 19]. In general, solar cell efficiency increases with the concentration ratio due to the logarithmic increase in V_{OC} until series resistance losses become high enough to cause the fill factor and the efficiency to decrease. Figure 3 shows that the triple junction solar cell efficiency increases with concentration of sunlight up to 1000 suns, indicating very low series resistance losses in the solar cell.

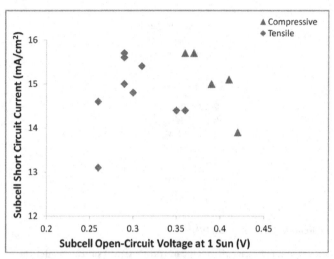

Figure 2: Short-circuit current and open-circuit voltage of low-Sb-GaInNAsSb subcells with compressive or tensile strain under the AM1.5D spectrum at 1 sun filtered by an optically thick GaAs subcell.

To continue to increase efficiency, it is desirable to have multijunction solar cells with more than three subcells. The optimal band gaps for the individual subcells change as the number of subcells increases; thus, a wide range of band gap tunability is essential to the realization of higher efficiencies. The band gap tunability of GaInNAsSb, coupled with the tunability of the AlInGaP and Al(In)GaAs alloys at higher band gap energies, allow for the optimal band gaps for multijunction solar cells with more than three subcells to be attained while maintaining lattice-matching. High quality GaInNAsSb subcells that are lattice-matched to GaAs and Ge have been achieved with band gaps across the range from 0.8 eV to 1.25 eV. Band gaps below 0.8 eV and as high as 1.4 eV are also possible. Figure 4 shows an example of the expected efficiencies for lattice-matched solar cells with three to six subcells with the materials and band gaps shown. Efficiencies of 46%, 48% and 50% are expected for the four, five and six junction solar cells. Additional designs with four to six subcells are possible using GaInNAsSb in the bottom subcells. These structures can be grown in upright or inverted configurations. Further, epitaxial lift-off can be used to remove the solar cell layers from the substrate, allowing for substrate reuse.

In addition to the structures shown above, the tunability of GaInNAsSb alloys in band gap, as well as in lattice constant, may make them useful for a wide range of additional solar cell and other optoelectronic devices, including lattice-mismatched or metamorphic solar cells, semiconductor bonded solar cells, and solar cell devices utilizing spectrum-splitting optics.

Efficiency vs. Concentration Ratio

HIPSS Data
Temperature: 25 ± 1°C

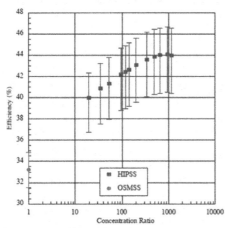

Figure 3: Efficiency as a function of concentration ratio for the world record InGaP/GaAs/GaInNAsSb solar cell, showing 44.0% efficiency at 947 suns.

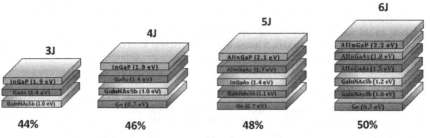

Figure 4: Expected efficiencies of lattice-matched solar cells using GaInNAsSb subcells.

CONCLUSIONS

Dilute nitride subcells with high short-circuit currents and open-circuit voltages have been achieved using molecular beam epitaxy, low Sb compositions and compressive strain. These subcells enabled new world records for solar cell efficiency of 43.5% (in 2011) and 44.0%

165

(in 2012) with the lattice-matched InGaP / GaAs / GaInNAsSb structure. The lattice-matched alloy systems AlInGaP, Al(In)GaAs, GaInNAsSb and Ge span a large fraction of the solar spectrum, enabling the optimal sets of band gaps to be achieved for multijunction solar cells with more than three subcells. A six junction solar cell with two GaInNAsSb subcells is expected to attain 50% efficiency.

ACKNOWLEDGMENTS

The authors thank the entire Solar Junction team for contributions to this work. We also thank Keith Emery and the PV Cell and Module Performance Group at the National Renewable Energy Laboratory for the efficiency measurements.

REFERENCES

1. W. Shockley and H. Queisser, J. Appl. Phys. **32,** 510 (1961).
2. A. Marti and G. Araujo, Sol. En. Mat. Sol. Cel. **43**, 203 (1996).
3. S. R. Kurtz, A. A. Allerman, E. D. Jones, J. M. Gee, J. J. Banas, and B. E. Hammons, Appl. Phys. Lett. **74**, 729 (1999).
4. K. Volz, D. Lackner, I. Nemeth, B. Kunert, W. Stolz, C. Baur, F. Dimroth, and A. W. Bett, J. Cryst. Growth **310**, 2222 (2008).
5. D. J. Friedman, J. F. Geisz, S. R. Kurtz, and J. M. Olson, J. Cryst. Growth **195**, 409 (1998).
6. D.J. Friedman and S.R. Kurtz, Prog. Photovolt: Res. Appl. **10**, 331 (2002).
7. J. F. Geisz, D. J. Friedman, J. M. Olson, S. R. Kurtz, and B. M. Keyes, J. Cryst. Growth **195**, 401 (1998).
8. W. Guter, J. Schöne, S. P. Philipps, M. Steiner, G. Siefer, A. Wekkeli, E. Welser, E. Oliva, A. W. Bett, and F. Dimroth, Appl. Phys. Lett. **94**, 223504 (2009).
9. R. R. King D. Bhusari, D. Larrabee, X.-Q. Liu, E. Rehder, K. Edmondson, H. Cotal, R. K. Jones, J. H. Ermer, C. M. Fetzer, D. C. Law and N. H. Karam, Prog. Photovolt: Res. Appl. **20**, 801 (2012).
10. S. Wojtczuk, P. Chiu, X. Zhang, D. Pulver, C. Harris, and M. Timmons, IEEE J. Photovolt. **2**, 371 (2012).
11. D. Wilt and M. Stan, Ind. Eng. Chem. Res. **51**, 11931 (2012).
12. H. Cotal, C. Fetzer, J. Boisvert, G. Kinsey, R. King, P. Hebert, H. Yoon and N. Karam, Energy and Environmental Science **2**, 174 (2009).
13. R. Jones, H. Yuen, T. Liu and P. Misra, US patent application number US20110232730 A1 (2011).
14. D. B. Jackrel, S. R. Bank, H. B. Yuen, M. A. Wistey, J. S. Harris, A. J. Ptak, S. W. Johnston, D. J. Friedman, and S. R. Kurtz, J. Appl. Phys. **101**,114916 (2007).
15. N. Miyashita, H. Oigawa and Y. Okada, Proc. 25th EU PVSEC, 946 (2010).
16. A.J. Ptak, D. J. Friedman, S. Kurtz, and R. C. Reedy, J. Appl. Phys. **98**, 094501 (2005).
17. T. W. Kim, T. J. Garrod, K. Kim, J. J. Lee, S. D. LaLumondiere, Y. Sin, W. T. Lotshaw, S. C. Moss, T. F. Kuech, R. Tatavarti, and L. J. Mawst, Appl. Phys. Lett. **100**, 121120 (2012).
18. M.A. Green, K. Emery, Y. Hishikawa, W. Warta and E.D. Dunlop, Prog. Photovolt: Res. Appl. **20**, 12 (2012).
19. M.A. Green, K. Emery, Y. Hishikawa, W. Warta and E.D. Dunlop, Prog. Photovolt: Res. Appl. **21**, 1 (2013).

Mater. Res. Soc. Symp. Proc. Vol. 1538 © 2013 Materials Research Society
DOI: 10.1557/opl.2013.670

Over 20% Efficiency Mechanically Stacked Multi-Junction Solar Cells Fabricated by Advanced Bonding Using Conductive Nanoparticle Alignments

Kikuo Makita, Hidenori Mizuno, Hironori Komaki, Takeyoshi Sugaya, Ryuji Oshima, Hajime Shibata, Koji Matsubara, and Shigeru Niki
National Institute of Advanced Industrial Science and Technology (AIST)
AIST Tsukuba, Central 2, Umezono 1-1-1, Tsukuba, Ibaraki, 305-8568, JAPAN

ABSTRACT

This paper shows a new semiconductor bonding technology for mechanically stacked multi-junction solar cells. Our strategy is the combination of conductive nanoparticle alignments and the van der Waals bonding technique. With this method, reasonably low bonding resistances and minimal optical absorption losses were simultaneously attained for the use of mechanically stacked solar cells. We examined a GaInP(Eg-1.89 eV)/GaAs (Eg-1.42 eV)/InGaAsP (Eg-1.15 eV) three-junction solar cell fabricated with this bonding method. As a result, the total efficiency of 22.5% was achieved, which was in good agreement with the theoretically predicted value. These results suggested that our bonding method is highly useful to fabricate high-efficiency mechanically stacked multi-junction solar cells.

INTRODUCTION

Multi-junction (MJ) solar cells have enabled very high efficiencies due to the effective utilization of the solar spectrum. These MJ solar cells are usually composed of complicated multi-layers on GaAs or Ge wafers using elaborative growth techniques. Therefore, these cells tend to be expensive and lack material flexibility. On the other hand, the MJ solar cells based on mechanical stacking have been considered as an alternative approach to produce MJ solar cells [1-3]. The mechanical stacking enables high efficiency and low-cost because of the flexible combinations of individually-processed cells. We have recently proposed an advanced bonding method using conductive nanoparticle alignments with a simple mechanical stacking process [4,5]. As an initial demonstration, a GaAs(Eg-1.42 eV) / InGaAsP(Eg-1.15 eV) two-junction solar cell was fabricated, and the efficiency of 11.8% was obtained [4]. This efficiency was limited by the small short-circuit current density of InGaAsP bottom cell, because E_g combination (1.42eV/1.15eV) does not meet the current matching. In this report, we design cell structures and cell combinations based on the current matching, and demonstrate a high-efficiency (over 20%) for GaInP (Eg-1.89 eV)/GaAs (Eg-1.42 eV)/InGaAsP (1.15 eV) three-junction solar cell.

DEVICE CONCEPT

Figure 1 shows the schematic drawing of the proposed cell structure, which possesses conductive nanoparticle alignments at the interface between top and bottom cells [4,5]. The conductive nanoparticle alignments were introduced on top of the bottom cell through the use of self-assembled block copolymer (polystylene-*block*-poly-2-vinylpyridine) templates. As conductive nanoparticle materials, Pd, Ag, Au, and Pt were available. Figure 2 shows the scanning electron microscopy (SEM) image of a typical Pd nanoparticle alignment on an InP substrate. Owing to the self-assembling behavior of the block copolymer, hexagonally patterned Pd domains were achieved. The domain size and spacing was 50 and 100 nm, respectively. The concentration of the Pd nanoparticles was estimated to be 1×10^{10} /cm^2, and the coverage area of the Pd nanoparticle was about 12%. As for the top cell, an epitaxially grown single or tandem cell was released from the growth substrate by epitaxial lift-off (ELO). Finally, two cells were stacked through van der Waals bonding technique [6]. Because of the formation of ohmic-contacts between semiconductors and Pd nanoparticles, two cells were interconnected electrically. In addition, light could pass through the interface without the absorption loss by Pd nanoparticles because of the extremely thin nanoparticle thickness (~10 nm) and the low surface coverage. Therefore, a reasonably low bonding resistance (<4 Ωcm^2) as well as a minimal interfacial optical loss (<2%) were possible [5].

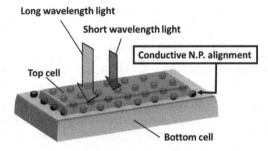

Figure 1. Schematic drawing of the proposed multi-junction solar cell. A conductive nanoparticle alignment exists at the interface of top and bottom cell.

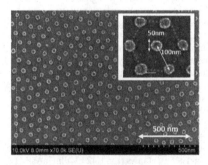

Figure 2. SEM image of a Pd nanoparticle alignment on an InP substrate. A hexagonally patterned Pd nanoparticles (size: 50 nm) was observed.

168

EXPERIMENT

Using this method, we fabricated a GaInP/GaAs/InGaAsP three-junction solar cell as shown in Fig. 3. As the top cell, a two-junction cell consisting of p-GaInP (Eg-1.89 eV) and p-GaAs (Eg-1.42 eV) absorbers connected with tunneling layers was used. The thicknesses of the GaInP and GaAs absorbers were designed to be less than 1000 nm to achieve the current matching with the bottom cell (1.15 eV). As for the bottom cell, a single cell with a p-InGaAsP (Eg-1.15 eV) absorption layer formed on an InP substrate was used. The thickness of the p-InGaAsP absorber was 2500 nm. These top and bottom cells were bonded through the Pd nanoparticle alignment. The photovoltaic performance was investigated under AM1.5 solar spectrum illumination (1 sun, 100 mW /cm^2). Figure 4 shows the current-voltage (J-V) characteristics of the fabricated three-junction cell with anti-reflection (AR) coating. It was revealed that the total efficiency was 22.5% with open-circuit voltage, short-circuit current density and fill factor of 2.86 V, 10.55 mA /cm^2 and 0.75, respectively.

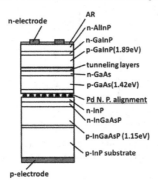

Figure 3. Schematic diagram of the fabricated GaInP (1.89 eV)/GaAs (1.42 eV)/InGaAsP (1.15 eV) three-junction solar cell. The top two-junction cell and the bottom cell were connected through the Pd nanoparticle alignment.

Fig. 5 shows the theoretically-estimated and experimentally-obtained J-V curves of the three-junction cell. The theoretical J-V curve for single junction cell appears to be the next relational expression [7]:

$$J = J_S(e^{q(V-JR_S)/kT} - 1) - J_L \qquad (1)$$

J_S is the saturation current. J_L is the constant source current due to the incident light. R_S is the series resistance including the cell resistance and the bonding resistance. In this case, each cell resistance was supposed to be 1Ωcm^2. The bonding resistance originated in Pd nanoparticle alignment was set to be 3 Ωcm^2. J-V curves of each cell were combined according to the Kirchhoff's Current Law. The short-circuit current density of tandem cell was adjusted to be experimental value of 10.55 mA/cm^2. Because this value was consistent with the short-circuit current density of the GaInP/GaAs top cell, it was considered to be electrically connected without current loss. By this means, theoretical J-V curve was obtained. The device parameters of each curve are also summarized in Fig.5. Although two J-V curves overlapped considerably, it

169

can be seen that a slightly higher bonding resistance was observed in the experimentally-obtained curve. Nevertheless, the experimental efficiency (22.5%) and open-circuit voltage (2.86V) were in good agreement with the theoretically predicted values (23.8% and 2.80V). The efficiency would be further improved by appropriate combinations of solar cells. According to our simulation, three junction solar cells with 1.89eV/1.42 eV/1.0 eV combination can achieve an efficiency of more than 30% due to the complete current matching. Furthermore, multi-junction solar cells with four-junction and five-junction are predicted to reach 40% efficiency. In such cases, this bonding method is highly useful to achieve high-efficiency mechanically stacked multi-junction solar cells.

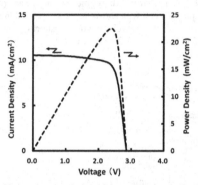

Figure 4. Current-voltage characteristics of three-junction cell. The total efficiency was 22.5% with the open-circuit voltage, short-circuit current density and fill factor of 2.86 V, 10.55 mA /cm^2 and 0.75, respectively.

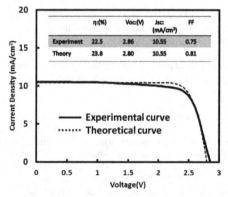

Figure 5. Experimental and theoretical current-voltage curves. The experimental efficiency was in good agreement with the theoretically predicted value.

CONCLUSIONS

In summary, we demonstrated a new bonding method using conductive Pd nanoparticle alignments to fabricate a mechanically stacked GaInP/GaAs/InGaAsP three-junction solar cell. A high-efficiency of 22.5% was achieved with this solar cell design, suggesting that the present method is promising to obtain high efficiency multi-junction solar cells.

ACKNOWLEDGMENTS

This work was supported by the New Energy and Industrial Technology Development Organization as a part of the Research and Development on Innovative Solar Cell Program. The authors thank Dr. T. Takamoto, K. Sasaki, N. Takahashi, and Dr. M. Togo for providing epitaxial wafers.

REFERENCES

1. K. Tanabe, A. F. Morral, H. A. Atwater, D. J. Aiken, and M. W. Wanlass, *Appl. Phys. Lett.*, **89**, 102106(2006).
2. T. Sameshima, J. Takenezawa, M. Hasumi, T. Koida, T. Kaneko, M. Karasawa, and M. Kondo, *Jpn. J. Appl. Phys.*, **50**, 052301 (2011)
3. K. Makita, H. Soya, H. Komaki, H. Sai, H. Tampo, T. Sugaya, S. Furue, S. Ishizuka, A. Yamada, A. Furukawa, H. Shibata, K. Matsubara, and S. Niki, *Technical Digest of the 21st International Photovoltaic Science and Engineering Conference*, 2B-4O-10 (2011).
4. K. Makita, H. Mizuno, H. Soya, H. Komaki, H. Sai, H. Tampo, T. Sugaya, R. Oshima, A. Furukawa, H. Shibata, K. Matsubara, and S. Niki, *Proceedings of the 27th European Photovoltaic Solar Energy Conference*, 78-80 (2012).
5. H. Mizuno, K. Makita, and K. Matsubara, *Appl. Phys. Lett.*, **101**, 191111(2012)
6. E. Yablonovitch, D. M. Hwang, T. J. Gmitter, L. T. Florez, and J. P. Harbison, *Appl. Phys. Lett.*, **56**, 2419 (1990).
7. S. M. Sze, "Physics of Semiconductor Devices", John Wiley & Sons, 640-653(1969)

Mater. Res. Soc. Symp. Proc. Vol. 1538 © 2013 Materials Research Society
DOI: 10.1557/opl.2013.1018

MoO₃ back contact for CuInSe₂-based thin film solar cells

Hamed Simchi[1,2], Brian E. McCandless[1], T. Meng[1], Jonathan H. Boyle[1,2], and William N. Shafarman[1,2]
[1]Institute of Energy Conversion, University of Delaware, Newark, DE 19716, U.S.A.
[2]Department of Materials Science and Engineering, University of Delaware, Newark, DE 19716, U.S.A.

ABSTRACT

MoO₃ films with a high work function (5.5 eV), high transparency, and a wide bandgap (3.0 - 3.4 eV) are a potential candidate for the primary back contact of Cu(InGa)Se₂ thin film solar cells. This may be advantageous to form ohmic contact in superstrate devices where the back contact will be deposited after the Cu(InGa)Se₂ layer and MoSe₂ layer doesn't form during Cu(InGa)Se₂ deposition. In addition, the MoO₃ may be incorporated in a transparent back contact in tandem or bifacial cells. In this study, MoO₃ films for use as a back contact for Cu(In,Ga)Se₂ thin film solar cells were prepared by reactive rf sputtering with O₂/(O₂+Ar) = 35%. The effect of post processing on the structural properties of the deposited films were investigated using x-ray diffraction and scanning electron microscopy. Annealing resulted in crystallization of the films to the α-MoO₃ phases at 400°C. Increasing the oxygen partial pressure had no significant effect on optical transmittance of the films, and bandgaps in the range of 2.6-2.9 eV and 3.1-3.4 eV were obtained for the as deposited and annealed films, respectively. Cu(In,Ga)Se₂ thin film solar cells prepared using an as-deposited Mo-MoO₃ back contact yielded an efficiency of >14% with V_{OC} = 647 (mV), J_{SC} = 28.4 (mA), and FF. = 78.1%. Cells with ITO-MoO₃ back contact showed an efficiency of ~12% with V_{OC} = 642 (mV), J_{SC} = 26.8 (mA), and FF. = 69.2%. The efficiency of cells with an annealed MoO₃ back contact was limited to 4%, showing a blocking diode behavior in the forward bias J-V curve. This may be caused by the presence of a barrier between the valence bands of the Cu(In,Ga)Se₂ and MoO₃, due to the higher bandgap of the annealed MoO₃ films. SEM cross section studies showed uniform coverage of the as-deposited MoO₃ layer and formation of voids for the annealed MoO₃ film. Structural orientation of the Cu(In,Ga)Se₂ absorber layer was also altered by the MoO₃ film and less-oriented films were observed for either cases.

INTRODUCTION

Cu(InGa)Se₂ solar cells are typically made in the substrate configuration in which the Mo back contact, Cu(InGa)Se₂ absorber layer, CdS buffer layer, high resistance ZnO, and ITO window layers are sequentially deposited on the glass substrate. The superstrate solar cell is a configuration where the light shines through the glass substrate. In this case, the back contact is formed after Cu(InGa)Se₂ deposition. Mo, the most widely used material for the base electrode in Cu(InGa)Se₂ thin film solar cells, creates a blocking contact with a barrier height of 0.8 eV relative to the CuInSe₂ [1], but the MoSe₂ layer which forms during Cu(InGa)Se₂ deposition gives ohmic characteristics [2]. The MoSe₂ layer will not form in the superstrate configuration case, as the back contact is deposited after the Cu(InGa)Se₂ layer, at a low temperature.

Models have shown that the barrier height at the back contact of the absorber shouldn't exceed 0.3 eV in order to not impede cell performance[3]. The Schottky model of the metal-semiconductor barrier[4] predicts that the barrier height (Φ_b) depends on the work function of the metal (Φ_m), the band gap (E_g) of the p-type semiconductor, and the electron affinity (χ) of the p-type semiconductor:

$$\Phi_b{}^p = E_g + \chi - \Phi_m \qquad (1)$$

which will be modified to the following equation if the primary contact is a semiconductor, with the Fermi level positioned at $E_f{}^{SC}$:

$$\Phi_b{}^p = E_g + \chi - E_f{}^{SC} \qquad (2)$$

Therefore, it's necessary to use a high work function metal contact or well-aligned buffer layer in order to reduce Φ_b.

Thus, MoO_3 films with a high work function (5.5 eV)[5] are proposed as a candidate for the primary back contact of $Cu(InGa)Se_2$ thin film solar cells. In particular, high transparency (>80% in the visible and near IR range) and wide bandgap (3.0-3.8 eV)[6,7] make MoO_3 a potential transparent back contact for superstrate and tandem devices. Its Fermi level and density of states can potentially be tuned by controlling the oxygen stoichiometry during the deposition and or post-processing.

In this study, MoO_3 films were prepared by reactive rf sputtering. The effects of MoO_3 back layer and post processing on the structural properties of the $Cu(In,Ga)Se_2$ films were studied using x-ray diffraction (XRD), and scanning electron microscopy (SEM). Evaluation of the MoO_3 electronic properties and back contact role in $Cu(In,Ga)Se_2$ thin film solar cells are also addressed.

EXPERIMENTAL DETAILS

Molybdenum oxide films were deposited on Mo/ITO-coated soda lime glass substrates by reactive rf sputtering of a 5cm diameter Mo target (99.95% purity). The sputtering gases were a mixture of argon (Ar) and oxygen (O2), with oxygen content O2/(O2+Ar) = 35%. The sputtering pressure was fixed at 10 mTorr and rf power was 250 W. For each run, the target was pre-sputtered for 30 minutes to clean and condition the surface. Films were deposited with two different thicknesses of 10 and 30 nm. Post-processing of the films was performed by annealing at 400 °C, for 1 h in an air atmosphere, in a 6 cm diameter quartz tube furnace.

Structural properties of the films were characterized by symmetric x-ray diffraction (XRD) with Cu $K\alpha$ radiation, and scanning electron microscopy (SEM).

$Cu(In_{1-x}Ga_x)Se_2$ films were deposited on top of the MoO_3 layer at 550 °C using multi-source elemental evaporation. Films were deposited with constant fluxes over time so that they have no intentional through-film composition gradients. All films have bulk compositions with group I/III ratio, Cu/(In+Ga) ≈ 0.9 and are 2 μm thick. Solar cells fabricated with the structure: Glass/(M or /ITO)/MoO_3/Cu(In,Ga)Se_2/CdS/ZnO:Al/Al-grid were characterized by current-voltage measurements under AM1.5G (100 mW/cm^2) illumination.

174

RESULTS AND DISCUSSION

MoO₃ is reported to have crystal structures of monoclinic (β) or orthorhombic (α) with the space groups of P21/c and Pbnm, respectively[8,9]. Similar to the MoSe₂, typically forms during Cu(In,Ga)Se₂ deposition on Mo, the fully oxidized α-MoO₃ phase has also the layered structure[10]. Each layer is composed of two corner-sharing octahedral nets that link by sharing edges along [001]. The adjacent layers along [010] are linked only by weak van der Waals forces[10].

We have recently studied the effects of oxygen partial pressure and post-deposition annealing on the structural, optical, and surface properties of the molybdenum oxide thin films prepared via RF reactive sputtering[11]. It was found that as deposited, 300 °C annealed, and 400-500 °C annealed samples have the amorphous, β-MoO₃ phase, and α-MoO₃ phase structures, respectively, independent of the oxygen partial pressure. High resolution XPS studies showed that Mo^{6+} (corresponding to MoO₃) is the dominant oxidation state present at the surface of the as-deposited films with a small contribution of the Mo^{5+} (Mo_4O_{11}) state due to oxygen deficiency. No contribution from any of lower oxidation states, such as Mo^{4+} was observed. The Mo^{6+} to Mo^{5+} ratio (MoO₃ to Mo_4O_{11}) didn't change noticeably with oxygen partial pressure, and increased with annealing at 400-500 °C. UV/Vis/NIR spectrophotometry revealed bandgap values in the range of 2.6-2.9 eV for as deposited and 300 °C annealed samples. However, samples annealed at 400 °C and 500 °C exhibited wider bandgaps within the range of 3.1-3.4 eV.

Figure 1 shows the XRD patterns of the CIGS films deposited on the Mo-MoO₃ and ITO-MoO₃ stacks. The MoO₃ layer has altered the structural orientation of the Cu(In,Ga)Se₂ absorber layer. As a result, films grown on MoO₃ are less oriented and (112) to (220)/(204) peak ratios reduced from ~10 for the control sample (Mo) to the range of ~3-4 for films deposited on MoO₃ independent of whether the MoO₃ is deposited on Mo or ITO.

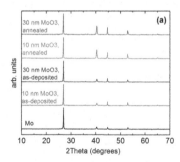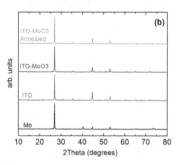

Figure 1. XRD pattern of the Cu(In,Ga)Se₂ deposited on a) Mo-MoO₃ stacks (10 and 30 nm), and b) ITO-MoO₃ (10 nm) stacks with different treatment.

SEM cross section images are shown in Fig. 2 and reveal a poor quality interface for the films deposited on annealed MoO₃. This might be caused by different expansion coefficient of annealed MoO₃ and Cu(In,Ga)Se₂ layers, which is less significant in the case of as deposited

175

MoO₃ with amorphous structure. However, no adhesion issue was observed on the complete cell structure. Large columnar grains were obtained in both cases.

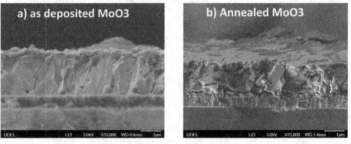

Figure 2. SEM cross-section of the cell made on the 30 nm MoO$_3$ back contact; a) as-deposited, and b) annealed at 400 °C.

Fig. 3a and Table 1 show the device properties of the cells made on the Mo-MoO$_3$ stack with different thicknesses and treatment conditions. It can be seen that the device made on the 10 nm MoO$_3$ back layer show properties comparable to the control sample with efficiency of ~14%. Cells made on the 30 nm MoO$_3$ back layer have also a regular diode shape but with lower JV parameters compared to the control sample (Mo contact). This might be due to a reduced concentration of Na diffusing from the soda lime glass substrate to the Cu(In,Ga)Se$_2$ layer passivating the surface and grain boundaries of the absorber layer[12]. The fill factor loss might also be partly due to higher series resistance of the cells made on MoO$_3$ stack. Series resistance values of 0.03, 0.06, and 0.3 Ω.cm were found for the cells made on the Mo, 10 nm MoO$_3$, and 30 nm MoO$_3$ layers, respectively.

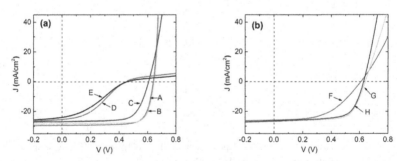

Figure 3. J-V profile of devices made on a) Mo and Mo-MoO$_3$ stacks and b) ITO and ITO-MoO$_3$ stacks.

Table 1. Cell properties of the devices made on different back contacts.

Sample	Back contact	MoO_3 thickness	MoO_3 anneal (°C)	Eff. (%)	V_{OC} (mV)	J_{SC} (mA/cm^2)	FF (%)
A	Mo	---	none	14.3	645	29.8	75.7
B	Mo+MoO$_3$	10 nm	none	14.4	647	28.4	78.1
C	Mo+MoO$_3$	30 nm	none	11.5	609	26.9	70.4
D	Mo+MoO$_3$	10 nm	400	4.2	453	25.5	37.8
E	Mo+MoO$_3$	30 nm	400	3.7	457	24.4	32.9
F	ITO	10 nm	none	8.0	614	26.2	50.2
G	ITO+MoO$_3$	10 nm	none	11.9	642	26.8	69.2
H	ITO+MoO$_3$	10 nm	400	11.3	629	26.6	67.3

The efficiency of cells with an annealed Mo-MoO$_3$ stack was limited to 4.2%, showing a blocking diode behavior in the forward J-V profile (Fig. 3a). This may be caused by the presence of a large barrier between the valence bands of the Cu(In,Ga)Se$_2$ and MoO$_3$, due to the higher bandgap of the annealed MoO$_3$ films, as noted previously.

Device properties of cells made on the ITO-MoO$_3$ stack are shown in Fig. 3b and also summarized in Table 1. It can be seen that MoO$_3$ layer has enhanced the performance of the cells with ITO back contact, with V_{OC} approaching cells on Mo. While the efficiency of the device made on the ITO was limited to 8%, mainly due to low fill factor, cells made on the 10 nm MoO$_3$ have efficiency of ~12% with V_{OC} = 642 (mV), J_{SC} = 26.8 (mA), and FF. = 69.2%. Annealing the MoO$_3$ layer did not adversely affect cell performance, likely due to different nature of MoO$_3$ layer grown on ITO, and similar cell performance was obtained.

CONCLUSIONS

Molybdenum oxide thin films were deposited on Mo and ITO coated glass substrates via rf reactive sputtering in O$_2$/(O$_2$+Ar) = 35%. The effect of MoO$_3$ thickness and post processing on the structural and device properties of the Cu(In,Ga)Se$_2$ thin film solar cells were investigated. It was found that as-deposited MoO$_3$ films have no significant effect on the structure of Cu(In,Ga)Se$_2$ absorber layers, but some voids at the MoO$_3$/Cu(InGa)Se$_2$ interface were observed in the case of MoO$_3$ films annealed at 400 °C. Cells made on Mo-MoO$_3$ back layers exhibited efficiency of ~14% and ~11% for the 10 and 30 nm thick MoO$_3$ back contacts, respectively. Lower efficiency of cells made on 30 nm MoO$_3$ layer might be due to limited diffusion of Na atoms from the glass substrate. Devices made on the ITO-MoO$_3$ transparent back layer has also showed efficiency of ~12%, presenting it as a promising back contact candidate for bifacial and tandem devices.

ACKNOWLEDGEMENTS

This material is based, in part, upon work supported by the Department of Energy under Award Number DE-EE0005317.

REFERENCES

[1] P.E. Russell, Applied Physics Letters **40**, 995 (1982).

[2] N. Kohara, S. Nishiwaki, Y. Hashimoto, T. Negami, and T. Wada, Solar Energy Materials and Solar Cells **67**, 209 (2001).

[3] R. Scheer and H.-W. Schock, *Chalcogenide Photovoltaics* (Wiley-VCH Verlag GmbH & Co. KGaA, Weinheim, Germany, 2011).

[4] D.K Schroder, *Contact Resistance and Schottky Barriers* (John Wiley & Sons Ltd., 2006), p. 128.

[5] N. Oka, H. Watanabe, Y. Sato, H. Yamaguchi, N. Ito, H. Tsuji, and Y. Shigesato, Journal of Vacuum Science & Technology A: Vacuum, Surfaces, and Films **28**, 886 (2010).

[6] V. Nirupama, M. Chandrasekhar, P. Radhika, B. Sreedhar, and S. Uthanna, Journal of Optoelectronics and Advanced materials **11**, 320 (2009).

[7] X. Fan, G. Fang, P. Qin, N. Sun, N. Liu, Q. Zheng, F. Cheng, L. Yuan, and X. Zhao, Journal of Physics D: Applied Physics **44**, 045101 (2011).

[8] ICDD DDView 4.8.3.4 using PDF-2/Release 2008 RDB 2.0804, The International Center for Diffraction Data, Newton Square, card number 01-089-1554.

[9] ICDD DDView 4.8.3.4 using PDF-2/Release 2008 RDB 2.0804, The International Center for Diffraction Data, Newton Square, card number 00-005-0508 (n.d.).

[10] R.L. Smith and G.S. Rohrer, Journal of Solid State Chemistry **124**, 104 (1996).

[11] H. Simchi, B.E. McCandless, T. Meng, J.H. Boyle, and W.N. Shafarman, Characterization of sputter deposited MoO_3 for solar cell application, Manuscript Under Preparation.

[12] S.-H. Wei, S.B. Zhang, and A. Zunger, Journal of Applied Physics **85**, 7214 (1999).

Mater. Res. Soc. Symp. Proc. Vol. 1538 © 2013 Materials Research Society
DOI: 10.1557/opl.2013.1002

Fabrication of Cu$_2$ZnSn(S,Se)$_4$ solar cells by printing and high-pressure sintering process

Feng Gao, Tsuyoshi Maeda, and Takahiro Wada
Department of Materials Chemistry, Ryukoku University, Seta, Otsu 520-2194, Japan

ABSTRACT

We fabricated Cu$_2$ZnSn(S$_x$Se$_{1-x}$)$_4$ (CZTSSe) solar cells by a printing and high-pressure sintering (PHS) process. First, the CZTSSe solid solution powders were synthesized by heating the elemental mixtures at 550°C for 5 h in an N$_2$ gas atmosphere. We fabricated CZTSSe films by a printing and high-pressure sintering (PHS) process. The obtained dense CZTSSe film was post-annealed at 550°C for 10 min under an N$_2$ +5% H$_2$S gas atmosphere. We fabricated CZTSSe solar cells with the device structure of Ag/ITO/i-ZnO/CdS/CZTSSe/Mo/soda-lime glass. The CZTSSe solar cell showed an efficiency of 2.1%, with V_{oc} of 272 mV, J_{sc} of 18.0 mA/cm^2 and FF of 0.44.

INTRODUCTION

After a Cu(In,Ga)Se$_2$ (CIGS) solar cell was achieved with a conversion efficiency of 20.3% [1], an increasing number of companies moved to commercialize CIGS PV modules. Therefore, substituting abundant elements for indium and gallium in CIGS has become an important issue because they are expensive rare metals. Cu$_2$ZnSnS$_4$ (CZTS) is anticipated to be an indium-free absorber material [2]. Katagiri et al. reported a CZTS solar cell with 6.7% efficiency fabricated by a combination of precursor layer sputtering and post-sulfurization [3]. An IBM group fabricated Cu$_2$ZnSn(S,Se)$_4$ (CZTSSe) solar cell with an efficiency of 11.1% by the hybrid coating process [4].

We studied the fabrication of CIGS films by a non-vacuum particulate-based deposition process. In such processes, solid particles are dispersed in a solvent to form an ink that can be coated onto a substrate [5]. We fabricated CIGS films by a combination of mechanochemical and screen printing/sintering processes [6, 7]. CIGS powder suitable for screen printing was prepared using a mechanochemical process (MCP) [8, 9]. First, we fabricated a CIGS solar cell with 2.8% efficiency [6]. Then we fabricated CIGS films by a mechanochemical process, wet bead-milling of CIGS powder, and screen printing and sintering processes, and obtained a CIGS solar cell with 3.1% efficiency [7]. Recently, we fabricated high-density CIGS films by a printing and high-pressure sintering (PHS) process and obtained a CIGS solar cell with 3.2% efficiency [10].

We prepared Cu deficient Cu$_{2(1-x)}$ZnSnSe$_4$ and characterized their crystal structures by XRD and XAFS [11]. Then, we characterized their optical properties. The band gaps of the CZTSSe solid solutions were determined from diffuse reflectance spectra of the powders and transmittance spectra of the films. The band gap (E$_g$) of the Cu$_2$ZnSn(S$_x$Se$_{1-x}$)$_4$ solid solution linearly increases from 1.05 eV for CZTSe (x=0.0) to 1.51 eV for CZTS (x=1.0) [12].

In this study, we fabricated CZTSSe films by PHS process [5] and CZTSSe solar cells with the device structure of Ag/ITO/i-ZnO/CdS/CZTSSe/Mo/soda-lime glass. A CdS buffer layer (100 nm) was formed by a conventional chemical bath deposition (CBD). The i-ZnO (100 nm), ITO (200 nm) layers and Ag grid (100 nm) were deposited by RF-sputtering. The performance of the solar cells was evaluated under standard AM1.5 illumination. The CZTSSe solar cell showed an efficiency of 2.1%, with V_{oc} of 272 mV, J_{sc} of 18.0 mA/cm^2 and FF of 0.44.

EXPERIMENT

Preparation of CZTSSe powder

Elemental powders such as Cu, Zn, Sn, S, and Se were weighed to give a molar ratio of $Cu_{1.9}Zn_{1.25}Sn(S_{0.4}Se_{0.6})_{4.5}$. The composition of $Cu_{1.9}Zn_{1.25}Sn(S_{0.4}Se_{0.6})_{4.5}$ was determined by reference to the conversion efficiency map of CZTS-based thin film solar cells reported by Katagiri et al. [13] and our previously reported paper [12]. The CZTSSe powder had a Cu-poor and Zn-rich composition. The CZTSSe powder was synthesized from elemental powders by planetary ball milling and post-heating at 550°C for 5 h in an N_2 gas atmosphere. In the case of $Cu(In,Ga)Se_2$, the CIGS powder could be synthesized from the elemental powders by planetary ball milling in an N_2 gas atmosphere without any additional heating [8]. However, we could not synthesize CZTSSe only by a mechanochemical process without additional heating. The ball-to-powder weight ratio was maintained at 5:1. The milling was conducted in a planetary ball mill (Fritsch premium line P-7) under a rotational speed of 800 rpm and a milling period of 20 min.

Fabrication of CZTSSe films

Particulate precursor ink was prepared by mixing the obtained $Cu_{1.9}Zn_{1.25}Sn(S_{0.4}Se_{0.6})_{4.5}$ powders with an an organic solvent such as ethylene glycol monophenyl ether. CZTSSe powder was mixed with an organic solvent by planetary ball milling (Fritsch premium line P-7) under a rotational speed of 750 rpm and a milling period of 5 h. The precursor CZTSSe layer was deposited on a soda-lime glass substrate by a screen-printing technique. The organic solvent was removed from the screen-printed CZTSSe films by heating at about 110°C in an N_2 gas atmosphere. The porous precursor layer was sintered into a dense polycrystalline film at 100°C for 30 min by high-pressure sintering. The pressure was as high as 6 MPa. The schematic of high-pressure sintering equipment is shown in Fig. 1 [10].The obtained dense CZTSSe film was post-annealed at 550°C for 10 min under an $N_2 + 5\%$ H_2S gas atmosphere. The annealing process was present with Se, S and Sn powder in the quartz box.

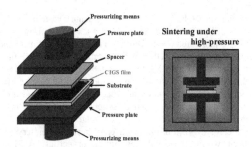

Figure 1 Schematic high-pressure sintering equipment [10].

Fabrication of CZTSSe solar cells

CZTSSe solar cells with our standard Ag/ITO/i-ZnO/CdS/CZTSSe/Mo/soda-lime glass structure solar cell with an area of about 0.34 cm^2 were fabricated. A CdS buffer layer (100 nm) was formed by a conventional chemical bath deposition (CBD). The i-ZnO (100 nm), ITO (200 nm) layers and Ag grid (100 nm) were deposited by RF-sputtering. The photocurrent density-voltage (J–V) characteristics were measured under the typical standard condition of a light power density of 1000 W/m^2 with the air mass 1.5G spectrum at 25 °C using a solar simulator (Wacom Electric WXS-50S-1.5).

RESULTS AND DISCUSSION

CZTSSe films

Figures 2 (a) and (b) show the surface SEM micrographs of the CZTSSe films deposited on Mo coated soda-lime glass substrate. Figures 2 (a) shows the surface of hot-pressed CZTSSe film and Fig. 2 (b) shows the surface of hot-pressed and post-annealed CZTSSe film. The CZTSSe surfaces of hot-pressed and post-annealed CZTSSe film are without large pores. The size of the grains is less than 1 μm, which is smaller than the CuInSe$_2$ (CIS) films [6]. In the previous study, the precursor ink of CIS was prepared by the mixing of the obtained Cu(In$_{0.95}$Ga$_{0.05}$)Se$_2$ powder with a small amount of sintering additive, CuSe. CIS grains grew by adding CuSe as a sintering additive.

(a)   (b)

Figure 2 Surface SEM micrographs of hot-pressed CZTSSe film (a) and hot-pressed and post-annealed CZTSSe film (b).

CZTSSe solar cells

Figure 3 shows the I–V characteristics of the CZTSSe solar cell with a nominal composition Cu$_{1.9}$Zn$_{1.25}$Sn(S$_{0.4}$Se$_{0.6}$)$_{4.5}$. The solar cell shows an efficiency (E_{ff}) of 2.1% with a V_{oc} of 272 mV, a J_{sc} of 18.0 mA/cm^2 and a fill factor (FF) of 0.44. The quantum efficiency of the solar cell with an efficiency of 2.1% is shown in Fig 4. The quantum efficiency at about 540 nm was as large as 0.63 but decreased with increasing wavelength. Our equipment for the EQE measurement (Bunkoukeiki Co. Ltd., SM-251) is only available up to 1100nm. The EQE onset between 900 and 1100 nm is so shallow because quality of the CZTSSe film obtained in this

181

study is not good. The obtained CZTSSe film is not dense and the grain size of CZTSSe film is small (less than 1 μm). However, secondary phases were not observed in the X-ray diffraction patterns and Raman spectroscopy of the films. In our future work, it is necessary to improve the film quality.

CONCLUSIONS

We fabricated $Cu_2ZnSn(S,Se)_4$ (CZTSSe) films by a printing and high-pressure sintering (PHS) technique. Then, we fabricated CZTSSe solar cells with the device structure of Ag/ITO/i-ZnO/CdS/CZTSSe/Mo/soda-lime glass. The performance of the solar cells was evaluated under standard AM1.5 illumination. The $I–V$ characteristics of some preliminary CZTSSe solar cells with an area of about 0.34 cm^2 were established. The best solar cell showed an efficiency of 2.1%, with V_{oc} of 272 mV, J_{sc} of 18.0 mA/cm^2 and FF of 0.44.

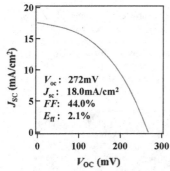 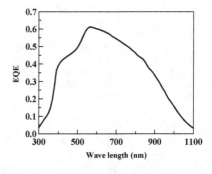

Figure 3 I–V characteristics of CZTSSe solar cells

Figure 4 Quantum efficiency of solar cell shown in Fig. 3 with efficiency of 2.1%.

ACKNOWLEDGMENTS

This work was supported by a grant based on the Advanced Low Carbon Technology Research and Development (ALCA) programs from Japan Science and Technology Agency (JST) under the Ministry of Education, Culture, Sports, Science and Technology in Japan.

REFERENCES

1. P. Jackson, D. Hariskos, E. Lotter, S. Paetel, R. Wuerz, and R. Menner, Prog. Photovolt. Res. Appl. 19, 894 (2011).
2. TM. Friedlmeier, N. Wieser, T. Walter, H. Dittrich, and HW. Schock, Proceedings of the 14th European Photovoltaic Specialists Conference, Barcelona, p. 1242 (1997).
3. H. Katagiri, K. Jimbo, S. Yamada, T. Kamimura, WS. Maw, T. Fukano, T. Ito and T. Motohiro, Applied Physics Express 1, 041201 (2008).
4. T. K. Todorov, J. Tang, S. Bag, O. Gunawan, T. Gokmen, Y. Zhu, and D. B. Mitzi, Adv. Energy Mater. 3, 34 (2013).

182

5. C. Eberspacher, C. Fredric, K. Pauls, and J. Serra, Thin Solid Films **387**, 18 (2001).
6. T. Wada, Y. Matsuo, S. Nomura, Y. Nakamura, A. Miyamura, Y. Chiba, A. Yamada and M. Konagai. Phys. Stat. Sol. (a) **203**, 2593 (2006).
7. J. Kubo, Y. Matsuo, T. Wada, A. Yamada, and M. Konagai, Thin-Film Compound Semiconductor Photovoltaics-2009, (Mater. Res. Soc. Symp. Proc. Vol. **1165**), 1165-M05-13.
8. T. Wada, H. Kinoshita, and S. Kawata, Thin Solid Films **431-432**, 11 (2003).
9. T. Wada and H. Kinoshita, J. Phys. Chem. Solids **66**, 1987 (2005).
10. T. Wada, J. Kubo, S. Yamazoe, A. Yamada, and M. Konagai, Proc. 25th European Photovoltaic Solar Energy Conference and Exhibition/5th World Conference on Photovoltaic Energy Conversion (Valencia), p. 3465 (2010).
11. F. Gao, S. Yamazoe, T. Maeda, and T. Wada, Jpn. J. Appl. Phys. **51**, 10NC28 (2012).
12. F. Gao, S. Yamazoe, T. Maeda, K. Nakanishi, and T. Wada, Jpn. J. Appl. Phys. **51**, 10NC29 (2012).
13. H. Katagiri, K. Jimbo, M. Tahara, H. Araki and K. Oishi, Thin-Film Compound Semiconductor Photovoltaics-2009, (Mater. Res. Soc. Symp. Proc. Vol. 1165), 1165-M04-01.

Mater. Res. Soc. Symp. Proc. Vol. 1538 © 2013 Materials Research Society
DOI: 10.1557/opl.2013.978

Cupric Oxide Thin Films for Photovoltaic Applications

Patrick J. M. Isherwood, Biancamaria Maniscalco, Fabiana Lisco, Piotr M. Kaminski, Jake W. Bowers and John M. Walls
CREST, School of Electrical, Electronic and Systems Engineering, Loughborough University, Loughborough, Leicestershire, LE11 3TU UK
Email: P.J.M Isherwood@lboro.ac.uk

ABSTRACT

Cupric oxide thin films were sputtered onto soda-lime glass slides from a single pre-formed ceramic target using a radio-frequency power supply. The effects of oxygen partial pressure and substrate temperature on the optical, electrical and structural properties of the films were studied. It was found that increasing temperature resulted in increased crystallinity and crystal size but also increased film resistivity. The most conductive films were those deposited at room temperature. Increasing oxygen partial pressure was found to reduce resistivity dramatically. This is thought to be due to higher charge carrier concentrations resulting from increased copper vacancies. Increasing oxygen partial pressure causes an increase in the optical band gap from a minimum of 0.8eV up to a maximum of 1.42eV. Oxygen-rich films display reduced crystallinity, becoming increasingly amorphous with increased oxygen content. These results show that the optical, electrical and structural properties of sputtered cupric oxide films can be controlled by alteration of the deposition environment.

INTRODUCTION

Transparent conductive oxides (TCOs) are wide-band gap semiconductors which when doped to a sufficiently high degree begin to show metal-like conductive behaviour [1]. They are easily formed, and common techniques such as chemical vapour deposition (CVD), magnetron sputtering and chemical bath deposition (CBD) are regularly employed [1]. They have good electrical properties, are relatively easy to produce and exhibit good chemical and thermal stability [1], [2]. They are used for a very wide range of purposes, including electrical contacts in computer screens, smartphone displays, LEDs and photovoltaic cells [1], [3]. TCOs in common usage are almost exclusively n-type. There are various p-type materials currently being investigated, but so far none have been developed with the same order of magnitude of electrical conductivity as n-type materials such as aluminium-doped zinc oxide (AZO) or tin-doped indium oxide (ITO) [4–6]. The most significant problem for p-type TCOs is that oxygen ions tend to localise holes (the majority charge carriers), preventing them from moving readily through the material [6]. There are a limited number of cationic species which are able to counteract this effect, notably copper, silver and gold [6–9]. Of these, copper is the most interesting largely because it is both relatively cheap and commonly available.

Cupric oxide (CuO) is one of two principal oxides of copper, and has the chemical formula CuO. It is an indirect band gap semiconductor, with reported experimental band gap values of 1.2 to 2eV [10], and theoretical values of around 1eV [11]. It is naturally p-type, even when not extrinsically doped. This is thought to be a result of copper vacancies and oxygen interstitials in the crystal structure [12]. So far it does not appear to have been successfully doped n-type, despite one or two attempts [12].

Whilst cuprous oxide is a well-researched source material for various of the potential p-type TCOs (e.g., [9]), there appears to have been relatively little work done on cupric oxide from this perspective. As an alternative p-type oxide, a better understanding of this material could allow for the design, development and improvement of new and existing p-type conductors, both transparent and otherwise. Given the reported band gap, it is also a candidate photoabsorber. This study provides an optical, electrical and structural characterisation of cupric oxide thin films, with a particular focus on the effects of substrate temperature and oxygen partial pressure during deposition.

EXPERIMENTAL

Cupric oxide was sputtered directly from a 3 inch diameter ceramic target onto 50 mm by 50 mm soda-lime glass slides using a radio frequency (RF) power supply. Power was kept at 120 Watts, and deposition time was two hours. Argon flow rate was kept at 7 standard cubic centimetres per minute (SCCM) and pressure was limited to 0.1333 Pa (1 millitorr). For the temperature experiments, oxygen input was kept at two SCCM of 100% oxygen, and for the oxygen content experiments, temperature was kept at 18 degrees Celsius. Temperature-variable depositions were run at 18, 50, 100, 150, 200, 250, 300, 350, 400 and 450 degrees Celsius. Oxygen-variable depositions were run using both a 1% oxygen in argon gas line and a 100% oxygen gas line. Depositions were run with oxygen partial pressures of 1.67×10^{-4}, 5.56×10^{-4}, 7.84×10^{-4}, 1.67×10^{-2}, 2.96×10^{-2} and 4.85×10^{-2} Pa. These correspond to oxygen flow rates of 0.01, 0.05, 0.1, 1, 2 and 4 SCCM respectively. Characterisation of the finished films involved conducting Hall tests on 1 cm square samples, spectrophotometer measurements to test the transmission and for use in band gap calculations, profilometer measurements on partially-etched samples to find the film thickness, SEM surface imaging and XRD analysis. Film depositions were carried out using an AJA International Orion 8HV magnetron sputtering system with an AJA 600 series radio frequency power supply. Hall measurements were taken using an Ecopia HMS 3000 system. Transmission measurements were taken using a Varian Cary 5000 spectrophotometer. The spectrum used was from 1400 nm to 200 nm, with some of the oxygen-deficient samples being measured from 2000 nm to 200 nm. Measurements were taken every 5 nm. Thickness measurements were done using an Ambios XP2 stylus profilometer. Measured resistivities were cross-checked using a four-point probe.

DISCUSSION

Temperature

It was found that increasing the substrate temperature during deposition increased the film resistivity (Figure 1). For most materials, it is normal for increasing substrate temperature to result in a more conductive film (e.g., [13]). This is thought to be due to the formation of better crystal structures at higher temperatures. For cupric oxide, it is thought that the low temperatures allow for an increase in film defects, particularly copper vacancies and oxygen interstitials. This would result in an increased carrier concentration, thereby reducing the resistivity. Unfortunately Hall effect measurements of both carrier concentration and mobility proved inconclusive. This was because many of the films were too resistive and beyond the capability of the equipment to measure accurately.

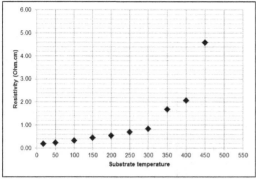

Figure 1. Resistivity against substrate temperature.

An increase in the number of defects would also have a detrimental effect on the crystallinity, with a sufficiently high defect density causing films to be amorphous. SEM photomicrographs of the film surfaces show that room-temperature films are very fine-grained, bordering on amorphous, with a gradual increase in crystallinity and crystal size with increasing temperature (Figure 2).

Figure 2. SEM photomicrographs showing the effects of substrate temperature on film crystallinity.

Oxygen content

Resistivity was found to increase dramatically with decreasing oxygen partial pressure (Figure 3). This change is thought to be due to a significant increase in the number of charge carriers with increasing oxygen content resulting from increased copper vacancies and oxygen interstitials. As with temperature-variable experiments, Hall measurements were inconclusive.

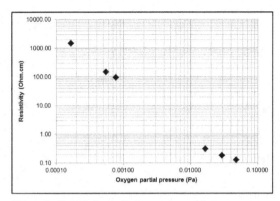

Figure 3. Resistivity against oxygen partial pressure.

At very high oxygen contents, the films appear to be very fine grained and almost amorphous (Figure 4). This suggests that the number of vacancies and other defects is sufficient to cause significant disruption of the crystal structure.

Whilst the crystal size for very low oxygen partial pressures and for high temperatures is similar, the crystal shape is very different. High temperature depositions produced crystals with a rounded appearance (Figure 2). Films with low oxygen content show a more defined plate-like structure (Figure 4). As the temperature-dependent experiments were all conducted with a relatively high oxygen partial pressure (2.96×10^{-2} Pa), the most likely explanation is that this change is due to oxygen content rather than temperature. It is likely that the two sets of films have the same basic crystal structure, but that the higher defect density due to increased oxygen content in the high temperature films causes some disruption of this structure. XRD analysis of these films was found to be inconclusive.

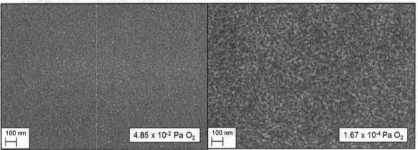

Figure 4. SEM photomicrographs showing the effects of oxygen partial pressure during deposition on film crystallinity.

The band gap was found to vary significantly with changing oxygen content, decreasing dramatically with reduced oxygen input (Figure 5). Both band gap changes and variation in resistivity appear to be the opposite of the behaviour shown by other amorphous oxides such as

188

zinc-doped indium oxide (IZO - [14]). The different behaviour for resistivity is clearly due to the difference in majority charge carrier between the two materials – IZO is n-type, whereas CuO is p-type. The band gap variation is significantly larger for CuO than it is for IZO, and appears to approximate an inverse exponential relationship to oxygen partial pressure.

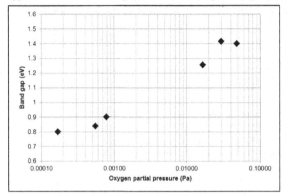

Figure 5. Band gap against oxygen partial pressure.

The observed dramatic changes in resistivity, crystal structure and band gap suggest that this material is structurally malleable, and as such, is a potential candidate for further research into mixed oxide semiconductors and mismatched alloys. Furthermore, the low resistivity values obtained without extrinsic doping show that CuO has potential as a p-type conductor. The variable band gap could yield uses in devices where absorption of a specific part of the visible and near-infrared spectrum is important, such as in multi-junction stacks. If it can be combined with other oxide materials in such a way as to increase the band gap further, it has promise as a p-type TCO.

CONCLUSIONS

It has been shown that by altering the temperature and oxygen partial pressure during deposition, the electrical, optical and structural properties of sputtered cupric oxide can be altered significantly. The greatest effects are shown by altering the oxygen partial pressure, with material band gaps changing from 0.8 to 1.42 eV and resistivity from highly insulating to around 0.1 Ohm.cm. The decrease in resistivity with increasing oxygen content is thought to be due to a greater number of copper vacancies and oxygen interstitials and a resulting increase in carrier concentration. As oxygen content increases, films also become increasingly fine-grained and possibly amorphous.

REFERENCES

[1]   D. S. Ginley and J. D. Perkins, "Transparent Conductors", *Handbook of Transparent Conductors*, ed. D. S. Ginely, H. Hosono and D. C. Paine. (London: Springer, 2010) p. 1-26.

[2]   D. S. Ginley, "Nanoparticle Derived Contacts for Photovoltaic Cells," *Electrochemical Society Proceedings*, vol. 99, no. 11, pp. 103-110, 1999.

[3]   C. G. Granqvist, "Transparent conductors as solar energy materials: A panoramic review," *Solar Energy Materials and Solar Cells*, vol. 91, no. 17, pp. 1529-1598, Oct. 2007.

[4]   G. Dong, M. Zhang, W. Lan, P. Dong, and H. Yan, "Structural and physical properties of Mg-doped CuAlO2 thin films," *Vacuum*, vol. 82, no. 11, pp. 1321-1324, Jun. 2008.

[5]   J. Tate et al., "Origin of p-type conduction in single-crystal CuAlO2," *Physical Review B*, vol. 80, no. 16, pp. 1-8, Oct. 2009.

[6]   H. Kawazoe, H. Yanagi, K. Ueda, and H. Hosono, "Transparent p-type conducting oxides: design and fabrication of pn heterojunctions," *MRS Bulletin*, vol. 25, no. 8, pp. 28–36, 2000.

[7]   D. O. Scanlon, K. G. Godinho, B. J. Morgan, and G. W. Watson, "Understanding conductivity anomalies in Cu(I)-based delafossite transparent conducting oxides: Theoretical insights.," *The Journal of chemical physics*, vol. 132, no. 2, p. 024707, Jan. 2010.

[8]   H. Yanagi, S.-ichiro Inoue, K. Ueda, H. Kawazoe, H. Hosono, and N. Hamada, "Electronic structure and optoelectronic properties of transparent p-type conducting CuAlO2," *Journal of Applied Physics*, vol. 88, no. 7, pp. 4159-4163, 2000.

[9]   B. J. Ingram et al., "Transport and Defect Mechanisms in Cuprous Delafossites. 1. Comparison of Hydrothermal and Standard Solid-State Synthesis in CuAlO2," *Chemistry of Materials*, vol. 16, no. 26, pp. 5616-5622, 2004.

[10]   L. Wang, K. Han, G. Song, X. Yang, and M. Tao, "Characterization of electro-deposited CuO as a low-cost material for high-efficiency solar cells," *Nature*, pp. 130-133, 2006.

[11]   D. Wu, Q. Zhang, and M. Tao, "LSDA+U study of cupric oxide: Electronic structure and native point defects," *Physical Review B*, vol. 73, no. 23, pp. 1-6, Jun. 2006.

[12]   Y. Peng, Z. Zhang, T. Viet Pham, Y. Zhao, P. Wu, and J. Wang, "Density functional theory analysis of dopants in cupric oxide," *Journal of Applied Physics*, vol. 111, no. 10, p. 103708, 2012.

[13]   J.-H. Lee, "Effects of substrate temperature on electrical and optical properties ITO films deposited by r.f. magnetron sputtering," *Journal of Electroceramics*, vol. 23, no. 2-4, pp. 554-558, Jul. 2008.

[14]   A. Leenheer, J. Perkins, M. van Hest, J. Berry, R. O'Hayre, and D. Ginley, "General mobility and carrier concentration relationship in transparent amorphous indium zinc oxide films," *Physical Review B*, vol. 77, no. 11, p. 115215, Mar. 2008.

Mater. Res. Soc. Symp. Proc. Vol. 1538 © 2013 Materials Research Society
DOI: 10.1557/opl.2013.1003

Spectral Calibrated and Confocal Photoluminescence of Cu2S Thin-Film Absorber

Hendrik Sträter[1], Rudolf Brüggemann[1], Sebastian Siol[2], Andreas Klein[2], Wolfram Jaegermann[2], Gottfried H. Bauer[1]
[1]Institut für Physik, Carl von Ossietzky Universität Oldenburg, D-26111 Oldenburg, Germany
[2]Fachbereich 11, Materialwissenschaft, Fachgebiet Oberflächenforschung, Technische Universität Darmstadt, D-64287 Darmstadt, Germany.

ABSTRACT

We have studied Cu2S absorber layers prepared by physical vapor deposition (PVD) by calibrated spectral photoluminescence (PL) and by confocal PL as function of temperature T and excitation fluxes to obtain the absolute PL-yield at an excitation flux equivalent to the AM1.5 spectrum and to calculate the splitting of the quasi-Fermi levels (QFL) $\mu = E_{f,n} - E_{f,p}$ and the absorption coefficient $\alpha(E)$, both in the temperature range $20\ K \leq T \leq 400\ K$. The PL-spectra reveal two peaks at $E_1 = 1.17\ eV$ and $E_2 = 1.3\ eV$, of which the low energy peak is only detectable at temperatures $T < 200\ K$. The samples show an impressive QFL-splitting of $\mu > 700$ meV at 300 K associated with a pseudo band gap of $E_g = 1.25\ eV$. The high energy peak shows an unexpected temperature behavior, namely an increase of the PL-yield with rising temperature at variance with the behavior of QFL-splitting that decreases with rising T from extrapolated $T = 0K$ value of $\mu = 1.3\ eV$. The PL-yield versus temperature will be discussed in terms of different defect states in the band gap. Our observations indicate that, contrary to common believe, it is not the PL-yield, but rather the QFL-splitting that is the comprehensive indicator of the quality of the excited state in an illuminated semiconductor. A further examination of the lateral variation of the opto-electronic properties by confocal PL shows a strong correlation between the QFL-splitting, the Urbach energy E_U and the optical band gap E_{opt}, respectively.

INTRODUCTION

Copper sulfide (Cu2S) is an abundant and non-toxic p-type semiconductor and thus a potential candidate for an absorber in thin film solar cells [1]. The Cu2S/CdS heterojunction was even the first reported thin film solar cell device [2], which later obtained efficiencies of $\eta \approx 10\%$ [3]. We present a detailed study of the excitation intensity and temperature dependent photoluminescence (PL) on the large and onto the microscopic scale. We use PL experiments as a contact-less technique to determine the radiative recombination of photo-generated excitation states within a semiconductor, which provides access to opto-electronic properties like the energy dependent absorption A(E), the splitting of quasi-Fermi levels $\mu = E_{f,n} - E_{f,p}$ and additionally information about defect levels within the band gap and the dynamics of transitions. Since the QFL-splitting can be interpreted as the limiting open-circuit voltage of a finally fabricated solar cell, $V_{oc} = \mu/e^-$, PL measurements allow analyzing the condition of an absorber material before it is used as a solar cell. According to Planck's generalized law, the emitted photon flux or PL-yield $Y_{PL}(E)$ at an energy E of an excited semiconductor with excitation energy larger than the band gap into the solid angle Ω can be written as [4]

$$Y_{PL}(E) = \frac{\Omega A(E) E^2}{4\pi^2 \hbar^3 c^3} \left[\exp\left(\frac{E-\mu}{kT}\right) - 1 \right]^{-1} \tag{1}$$

with the speed of light in vacuum c and Boltzmann constant k. With the exactly determined photon flux $Y_{PL}(E)$ one may calculate μ by fitting in the high energy regime, where $A(E) \approx 1$.

EXPERIMENT

In our work we focus on Cu_2S samples prepared at the TU Darmstadt by PVD with a substrate temperature of $T_{sub} = 90°C$ during the deposition process and thicknesses of about d = 500 nm. The films have a stoichiometry of Cu:S=2:1 and show a monoclinic structure. In our paper we focus on the sample with the best electro-optical properties (highest absolute QFL-splitting and largest PL-yield). Details of the deposition process can be found in [5].

To determine the absolute quantitative PL flux two variants of the same PL setup were used, which has been calibrated with a tungsten filament lamp. The samples were illuminated with a stationary 532 nm Nd:YAG laser with intensities of 10^{-3}-10^1 times the AM1.5 solar spectrum equivalent. For cooling the sample down to 12 K a closed-cycle cryostat was used and the sample was heated against the cooling up to 400 K. To collect the emitted PL we used either a calibrated monochromator with attached Si-diode for calibrated measurements or a reflective collimator, which feds the PL radiation into a glass fiber and an optical multichannel analyzer (OMA) with either a 512x1 pixel InGaAs CCD line or a 1340×100 pixel Si CCD array, both detectors cooled with LN2. For the confocal PL measurement we used a confocal microscope (WITEC alpha SNOM) and with lateral resolution $\Delta x \approx 0.9$ μm and excitation at 532 nm. The PL signal was also coupled into an OMA with a 512x1 pixel CCD line.

RESULTS AND DISCUSSION

Temperature-dependent photoluminescence

The T-dependent PL-spectra in Fig. 1a) reveal two different peaks from T = 40 K to T = 200 K with different temperature behavior. While the low-energy (LE) peak, denoted as #1 in the following, at $E_1 = 1.17$ eV vanishes with increasing T, the high energy (HE) peak, denoted as #2, starts to grow at $E_2 = 1.29$ eV and undergoes a red-shift for T > 120 K while peak #1 undergoes a small blue-shift. The PL-peak E_2 lies at an energetic position of 1.22 eV at room temperature. In contradiction to common PL-measurements the PL-yield of peak #2 increases with increasing temperature up to one order of magnitude between T = 30 K and T = 340 K (Fig. 2a) as well as the integrated PL-yield of the whole spectrum (Fig. 1b).The QFL-splitting in Fig. 1c) decreases with increasing T and shows an astonishing value of $\mu = 710$ meV at room temperature and an excitation intensity equivalent to the AM1.5 spectrum. Please be aware, that there is a small shift between the set cryostat temperature and the temperature from the fit due to noisy data and an insufficient spectral resolution in the high energy regime. The QFL-splitting extrapolates to T = 0 K with an energy of $E_0 = 1.30$ eV, which can be interpreted as the band gap at T = 0 K.

There is a slight blue-shift of LE peak #1 while the HE peak #2 undergoes a small blue-shift with increasing T up to T=150 K and undergoes a red-shift for higher T. Interestingly, the red-shift of peak #2 begins when peak #1 is not visible anymore. The temperature behavior of peak #2 can be explained by the theory of negative thermal quenching [7]. Due to thermal activated emitting of holes/electrons into the valence/conduction band, the PL-yield is increased. The decrease for T > 350 K in Fig. 2a) can be attributed to a phase transition of the Cu_2S [6].

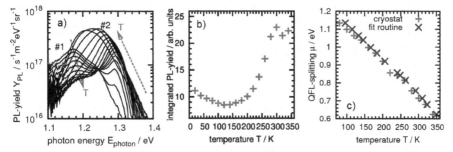

Figure 1: a) PL-spectra in the range 20 K ≤ T ≤ 340 K. Two peaks with different temperature behavior can be determined. b) The integrated PL-yield increases with increasing T while the QFL-splitting in c) decreases with increasing T and has its maximum of μ = 1.3 eV at T = 0 K.

The PL-yield as function of T can then be generally described by: [7]

$$\frac{Y_{PL}(T)}{Y_{PL,0}} = \frac{1 + T^{3/2} \sum_{i=1}^{m} C_i \exp\left(-E_{act,i}/kT\right)}{1 + T^{3/2} \sum_{j=1}^{n} D_j \exp\left(-E_{act,j}/kT\right)}$$ (2)

with the activation energy $E_{act,i}$ of each donor or the acceptor state, m the number of states with negative thermal quenching, n the number of states with thermal quenching, C_i, D_i describing the transition probability, and $Y_{PL,0}$ a fitting parameter. If the transition of carriers evolves into another defect instead of the valence or conduction band, the term $T^{3/2}$ can be omitted.

For the LE peak #1 we can extract two activation energies $E_{act,1}$ = 0.3 meV and $E_{act,2}$ = 11 meV with a fit of Eq.(2) with m = 0, n = 2 and $E_{act,2}$ with a $T^{3/2}$ dependence, see Fig. 3a). With increasing temperature, peak #1 undergoes a blue-shift of the order of kT/2 for T < 20 K and a blue-shift of the order of 1.75kT for higher T, indicating a free-to-bound transition for T < 20 K [8], which changes into a DA pair transition [9].

For the HE peak #2 the best fit of Eq.(2) reveals one level with negative thermal quenching (m = 1) and one level with positive thermal quenching (Fig.3b). The activation energies are $E_{act,1}$ = 7 meV and $E_{act,2}$ = 53 meV. The increase of the PL-yield is a result of the emission of carriers from a level 7 meV below the conduction band (or 7 meV above the valence band). The PL peak position changes with 0.9 kT for T < 100 K, indicating also a DA pair transition (Fig.3c). We can speculate, that $E_{act,1}$ and $E_{act,2}$ describe the same defect level and the mismatch is a result from the fitting routine. The difference of the peak maxima ΔE=E_2-E_1 is too large to be explained with the calculated activation energies indicating a third donor or acceptor level deep in the band gap, which forms the observed DA pair with one of the defect states with activation energies extracted from the fit of the $Y_{PL}(T)$ behavior of peak #1. Interestingly, in this case the negative thermal quenching occurs at much higher temperatures as it has been observed before [7] and might be the only reason, why the Cu_2S sample shows such a large PL-yield and QFL-splitting at room temperature.

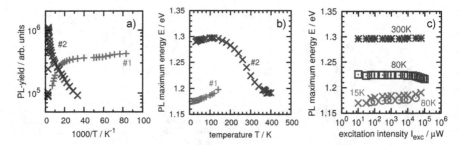

Figure 2: a) Temperature dependence of PL-yield and PL-peak position of both peaks. b) Peak #1 undergoes a blue-shift with increasing temperature while peak #2 undergoes at first a small blue-shift and at higher T a strong red-shift. c) With increasing excitation intensity the position of peak #1 shifts to higher energy (15 K, 80 K) while peak #2 does not change its position (80 K, 300 K) unless for very high intensities, which leads to a small red-shift due to a heating of the film.

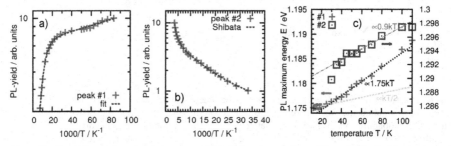

Figure 3: Fit of the temperature behavior with Eq.2 for a) peak #1 and b) peak #2, details are given in the text. c) Temperature dependence of PL peak position with varying slope of kT for both peaks.

Excitation-dependent photoluminescence

In general, the PL-yield as function of excitation intensity I_{exc} follows a power law [10]

$$Y_{PL}(I_{exc}) \propto I_{exc}^{\gamma} . \tag{3}$$

The exponent γ might give a hint to the underlying recombination process of the transitions, although γ is not necessarily constant over a large intensity range and can differ with T. For free-to-bound and DA pair recombination transitions one would expect $\gamma < 1$ and $\gamma = 1$ for band-to-band transitions, while for band-to-band recombination at high excitation energies and for excitons $\gamma \leq 2$ is expected [10,11].

From the PL-yield as function of I_{exc} we obtain an exponent $\gamma \approx 1$ at T = 15 K for peak #1, indicating a free-to-bound [11] transition. Interestingly, γ increases with increasing excitation energy at T = 80 K, what might be an effect of the concurring band-to-band transition, which is not visible for T = 15 K. For peak #2 γ increases from 0.8 up to 1 with increasing intensity at T = 80 K and γ ≈ 1 for T = 120 K, 200 K and 300 K over a large intensity regime. The excitation intensity dependent PL measurement shows a small blue-shift for the LE peak #1 with β = 3 meV/decade for a temperature of T = 15 K as well as for T = 80 K, indicating a weak compensated DA pair transition [11]. Peak #2 does not change its position with varying intensity. The observed energetic red-shift for very high intensities is moreover related to a rising temperature of the film due to the illumination.

Lateral variation of opto-electronic properties

The maps of the lateral variation of the opto-electronic properties in Fig. 4 show that pixels with a high integrated PL-yield do not necessarily show a higher QFL-splitting (see pixel 2). It is noted that $\Delta\mu$ denotes the variation from the mean μ. A higher μ is moreover connected to smaller Urbach energy [12] E_U and a higher E_{opt} (defined as energy at which $A(E)=\exp(-1)$) (see pixel 1). Regions with a smaller μ show a small PL-yield, a large E_U and a significant smaller band gap (pixel 6).

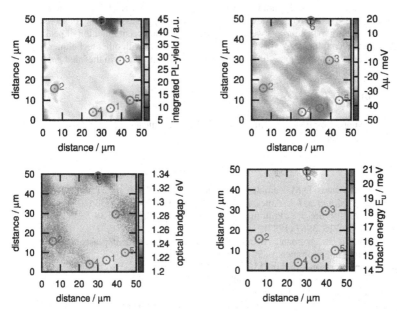

Figure 4: Lateral variation of opto-electronic properties of a 50 μm × 50 μm large section with marked pixels.

The temperatures T_{fit} obtained from the fitting routine of the high-energy wing of the PL-spectrum is also increased, what is probably an effect of a higher subgap-absorption and a larger (non-radiative) thermalization, which also reduces the radiative PL-emission. As T_{fit} of pixels (1) - (5) are nearly the same we can exclude a wrong fitting procedure, whereas T_{fit} of pixel 6 is nearly 100 K larger. This could be an effect of the larger defect absorption, which results in a higher E_U. The full-width-half-maximum (FWHM) of the $\Delta\mu$ distribution is 11 meV and the FWHM of the E_{opt} distribution is 30 meV.

CONCLUSIONS

Temperature and excitation intensity dependent PL measurements of a thin film Cu_2S layer show both an astonishing high QFL-splitting of $\mu = 710$ meV at room temperature and AM1.5 condition indicating a high quality of the semiconductor regarding its potential as solar cell absorber material. The temperature behavior shows a peak with increasing PL-yield as the temperature increases. We can explain the temperature behavior of this peak with the model of negative thermal quenching due to a band-edge near defect level, which emits carriers into the band with increasing T and acts as a thermal activated donor. At this point we want to emphasize that instead of the PL-yield it is the QFL-splitting which indicates the "quality" of the photo-excited state and that a look at the PL-yield only may be misleading if different samples or conditions have to be compared. In summary, we determined several defect states in the band gap and allocated these states to common recombination models. We assigned the observed temperature- and spectral behavior to a weak compensated DA pair recombination of peak #1 and a band-to-band recombination of peak #2. Further experiments with other techniques (e.g. DLTS) would be a useful tool to pinpoint the estimated defects absolutely in the band gap.

ACKNOWLEDGMENTS

Financial support by BMBF contract 03SF0358 and technical support by Peter Pargmann are gratefully acknowledged.

REFERENCES

1. C. Wadia, A.P. Alivisatos, D.M. Kammen, Environ. Sci. Technol. **43**, 2072 (2009).
2. D.C. Reynolds, B.M. Leies, L.L. Antes, R.E. Marburger, Phys. Rev. **96**, 533 (1954).
3. R. B. Hall, R. W. Birkmire, J. E. Phillips, J. D. Meakin, Appl. Phys. Lett. **38**, 925 (1981).
4. P. Würfel, J. Phys. C **15**, 3967 (1982).
5. S. Siol, H. Sträter, R. Brüggemann, A.Klein, G.H. Bauer, W. Jaegermann presented at the 2013 MRS Spring Meeting in San Francisco, 2013 (unpublished).
6. B.A. Mansour, phys. status solidi (a) **136**, 153 (1993).
7. H. Shibata, Jpn. J. Appl. Phys. **37**, 550 (1998).
8. W.P. Dumke, Phys. Rev. **132**, 1998 (1963).
9. J.J. Pankove, *Optical Processes in Semiconductors* (Dover, New York, 1972), ch. 6.
10. T. Schmidt, K. Lischka, W. Zulehner, Phys. Rev. B **45**, 8989 (1992).
11. S. Siebentritt in *Wide-Gap Chalcopyrites*, edited by S. Siebentritt, U. Rau (Springer-Verlag, Berlin Heidelberg, 2006), ch. 7.
12. M.V. Kurik, phys. status solidi (a) **8**, 9 (1971).

Mater. Res. Soc. Symp. Proc. Vol. 1538 © 2013 Materials Research Society
DOI: 10.1557/opl.2013.1025

CuO and Cu$_2$O Nanoparticles for Thin Film Photovoltaics

Jan Flohre[1], Maurice Nuys[1], Christine Leidinger[1], Florian Köhler[1] and Reinhard Carius[1]
[1]Institute of Energy and Climate Research 5 -Photovoltaics-, Forschungszentrum Jülich GmbH, D-52425 Jülich Germany

ABSTRACT

Laser annealing experiments on commercially available phase pure tenorite (CuO) nano-particles (NPs) were performed in air and nitrogen atmosphere to improve the structural and electronic properties, with respect to their suitability for photovoltaic applications. The particles exhibit size variations from about 30 nm to 100 nm. The influence of the thermal treatment is investigated by photoluminescence (PL) and Raman spectroscopy. Annealing of the particles in air by a laser treatment improved the material quality by defect reduction. Additional laser an-nealing in N$_2$ atmosphere leads to a phase transition of the NPs from tenorite to cuprite (Cu$_2$O). Due to the low partial oxygen pressure, the transition is initiated at about 1/3 of the maximum laser power used for the series in air, which is indicated by a drastic increase of the band edge emission from Cu$_2$O. However, annealing with higher laser power leads to strong defect lumi-nescence, which originates from copper and oxygen vacancies. A weak remaining tenorite band edge emission shows an incomplete phase transition.

INTRODUCTION

Cost effective solar cells with high efficiency based on abundant, non-toxic materials is the long term target of present research and development. Multijunction, material saving, thin film solar cells, including nanostructures or NPs, is considered as an important option for future solar cell technologies. Tenorite and cuprite fulfill many of these requirements, e.g. the suitable band gap energies of about 1.3 eV to 1.5 eV and 2.1 eV [1, 2]. In previous studies we showed that thermal annealing in an oven in nitrogen atmosphere improves the material quality of the tenorite NPs up to a temperature of 700°C and leads to a phase transition to cuprite after anneal-ing at 800°C [3]. In this work, commercially available tenorite nanoparticles are investigated and modified by laser annealing in order to improve their electronic and structural properties for use as active absorber material in solar cells. Laser annealing has the advantage that the heat pene-tration into the material is locally, so that a thermal treatment can be performed on non-heat re-sistant substrates, which is important for the use in thin film technology. The impact of the laser annealing on the material properties is investigated by Raman and PL spectroscopy. Since a cor-relation between cell efficiency and PL at room temperature is given by the theory of detailed balance [4], PL is a powerful method to investigate absorber materials for photovoltaic devices.

EXPERIMENT

Commercially available tenorite NPs (IoLiTec GmbH; Germany) were studied. The nominal diameter of the particles is 40 nm to 80 nm. Preliminary TEM investigations showed that the particles exhibit size variations from about 30 nm to 150 nm [3]. The NPs were dispersed in deionized water by treatment in an ultrasonic bath. Furthermore, a small amount of this solvent was dropped onto a quartz substrate and allowed to dry in air. The NPs form larger agglomerates on the substrate leading to an inhomogeneous coverage. The structural and electronic properties were characterized by a micro Raman and photoluminescence setup. Here, a microscope is used to focus a 532 nm solid-state laser to a spot size diameter of about 1 μm. Moreover, the same configuration is used for Raman and PL measurements, so that the corresponding signals arise from the same NP ensemble. The PL is detected by a Si detector and an additional InGaAs detector leading to a spectral range of about 0.8 eV to 2.3 eV. The PL signals are corrected for the spectral response of the experimental setup. In addition, samples can be probed in air and in nitrogen atmosphere. For the thermal treatment, the laser power of the solid-state laser is stepwise increased, and the particles are annealed for 10 minutes. Afterwards, PL and Raman measurements at the annealed positions were performed with the initial low laser power.

Since the phonon energy is depending on the temperature, we used the temperature induced blue-shift of the Raman lines to estimate the temperature within the laser focus. As a reference, temperature dependence of tenorite Raman lines of NPs from Xu et al. [8] is used. An estimation of the temperature by the ratio of the Stokes and anti-Stokes Raman scattering intensity was also tested but did not work due to a strong background signal, likely caused by PL.

The samples coverage and the nano-material are inhomogeneous. For this reason the investigated and treated area on the samples is very small (1 μm). But then it is not possible to perform macroscopic absorption measurements on the treated samples, e.g. by photo-thermal deflection spectroscopy (PDS). Therefore we compare the results with oven annealing experiments where investigations by PDS were feasible.

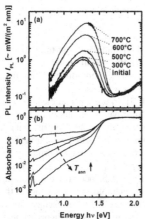 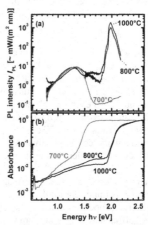

Figure 1. Representative PL spectra (a) and PDS signal (b) of tenorite NPs stepwise oven annealed up to a temperature of 700°C taken from [3].

Figure 2. Representative PL spectra (a) and PDS signal (b) of tenorite and cuprite NPs stepwise oven annealed up to a temperature of 1000°C taken from [3].

DISCUSSION

Oven annealing in N₂

In our previous work, oven annealing experiments were performed with similar tenorite NPs in nitrogen atmosphere. In figure 1a-b the PL and PDS data of the annealing series up to 700°C is shown. The band edge emission of the tenorite NPs increases with increasing annealing temperature by about one order of magnitude due to defect reduction. This can be concluded from the decreasing sub gap absorption in the PDS signal [3]. Figure 2a-b shows PL and PDS data after the 800°C and 1000°C annealing step. Strong band edge emission from cuprite is observed at 2 eV as well as less intense defect luminescence. The absorption edge in the PDS data is shifted to the position of the cuprite band gap energy, and no tenorite signal is observed after the 1000°C temperature treatment [3]. It should be noted that the NPs seems to be stable as all PL measurements were performed in air and no change of the intensity was observed.

According to the stability diagram of the Cu-O system in figure 3 [5, 6], the phase transition from tenorite to cuprite occurs at about 1000°C in air. Since we observed the phase transition between 700°C to 800°C, the partial oxygen pressure must be between -2 mm_{Hg} to 0 mm_{Hg}.

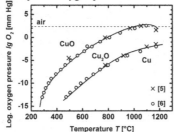

Figure 3. Cu-O stability diagram [5, 6].

Laser annealing in air

In figure 4 we present Raman spectra of the laser annealed tenorite NPs in air. The three most intense peaks of the untreated NPs at 296 cm^{-1}, 344 cm^{-1} and 629 cm^{-1} can be attributed to tenorite Raman modes [7]. All other weak scattering intensities at 320 cm^{-1}, 384 cm^{-1}, 438 cm^{-1}, and 486 cm^{-1} as well as the edge at 500 cm^{-1} are due to the quartz substrate. With increasing laser power P_A the full width at half maximum (FWHM) of the tenorite modes decreases. Likewise, the intensity of these modes is increasing. For example, the FWHM of the peak at 296 cm^{-1} decreases from about 19 cm^{-1} to 15 cm^{-1}, while the intensity is increasing by about 30 %. The reduction of the FWHM and the increasing Raman intensity indicates an improved microstructure, e.g. by reduction of the density of structural defects and a release of stress.

In figure 5, the corresponding room temperature PL spectra are pictured, which are taken from the same spot as the Raman spectra measured in figure 4. The PL spectra show the tenorite band edge emission about 1.28 eV, which is increased by a factor of three after the laser annealing process. In contrast, the PL signal at energies >1.7 eV decreases by a factor of two for the high temperature treatment. We attribute the increase of the PL signal to a reduction of defects due to a relaxation of the structure caused by the temperature treatment. This leads to a reduction of non-radiative recombination and higher PL intensity. The decreasing PL for energies >1.7 eV is currently not understood.

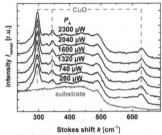

Figure 4. Raman spectra of stepwise laser annealed copper oxide NPs in air and Raman spectrum of the substrate. The CuO Raman modes are marked.

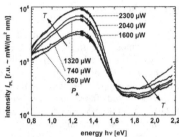

Figure 5. PL spectra of stepwise laser annealed copper oxide NPs in air.

The temperature during the annealing process, estimated by the induced blue-shift of the Raman mode 296 cm^{-1}, is shown in figure 8. The temperature is nearly constantly increasing from 20°C at 260 µW laser power to 590°C at 2040 µW but saturates for higher annealing power. This value is in good agreement with the increasing PL signal because also oven annealing at 600°C showed a PL increase about three times.

Laser annealing in N$_2$

In figure 6, Raman spectra of the laser annealing series measured in N$_2$ atmosphere on tenorite NPs are presented. Analogous to the Raman spectra of the laser annealing series in air, one can find the typical tenorite Raman modes for the first three temperature treatments, as well as the discussed quartz modes. Besides, no cuprite Raman modes are observed. After the laser treatment at 915 µW and above, additional peaks appear at 413 cm^{-1}, and 487 cm^{-1}. Furthermore, the intensity of the peak around 628 cm^{-1} increases as well as the Raman signal towards low wave numbers. All observed peaks can be attributed to cuprite [9]. In particular, the increasing signal towards low wave numbers is due to a strong cuprite Raman mode at 219 cm^{-1}. Moreover, stronger laser annealing leads to a narrowing of the mentioned peaks. As intensity of the tenorite modes are decreasing and the cuprite modes are increasing with increasing laser annealing power, it is evident that tenorite is transformed into cuprite.

The corresponding room temperature PL spectra are shown in figure 7. In the spectra, which were measured after annealing with 225 µW to 490 µW, tenorite band edge emission at 1.28 eV dominates. This is in agreement to the observed Raman signals in figure 6. However, for energies >1.5 eV the PL intensity increases, indicating a realignment of the electronic structure. Furthermore, annealing with 915 µW increases the PL drastically and a peak at about 2 eV is observed. While the optical band gap of cuprite has a value of 2.1 eV and the corresponding Raman spectrum shows cuprite Raman modes, we attribute the PL signal at 2 eV to the band edge emission from cuprite. Annealing with higher laser power leads to the formation of a fine structure, increases the band edge emission by two orders of magnitude, and shifts the peak from 1.92 eV to 1.94 eV. This peak shift can be explained by a shift of the band gap due to the modification of the vacancy concentration and stress variation within the nano-crystal. The fine structure is attributed to excitonic transitions in the cuprite nano-crystal [10]. Additionally, two peaks can be observed at 1.27 eV and 1.61 eV which can be due to remaining and improved tenorite band edge emission as well as defect luminescence coming from singly charged copper vacancies and doubly charged oxygen vacancies, respectively [11]. Compared to the oven annealing series in N_2, where the cuprite band edge emission is increased by four orders of magnitude, three orders of magnitude increase is achieved by laser annealing in N_2. The high defect concentration might be due to the rapid cooling process by the laser. For this reason, the equilibrium state of the atoms is not achieved. Furthermore, the annealing time might have an impact to the material modification. Further studies are needed to investigate this issue.

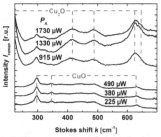

Figure 6. Raman spectra of step wise laser annealed copper oxide NPs in nitrogen. The CuO and Cu_2O Raman modes are marked.

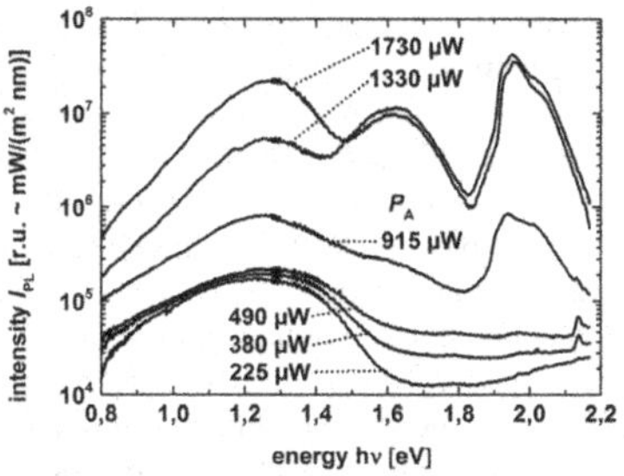

Figure 7. PL spectra of stepwise laser annealed copper oxide NPs in nitrogen.

After the first laser annealing step, an emission peak occurs at 2.13 eV (1550 rel cm^{-1}) and it can be found in the spectra until the laser annealing with 915 µW. We identified this peak to be a Raman O_2 vibration by performing laser annealing experiments also with a 488 nm argon ion laser [12]. Similarly, a peak occurs at 1550 rel cm^{-1}. Due to the phase transition, diffusing oxygen might be captured in cavities within the crystal.

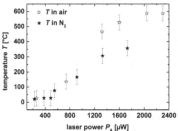

Figure 8. Temperatures within the laser focus estimated by the temperature induced Stokes shift of the laser annealing series in air (open stars), and in N_2 (filled stars).

The temperatures within the laser focus estimated by using the 296 cm^{-1} tenorite line rise from 20°C to 360°C and are presented in figure 8. Since the NPs are randomly spread on the sample surface, the temperature distribution within the laser focus might be very inhomogeneous, so that separated particles might be very hot, while bigger agglomerates are quite cold. Furthermore, for this estimation, only the tenorite particles were taken into account. The cuprite NPs might be hotter, since their phase has changed. For this reason, the estimated temperatures at high laser power are higher for the annealing series in air. Nevertheless, as the phase transition temperature depends on the partial oxygen pressure (see figure 3), which is unknown for this experiment, it is not feasible to calculate the exact temperature.

CONCLUSIONS

We showed that an improvement of the tenorite NPs can be achieved by laser annealing in air, which allows a temperature treatment on non-heat resistant substrates. The PL band edge emission was increased by a factor of three. The temperature estimation within the laser focus turned out to be difficult. A reason for this could be an inhomogeneous heat distribution within the laser focus with hot and cold NPs. A phase transformation from tenorite to cuprite was induced by laser annealing in nitrogen atmosphere due to the low partial oxygen pressure. By increasing the laser power a larger fraction of the material was transformed into the cuprite phase. Strong cuprite band edge emission was observed as well as remaining tenorite band edge emission. In contrast to previous oven annealing experiments, strong defect luminescence arising from copper and oxygen vacancies was detected. The defect luminescence is related to the rapid cooling process by the laser treatment. In conclusion, we showed that laser annealing is an appropriate method to modify and improve the material quality of CuO and Cu_2O NPs on specific positions at different temperatures in air and nitrogen atmosphere. The low heat penetration depth makes this method suitable for use in thin film photovoltaic applications.

ACKNOWLEDGMENTS

This work was partially funded by the German Federal Ministery of Education and Research (BMBF) project nr. 03SF0402A „NADNuM".

REFERENCES

1. F.P. Koffyberg, and F.A. Benko, Journal of Applied Physics 53(2), 1173 (1982)
2. A. Roos, and B. Karlsson, Solar Energy Materials 7, 467 (1983)
3. M. Nuys, J. Flohre, C. Leidinger, F. Köhler, and R. Carius, Mat. Res. Soc. Symp. Proc., 1494, (2013)
4. U. Rau, Phys. Rev. B 76, 085303 (2007)
5. M. O'Keeffe, and W. J. Moore, The Journal of Chemical Physics 36(11), 3009 (1962)
6. J. R. Coughlin, U.S. Bur. Mines Bull. 20, 542 (1954)
7. J. Chrzanowski, and J.C. Irwin, Solid State Communications 70, 11 (1989)
8. J.F. Xu, W. Ji, Z.X. Shen, S.H. Tang, X.R. Ye, D.Z. Jia, and X.Q. Xin, Mat. Res. Soc. Symp. Proc. 571, 229 (2000)
9. J. Reydellet, Phys. Stat. Sol. (b) 52, 175 (1972)
10. D.W. Snoke, A.J. Shields, and M. Cardona, Phys. Rev. B 45, 693 (1992)
11. T. Ito, and T. Masumi, J. Phys. Soc. Jpn. 66, 2185 (1997)
12. J. Cabannes, and A. Rousset, J. Phys. Radium 1(6), 210 (1940)

Mater. Res. Soc. Symp. Proc. Vol. 1538 © 2013 Materials Research Society
DOI: 10.1557/opl.2013.980

OPTIMISING THE PARAMETERS FOR THE SYNTHESIS OF CuIn-NANOPARTICLES BY CHEMICAL REDUCTION METHOD FOR CHALCOPYRITE THIN FILM PRECURSORS

Matthias Schuster[1], Stefan A. Möckel[1], Rachmat Adhi Wibowo[2], Rainer Hock[2], Peter J. Wellmann[1]

[1]Department Materials Science, Chair Materials for Electronics and Energy Technology, Friedrich-Alexander-University of Erlangen-Nürnberg, Martensstr. 7, 91058 Erlangen, Germany;
[2]Chair for Crystallography and Structure Physics, Friedrich-Alexander-University of Erlangen-Nürnberg, Staudtstr. 3, 91058 Erlangen, Germany.

ABSTRACT

Roll-to-roll deposition techniques for the fabrication of chalcopyrite solar cells are of major interest and are a promising alternative to state of the art vacuum processes. However, for roll-to-roll processes the preparation of precursor materials like nanoparticle inks is a crucial point. In this work a study on the preparation technique of copper-indium intermetallic nanoparticles was conducted. The preparation of the nanoparticles is based on the chemical reduction of copper and indium cations with sodium borohydride. Different parameters are discussed regarding their influence on (1) size and shape of the nanoparticles, (2) Cu/In ratio within the synthesised nanoparticles and (3) yield of the synthesis. Results show a strong dependency of the Cu/In ratio of the nanoparticles and the yield of the synthesis on the synthesis parameters. The influence of different parameters like (a) the ratio of metal cations to BH_4^- anions, (b) the Cu^{2+}/In^{3+} cation ratio within the precursor solution and (c) the dropping rate of the copper-indium precursor solution are discussed. The Cu/In ratio within the nanoparticles can mainly be controlled by the Cu^{2+}/In^{3+} cation ratio and the dropping rate of the copper-indium precursor solution. The yield of the synthesis shows saturation behaviour depending on the ratio of metal cations to BH_4^- anions. Shape and size of the nanoparticles are independent of the varied parameters.

INTRODUCTION

Chalcopyrite solar cells are a promising alternative to conventional silicon solar cells. Due to their high absorption coefficient layer thickness and thereby material usage can be reduced. Solar cells with an efficiency up to 20.3% [1] can be achieved by well-established fabrication processes like co-evaporation. However, to reduce fabrication costs non-vacuum processes are investigated [2].
Recently a lot of research was conducted on preparation methods using bimetallic CuIn nanoparticles for roll-to-roll processing of copper indium diselenide (CISe) solar cell absorbers [3–7]. Efficiencies of cells processed by this route range from 1.43% [3] to 7% [4]. However, for CISe thin-film solar cells the ratio of elements within the absorber layer is a critical point. Champion solar cells, which also incorporate gallium, were fabricated with a Cu-depleted composition [8]. The main reason for this composition is to avoid the formation of conductive copper-selenides which may shunt the solar cell.

To avoid etching with KCN to remove these copper-selenides [4] a good control of Cu/In ratio in synthesised nanoparticles is essential. In the following a study on the influence of different synthesis parameters on nanoparticle characteristics is presented. Parameters like (a) the ratio of metal cations to BH_4^- anions, (b) the Cu^{2+}/In^{3+} cation ratio within the precursor solution and (c) the dropping rate of the copper-indium precursor solution are varied. The influence of these parameters on (1) size and shape of the nanoparticles, (2) Cu/In ratio within the synthesised nanoparticles and (3) yield of the synthesis are discussed.

EXPERIMENT

Copper (II) chloride (Sigma Aldrich 97%), indium (III) chloride (Alfa Aeser 99.99% anhydrous), sodium borohydride (Sigma Aldrich 99%) and tetraethylene glycol (TEG) (Sigma Aldrich 99%) were used as received without any further purification.

All samples were prepared under ambient conditions at room temperature. Two different solutions were prepared. Copper (II) chloride and indium (III) chloride were dissolved in TEG with standard ratio $Cu^{2+}/In^{3+} = 0.83$ while for some experiments this ratio was varied between 0.67 and 1.12. In a standard process 1 mmol $CuCl_2$ and 1.2 mmol $InCl_3$ were dissolved in 15 ml TEG. For the variation of Cu^{2+}/In^{3+} the quantities for both chlorides had to be adjusted. The overall quantity for both chlorides in these experiments varied from 2.1 to 2.3.

A 1M sodium borohydride solution in TEG was prepared in a separate baker. The standard ratio of $BH_4^-/(Cu^{2+} + In^{3+})$ was set to 7 to ensure a sufficient reducing power and overcome the alcoholysis [9,10]. For some experiments this ratio was varied between 7 and 2.

Samples were prepared with two different setups. In the first setup the Cu^{2+}/In^{3+} precursor solution was dropped into the BH_4^- reducing agent solution. During the dropping the reducing agent solution was stirred vigorously. In the second setup the precursor solution and the reducing agent solution were dropped together simultaneously in a third baker containing vigorously stirred TEG. The BH_4^- solution was not stirred in this case. An *Ismatec REGLO Digital MS-4* peristaltic pump was used to control the dropping rates for both setups. The concentrations of both, the Cu^{2+}/In^{3+} precursor solution and the BH_4^- reducing agent solution were adjusted to keep the flow rate ratios like the respective $BH_4^-/(Cu^{2+} + In^{3+})$ ratios of the reaction. Therefore the dropping rate is standardised to the concentration and displayed as mmol/min. For both setups black precipitates were observed immediately after the contact of both solutions.

To stop the alcoholysis with TEG briefly after finishing the reaction 5 ml Acetone were added. The reaction of acetone with $NaBH_4$ is well reported [11]. Nanoparticles were isolated and washed with 2-propanol and dried in a vacuum desiccator at 40°C. The yield of each synthesis was determined by weighing the dry powder. After the synthesis the nanoparticles were stored under ambient conditions.

For phase analysis X-ray diffraction was performed using Cu Kα radiation. Morphology and size were determined using a *Zeiss field emission scanning electron microscopy* (FESEM) *ultra 55* with accelerating voltage of 10 kV. Chemical composition was analysed by energy dispersive X-ray spectroscopy (EDX) with accelerating voltage of 20 kV and Si doped with Li detector without using a standard on a *Jeol JSM 6400 Scanning Electron Microscope*. For each sample 5 measures were conducted. The mean value and standard deviation were determined.

RESULT AND DISCUSSION

To study the morphology and size of the CuIn nanoparticles these were spread on a carbon tape and characterised by FESEM at 10 kV. Particles in the range of 50 to 100 nm of were determined. All samples showed flake like nanoparticles as Figure 1 shows. The porose structure of the layer is attributed to the solvent free preparation method which did not aim at good layer quality. All samples showed a similar morphology, independent on the process parameters. Figure 1 shows three different samples synthesised with different dropping rates of 0.09 mmol/min, 0.18 mmol/min and 0.60 mmol/min.

Figure 1: FESEM-picture of CuIn nanoparticles synthesised with a dropping rate of 0.09 mmol/min (left), 0.18 mmol/min (middle) and 0.60 mmol/min (right)

a) Variation of BH_4^- / $(Cu^{2+} + In^{3+})$

Figure 2 shows the Cu/In ratio of the synthesised nanoparticles and the yield of the synthesis versus the $BH_4^-/(Cu^{2+} + In^{3+})$ ratio. Standard conditions at $Cu^{2+}/In^{3+} = 0.833$ and 0.36 mmol/min were used or the Cu^{2+}/In^{3+} ratio in the precursor solution and the dropping rate respectively. Cu^{2+}/In^{3+} precursor solution was dropped into BH_4^- solution.

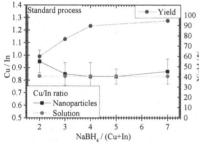 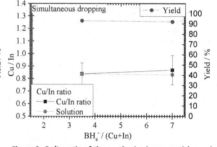

Figure 2: Cu/In ratio of the synthesised nanoparticles and yield of the synthesis is plotted against BH_4^- / $(Cu^{2+} + In^{3+})$ ratio. With increasing BH_4^- amount the Cu/In ratio of the nanoparticles converges with the Cu^{2+}/In^{3+} ratio in the precursor solution. The yield shows saturation behaviour with increasing BH_4^- amount. Cu^{2+}/In^{3+} ratio was kept constant at 0.833 and dropping rate at 0.36 mmol/min.

Figure 3: Cu/In ratio of the synthesised nanoparticles and yield of the synthesis is plotted against BH_4^- / $(Cu^{2+} + In^{3+})$ ratio of samples produced by simultaneous dropping of $(Cu^{2+} + In^{3+})$- and BH_4^- solution. Cu/In ratio in the nanoparticles and yield are seemingly independent on the BH_4^- amount. Cu^{2+}/In^{3+} ratio was kept constant at 0.833 and dropping rate at 0.36 mmol/min.

With increasing BH_4^- amount the Cu/In ratio converges with the Cu^{2+}/In^{3+} ratio in the precursor solution. When the BH_4^- amount is lowered the influence of the effect of alcoholysis becomes more important. The standard reductive potential for copper $E^0_{Cu^{2+}/Cu} = 0,345\,eV$ is significantly higher than for indium $E^0_{In^{3+}/In} = -0,338\,eV$ [3]. Hence the reduction of Cu^{2+} is

205

preferred comparing the reduction of In^{3+} and the Cu/In ratio in the nanoparticles shifts to copper-rich composition.

Figure 3 shows the result of the BH_4^- variation for the samples produced by simultaneous dropping of both, $(Cu^{2+} + In^{3+})$ precursor solution and BH_4^- solution. Both Cu/In ratio and yield are independent on the amount of BH_4^-. This is attributed to the handling of the BH_4^- solution. During the synthesis performed with simultaneously dropped solutions the BH_4^- solution was not stirred. Hence the alcoholysis did not happen as fast as it happened in the stirred solution in standard process. The reducing power of the solution did not decrease significantly and the whole amount of Cu^{2+} and In^{3+} cations can be reduced.

b) Variation of Cu^{2+}/In^{3+}

Figure 4 shows the Cu/In ratio of the nanoparticles versus the Cu^{2+}/In^{3+} ratio in the precursor solution. Other synthesis parameters were standard like $BH^{4-}/(Cu^{2+} + In^{3+}) = 7$ and dropping rate of 0.36 mmol/min. The Cu/In ratio of the nanoparticles follows the Cu^{2+}/In^{3+} ratio of the precursor solution. A linear fit shows a behaviour of Cu/In $= 0.005 + 1.02 (Cu^{2+}/In^{3+})$.

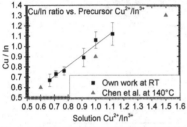 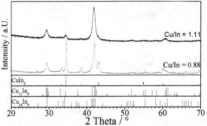

Figure 4: Cu/In ratio of the synthesised nanoparticles is plotted against Cu^{2+}/In^{3+} ratio of the precursor solution. Here all samples were prepared using standard synthesis conditions with $BH_4^-/(Cu^{2+}+In^{3+}) = 7$ and dropping rate of 0.36 mmol/min.

Figure 5: XRD pattern of a copper poor and copper rich sample. The copper poor sample shows a significant amount of the copper poor phase $CuIn_2$ while in the copper rich sample mainly copper rich phases like $Cu_{11}In_9$ and $Cu_{16}In_9$ can be observed.

These results differ from the results of Chen et al. [3]. From his results a linear behaviour with a slope $= 0.77$ can be determined, indicating a preferred reduction of copper. Also the preparation method differs. Chen et al. [3] worked at 140°C and a different $BH_4^-/(Cu^{2+} + In^{3+})$ ratio. For standard synthesis the ratio was about 2.65 [3] while for different Cu^{2+}/In^{3+} ratios the amount of $CuCl_2$ was decreased or increased, respectively. Other reaction conditions were kept constant [3]. This leads to different $BH_4^-/(Cu^{2+} + In^{3+})$ ratios reaching from 2.12 to 3.31. As shown above, the $BH_4^-/(Cu^{2+} + In^{3+})$ is a critical point. The ratio where the composition of nanoparticles is the same as the composition of the precursor solution is between 3 and 4. This is in good agreement with the results of Chen et al. [3]. In the work of Chen et al. [3] nanoparticles from a copper poor syntheses with $BH_4^-/(Cu^{2+} + In^{3+})$ ratio of 3.31 conform to the $Cu^{2+}In^{3+}$ ratio of the precursor solution. With decreasing $BH_4^-/(Cu^{2+} + In^{3+})$ ratio the Cu/In ratio of the nanoparticles diverges from the Cu^{2+}/In^{3+} ratio.

In Figure 5 characteristic XRD pattern of two samples, copper poor and copper rich, are presented. A mixture of three phases was observed. $Cu_{11}In_9$ and $Cu_{16}In_9$ which are both part of the thermodynamic phase diagram as published by Bolcavage et al. [12] but hardly distinguishable from each other by XRD. Also $CuIn_2$ was observed by XRD characterisation. Although this phase does not appear in the thermodynamic phase diagram it is well reported [12–14].

It is assumed, that these three phases are formed in the nanoparticles. Depending on the chemical composition the amount of each phase shifts. In copper rich samples mainly $Cu_{11}In_9$ and $Cu_{16}In_9$ are observed. In copper poor samples strong reflections of $CuIn_2$ can be observed as well. No elemental copper or indium nor crystalline oxides of these metals were observed for both samples. Hence the chemical composition has to be covered by intermetallic phases from copper and indium. If the amount of Cu/In ratio decreases below 1.22 (55 at.% Cu), which is the composition of $Cu_{11}In_9$ [12], $CuIn_2$ phase has to arise. With decreasing amount of copper the amount of $CuIn_2$ has to increase.

c) Variation of dropping rate

In Figure 6 and Figure 7 Cu/In ratio of synthesised nanoparticles and yield of the synthesis is shown versus dropping rate during the synthesis for particles produced by standard process and simultaneously dropped solutions respectively. For both setups the behaviour is similar to the variation of $BH_4^-/(Cu^{2+} + In^{3+})$ ratio, but the effects are less pronounced.

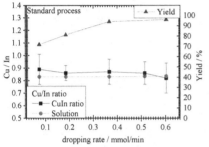 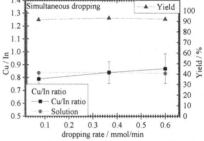

Figure 6: Cu/In ratio of the synthesised nanoparticles and yield of the synthesis is plotted against dropping rate during the synthesis. With increasing dropping rate the Cu/In ratio in the nanoparticles decreases and converges with the Cu^{2+}/In^{3+} ratio of the precursor solution. The yield of the synthesis shows saturation behaviour with increasing dropping rate. Cu^{2+}/In^{3+} ratio was kept constant at 0.833 and $BH_4^-/(Cu^{2+} + In^{3+})$ at 7.

Figure 7: Cu/In ratio of the synthesised nanoparticles and yield of the synthesis is plotted against dropping rate during the synthesis samples produced by simultaneous dropping of $(Cu^{2+} + In^{3+})$- and BH_4^- solution. Yield of the synthesis is independent on the dropping rate. Cu/In ratio of the nanoparticles shows a slight trend to higher copper content with increasing dropping rate. Cu^{2+}/In^{3+} ratio was kept constant at 0.833 and $BH_4^-/(Cu^{2+} + In^{3+})$ at 7.

For the standard process the yield shows saturation behaviour while the Cu/In ratio of the nanoparticles shows a slight tendency to converge with the Cu^{2+}/In^{3+} ratio of the precursor solution. With increasing dropping rate also the whole reaction time is decreased. The time between solving $NaBH_4$ and stopping the synthesis by adding acetone was roughly monitored and varied between 12 min and 38 min depending on the dropping rate. Typical synthesis time with standard dropping rate was about 14 min to 19 min. During this time alcoholysis affected the reducing power of BH_4^- solution. Stirring increased this effect. Hence the samples synthesised with low dropping rate show a more copper rich composition than samples produced with high dropping rates.

Particles synthesised by simultaneous dropping of both solutions show an unexpected slight trend to copper poor composition with increasing dropping rate. This could be attributed to a kinetic effect caused by different $BH_4^-/(Cu^{2+}+In^{3+})$ ratios at the beginning of the synthesis between both setups. For simultaneous dropping the $BH_4^-/(Cu^{2+} + In^{3+})$ within the reaction baker changes slower compared to standard process. For the standard process a small drop of precursor solution is dropped into the complete BH_4^- solution. At the beginning the $BH_4^-/(Cu^{2+} + In^{3+})$ is high while at the end by adding Cu^{2+} and In^{3+} cations and alcoholysis this ratio decreases. With increasing dropping rate at simultaneous dropping the equilibrium may not be reached as fast as Cu^{2+} is supplied. Hence Cu^{2+} is reduced preferentially due to its lower reductive potential [3] and copper rich particles are formed.

CONCLUSION

CuIn nanoparticles were synthesised via chemical reduction method. Process parameters like $BH_4^- / (Cu^{2+} + In^{3+})$ ratio, Cu^{2+}/In^{3+} ratio and dropping rate were varied while the influence on size and shape of the nanoparticles, Cu/In ratio of nanoparticles and yield of the synthesis was examined. $BH_4^- / (Cu^{2+} + In^{3+})$ ratio was found to be a crucial parameter of the synthesis. With sufficient amount of BH_4^- yield can be optimised and Cu/In ratio of precipitated nanoparticles is conform to the Cu^{2+}/In^{3+} ratio. The key point to provide sufficient BH_4^- is to control the effect of alcoholysis. The method of dropping of $Cu^{2+} + In^{3+}$ precursor solution into a stirred BH_4^- solution needs a high amount of BH_4^- as stirring increases alcoholysis. The role of alcoholysis is less important if $Cu^{2+} + In^{3+}$ precursor solution and BH_4^- reducing agent solution are dropped simultaneously together from not stirred solutions.

ACKNOWLEDGMENTS

The authors would like to thank Deutsche Forschungsgemeinschaft under contract number GRK 1161 for financial support.

REFERENCES

1   P. Jackson, D. Hariskos, E. Lotter, S. Paetel, R. Wuerz, R. Menner, W. Wischmann, M. Powalla, Prog. Photovolt: Res. Appl., 19, 894, (2011).
2   C.J. Hibberd, E. Chassaing, W. Liu, D.B. Mitzi, D. Lincot, a. N. Tiwari, Progress in Photovoltaics: Research and Applications, 18, 434, (2010).
3   G. Chen, L. Wang, X. Sheng, D. Yang, J Mater Sci: Mater Electron, 22, 1124, (2011).
4   C. Kind, C. Feldmann, A. Quintilla, E. Ahlswede, Chemistry Letters, 23, 5269, (2011).
5   J. Chang, J.H. Lee, J.-H. Cha, D.-Y. Jung, G. Choi, G. Kim, Thin Solid Films, 519, 2176, (2011).
6   S.-H. Liu, F.-S. Chen, C.-H. Lu, Journal of Alloys and Compounds, 517, 14, (2012).
7   S.-H. Liu, F.-S. Chen, C.-H. Lu, Chemistry Letters, 39, 1333, (2010).
8   I. Repins, M.A. Contreras, B. Egaas, C. Dehart, J. Scharf, C.L. Perkins, 235, (2008).
9   J. Zhang, T. Fisher, J. Gore, D. Hazra, P. Ramachandran, International Journal of Hydrogen Energy, 31, 2292, (2006).
10   V. Dalla, J.P. Catteau, P. Pale, Tetrahedron Letters, 40, 5193, (1999).
11   H.C. Brown, E.J. Mead, B.C. Subba Rao, Journal of the American Chemical Society 77, 77, 6209, (1955).
12   A. Bolcavage, S.V. Chen, C.R. Kao, Y.A. C, Journal of Phase Equilibria, 14, 14, (1993).
13   G.S. W. Keppner, T. Klas, W. Körner, R. Wesche, Physical Review Letters, 45, 2371, (1985).
14   W. Keppner, R. Wesche, T. Klas, J. Voigt, G. Schatz, Thin Solid Films, 143, 201, (1986).

Mater. Res. Soc. Symp. Proc. Vol. 1538 © 2013 Materials Research Society
DOI: 10.1557/opl.2013.982

Moisture Resistant Ga-Doped ZnO Films with Highly Transparent Conductivity for Use in Window Layers of Thin-Film Solar Cells

H. -P. Song[1], H. Makino[1], S. Kishimoto[1, 2] and T. Yamamoto[1]
[1]Materials Design Center, Research Institute, Kochi University of Technology, Kami-City, Kochi 782-8502, Japan
[2]Department of Mechanical Engineering, Kochi National College of Technology, 200-1 Otsu Monobe Nankoku-City, Kochi 783-8508, Japan

ABSTRACT

Highly transparent conductive Ga-doped ZnO (GZO) films are one of the promising transparent conductive oxide (TCO) films for use in electrodes of flat display panels and window layers of thin film solar cells. For the ZnO-based TCO films, the stability to damp-heat environment is a crucial issue for practical applications. We will report moisture resistant GZO codoped with indium films (GZO:In) on the basis of analysis of data obtained a damp-heat test for solar cells (85°C and 85% relative humidity for 1000 hours).

We used ZnO sintered targets with contents of 3 wt% Ga_2O_3 and 0.25 wt% In_2O_3 to grow GZO:In films in ion plating with direct current arc-discharge system. GZO:In films with different thicknesses (0.1-1 µm) were deposited on glass substrates at 200°C under the O_2 flow rate of 15 sccm. As the film thickness increased from 0.1 to 1 µm, the resistivity and sheet resistance decreased from 4.3 µΩm to 2.6 µΩm and from 42.7 Ω/Sq. to 2.6 Ω/Sq., respectively. And the average optical transmittance (T_{av}) in the range from 0.4 to 1 µm decreased from ~ 86% to ~ 75%. The GZO:In film with a thickness of ~300 nm had a low sheet resistance of 10.5 Ω/Sq. and a T_{av} of 82.5%. After 1000 hours damp-heat (DH) test under 85°C and 85% relative humidity, the relative change of sheet resistance is 3.4% with a Hall mobility of 26.4 cm^2/V.s and a T_{av} of 82.7% after test. The film thicker than 300 nm has a sheet resistance lower than 10 Ω/Sq. and a relative change of resistance of ~3% after DH test.

INTRODUCTION

ZnO-based transparent conductive oxide (TCO) films such as Al-doped or Ga-doped ZnO (AZO or GZO) films have been widely used in many applications including photovoltaic (PV) solar cells in recent years due to many advantages such as low material cost, abundant storage, and eco-friendly (non-toxicity) [1-5]. Moreover, compared to SnO_2 or In_2O_3-based films, ZnO-based TCO films are more durable to reductive hydrogen plasma which is commonly used to fabricate the amorphous (α-Si: H) or microcrystalline silicon (µ-Si: H) type solar cells [6]. High reliability is a crucial issue for PV solar modules which are settled outdoors. Hydrothermal operating environment is inevitable. Moisture ingress from the environment will attack the device components and cause performance degeneration or even the failure of solar module. To date large amount of research works have been performed to assess the stability of TCO materials used in PV solar cells [7-10].

Many reports show that the properties of ZnO-based TCO films were easily degenerated by a standard damp heat (DH) test (85°C and 85% relative humidity) for solar cell application. And the degeneration of ZnO-based TCO materials in PV modules is always the dominating failure factor. The issue on ZnO-based TCO films must be resolved for the application. Recently we have achieved high humidity resistant 100-nm-thick GZO:In (Ga_2O_3 content of 3 wt% and

In$_2$O$_3$, 0.75 wt%) films [11]. For GZO:In film deposited at an OFR of 15 sccm, the relative change of ρ was 6.0 % with ρ values of 5.0/5.3 $\mu\Omega$m before/after a 500 h humidity test (the standard used for liquid crystal display: 60 $^\circ$C and 95 % relative humidity). Note that the GZO (Ga$_2$O$_3$ content of 3 wt%) film deposited under the same condition had a relative change in ρ of 33 % with ρ values of 3.7/4.9 $\mu\Omega$m before/after the humidity test as a reference. The results revealed that the co-doping technique of both Ga and In species provides high-humidity-resistant ZnO-based TCO films. In this study, we will report GZO films codoped with indium as the suggested TCO films that will be adequate for the application as the transparent conductors in solar cells.

EXPERIMENT

GZO:In films with different thicknesses (0.1-1 μm) were deposited on glass substrates at 200 $^\circ$C by ion plating with dc arc discharge. We used sintered ZnO tablets with contents of 3 wt% Ga$_2$O$_3$ and 0.25 wt% In$_2$O$_3$ as resources to deposit the films. Details of the deposition conditions have been reported in our previous works [2]. During the growth, the flow rate of argon (Ar) gas was fixed to 140 sccm (sccm denotes standard cubic centimeters per minute). Oxygen (O$_2$) gas was introduced into the deposition chamber to compensate for the oxygen deficiency. O$_2$ gas flow rate (OFR) is a critical growth parameter: OFR determine (1) the deposition rate; (2) electrical and optical properties and (3) film stability to moisture [11,12]. Through a series of study, we found that OFR=15 sccm is the optimized growth condition to achieve the GZO:In TCO films with excellent performance. With further increasing OFR, the stability is only slightly enhanced, whereas the ρ values increase rapidly. It is worth mentioning that the OFR=15 sccm process with a high deposition rate of ~180 nm/min at a substrate temperature of 200 $^\circ$C has the cost merit for the production of thick window layers in solar cells.

The crystalline quality of the GZO films was characterized by X-ray diffraction (XRD) measurements (Rigaku, ATX-G). The full width at half maximum (FWHM) values for a (0002) peak is estimated from its rocking curve. The lateral size of grains perpendicular to the substrates was estimated by analyzing the in-plane XRD measurement data. The electrical properties was determined by Hall effect measurements at a room temperature using the van der Pauw method. Optical transmittance and reflection spectra were measured in the wavelength ranging from 200 to 2500 nm using a grating spectrometer (Hitachi, U4100) with 5° incidence angle. The surface morphology together with the root-mean-square (RMS) surface roughness was characterized by atomic force microscopy (AFM: JEOL JSPM-5200). Damp heat (DH) test was performed under 85°C and 85% relative humidity for 1000 h, which is in accordance with IEC61646 standard.

RESULTS and DISCUSSION

Table I lists all the GZO:In samples with various properties. T_{av} denotes the averaged optical transmittance in the wavelength from 400 to 1000 nm. Figure of merit is defined as the ratio of the electrical conductivity σ to the visible absorption coefficient α at specific wavelength (500 nm herein):

$$\frac{\sigma}{\alpha} = -\frac{1}{Rs.\ln(T + R)} \qquad (1)$$

where R_s is the sheet resistance, T and R is the optical transmittance and reflectance of films, respectively. Relative change of R_s (ΔR_s) in unit of % was calculated by $(R_{sa} - R_{sb})/R_{sb}$ where R_{sa} and R_{sb} are R_s after and before the DH test, respectively.

Table I. Sample list with numbers, thickness and the typical properties.

Number (GZO:In)	Thickness (nm)	R_s (Ω/Sq.)	T_{av} (%)	Figures of Merit (Ω^{-1})	Hall mobility (cm^2/V.s)	ΔR_s (%)
A1	100.6	42.7	86.5	2.4	24.6	22.2
A2	198.0	18.2	86.4	3.9	26.5	6.7
A3	301.2	10.5	82.5	4.5	27.2	3.4
A4	400.0	7.9	81.5	5.3	29.8	2.7
A5	491.2	6.0	80.0	6.6	31.3	2.5
A6	693.0	4.1	77.3	7.8	33.4	1.8
A7	994.0	2.6	74.4	9.6	36.6	2.5

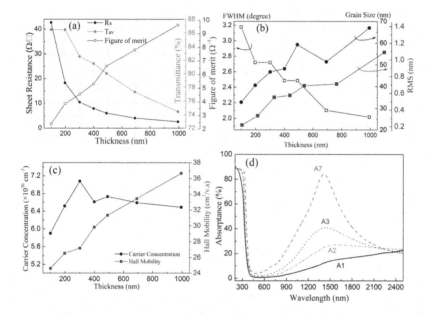

Figure 1 The properties of as-grown films: (a) the sheet resistance, averaged optical transmittance and figure of merit at a wavelength of 500 nm; Green dash line indicates R_s=10 Ω/Sq; (b) the FWHM values for the (0002) peak, the lateral grain size and RMS surface roughness; (c) carrier concentration and Hall mobility as a function of thickness. (d) the absorption spectra of representative samples (A1, A2, A3, A7) calculated by (100-T-R), as a function of wavelength.

Figure 1 shows thickness dependent of GZO:In properties. The R_s decreases together with an increase in the figure of merit as thickness increases, as shown in figure 1(a). Figure 1(b) indicates that this is mainly due to the improvement of microstructure quality of GZO:In films. Both the improvement of c-axis alignment between the grains and the enlargement of the lateral grain size will reduce the contribution of the grain boundary scattering to the carrier transport in the films [12]. We confirmed an increase in Hall mobility with increasing thickness, as shown in figure 1(c).

Despite the advantages mentioned above due to the increase of film thickness, we cannot increase the film thickness freely to get the best TCO film for using in solar cells, because the absorption in GZO films will also increase, especially the plasma absorption in the near-infrared region shown in figure 1(d). The plasma wavelength in TCO films is given by:

$$\lambda_p = 2\pi c \sqrt{\frac{\varepsilon_0 m_e^*}{N_e e^2}} \tag{2}$$

where c is velocity of light in vacuum, m_e^* is an effective mass of TCO films, and ε_0 is the vacuum permittivity, e is electron charge and N_e is the carrier concentration. The GZO:In films can have an effective mass about 0.3-0.4 m_0 (m_0 is the electron rest mass). For TCO films as the window layers in solar cells, the values of λ_p should be longer than 1.2 μm to enhance the transmittance of infrared part of the solar spectrum. An effective resolution to the issue above is that N_e should be lower, as indicated from eq. (2). Figure 1(c) shows that GZO:In films with thicknesses of more than 400 nm have high N_e. More development of a process technique together with optimized contents of n-type dopants, Ga and In, to reduce N_e together with an enhancement of Hall mobility should be needed. Considering the requirements from low ρ and high optical transparency, 300~500-nm-thick GZO:In films with the R_s of less than 10 Ω/Sq. and T_{av} above 80% will be a beter target for use in the window layers in solar cells.

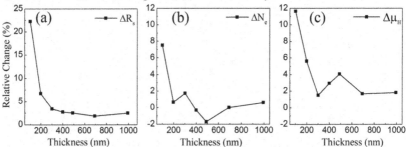

Figure 2. The relative change of (a) sheet resistance; (b) carrier concentration; (c) Hall mobility of GZO:In films with various thicknesses after 1000 h DH test.

To evaluate the moisture stability of GZO:In films after the DH test, the relative changes of sheet resistance, carrier concentration and Hall mobility are defined as: $\Delta R_s = (R_{sa}-R_{sb})/R_{sb}$, $\Delta N_e = -(N_{ea}-N_{eb})/N_{eb}$ and $\Delta\mu_H = -(\mu_{Ha}-\mu_{Hb})/\mu_{Hb}$, where R_{sb}, N_{eb}, μ_{Hb} and R_{sa}, N_{ea}, μ_{Ha} are the electrical sheet resistance, carrier concentration and Hall mobility of the samples before and after the DH test, respectively. Figure 2 shows characteristic of thickness dependent moisture stability performance in electrical properties: (1) ΔR_s (=22.2 %) for A1 with a thickness of 100.6 nm is much higher than those of thicker films (see figure 2(a)); (2) with increasing thicknesses up to 300 nm, ΔR_s decreases rapidly, followed by very slowly decrease. Note that we obtained moisture resistant GZO:In films of thicker than 300 nm. ΔR_s of them is just about 3.0 % (see figure 2(a)); (3) $\Delta\mu_H$ is a dominant factor which limits ΔR_s (see figures 2(b) and 2(c)). For the samples with $\Delta N_e \sim 0$, no chemisorbed species but physisorbed species probably not only in in-grains but also at grain boundaries of polycrystalline GZO:In films after the DH test affect the behavior of $\Delta\mu_H$. More study about what happens in in-grains and/or at grain boundaries during the DH test will be needed.

In many reported results, obvious changes in optical spectra and XRD patterns happened after the DH test under the standard DH test condition for solar cells. Note that for the GZO:In films with thicknesses of more than 300nm, as shown in figure 3(a), there was little change in the optical spectra in the wavelength from 400 to 1500 nm after DH test compared with those before the DH test. The out-of-plane ($2\theta/.\omega$ range: $30\text{-}40^{\circ}$) and in-plane ($2\theta_\chi/\square$ range: $20\text{-}120^{\circ}$) XRD patterns, moreover, showed that there was no additional peak after the DH test. This proved both the optical and structural properties of the GZO:In films developed in this study are very stable to the moisture.

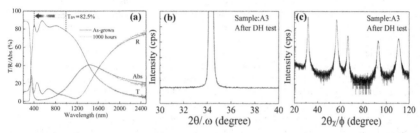

Figure 3. (a) The transmittance, reflectance and absorption spectra before and after the DH test; (b) out-of-plane and (c) in-plane XRD spectra after DH test. The sample in this figure is A3 with a thickness of 301.2 nm (see Table I).

CONCLUSIONS

Highly transparent conductive GZO:In films with different thicknesses were grown on glass substrates at 200 °C by ion plating with dc arc discharge. The structural and electrical properties of the GZO:In films were improved as the film thickness increased, while the optical transmittance in visible to infrared range was decreased. We found that thicker films have better performance on the moisture stability. For the sample with a thickness ~300 nm, the relative change of sheet resistance remained just about 3%, with a R_s of 11 Ω/Sq. and a T_{av} of 82.7% after the DH test. Thicker GZO:In films become very stable in electrical, optical and structural properties to the moisture. The high-performance GZO:In films in this study showed a great potential for use in the window layers for thin film solar cell devices.

ACKNOWLEDGMENTS
This work has been supported by Japan Society for the promotion of science JSPS (Kakenhi Grant Number 30320120), Basic Research A, the title is High performance ZnO-based hydrogen gas sensor.

REFERENCES

[1] T. Minami, MRS Bull. **25**, 38 (2000).
[2] N. Yamamoto, T. Yamada, H. Makino and T. Yamamoto, J. Electrochem. Soc. **157**, J13 (2010).
[3] T. Minami and T. Miyata, Thin Solid Films **517**, 1474 (2008).
[4] M. Oh, D. Hwang, D. Seong, H. Hwang and S. Park, E. D. Kim, J. Electrochem. Soc. **155**, D599 (2008).
[5] K. Ellmer, J. Phys. D: Appl. Phys. **34**, 3097 (2001).
[6] B. Yan, J. Yang and S. Guha, J. Vac. Sci. Technol. **A30**, 04D108 (2012).
[7] F. J. Pern, S. H. Glick, X. Li, C. DeHart, T. Gennett, M. Contreras and T. Gessert in Stability of TCO window layers for thin-film CIGS solar cells upon damp heat exposures: part III, edited by N. G. Dhere, J. H. Wohlgemuth and D. T. Ton, (SPIE Proc. **7412**, San Diego, CA, 2009) pp. 74120K/1-74120K/12.
[8] R. Sundaramoorthy, F. J. Pern, C. DeHart, T. Gennett, F. Y. Meng, M. Contreras and T. Gessert in Stability of TCO window layers for thin-film CIGS solar cells upon damp heat exposures: part II, edited by N. G. Dhere, J. H. Wohlgemuth and D. T. Ton, (SPIE Proc. **7412**, San Diego, CA, 2009) pp. 74120J/1-74120J/12.
[9] W. Lin, R. Ma, J. Xue and B. Kang, Sol. Energy Mater. Sol. Cells, **91**, 1902 (2007).
[10] J. N. Duenow, T. A. Gessert, D. M. Wood, B. Egaas, R. Noufi and T. J. Coutts in Investigation of ZnO:Al Doping Level and Deposition Temperature Effects on CIGS Solar Cell Performance, edited by K. Durose, T. Gessert, C. Heske, S. Marsillac and T. Wada, (Mater. Res. Soc. Proc. **1012**, San Francisco, CA, 2007) pp. Y01-Y08.
[11] H. Song, H. Makino and T. Yamamoto, submitted to Japanese patent.
[12] Y. Sato, H. Makino, N. Yamamoto and T. Yamamoto, Thin Solid Films **520**, 1395 (2011).
[13] T. Yamamoto, H. Song and H. Makino, Phys. Status Solidi **C** (2013) (in press).
[14] H. Zhu, J. Hüpkes, E. Bunte and S.M. Huang, Appl. Surf. Sci. **261**, 268 (2012).
[15] Y. Wang, X. Zhang, L. Bai, Q. Huang, C. Wei and Y. Zhao, Appl. Phys. Lett. **100**, 263508 (2012).

Mater. Res. Soc. Symp. Proc. Vol. 1538 © 2013 Materials Research Society
DOI: 10.1557/opl.2013.1070

Effective electrochemical n-type doping of ZnO thin films for photovoltaic window applications

B. Marí-Soucase[1,*], P. Cembrero-Coca[1], M. Mollar[1], M. E. Calixto[2]

[1]Física Aplicada-IDF, Universitat Politècnica de València, València, Spain.

[2]Instituto de Física, Benemérita Universidad Autónoma de Puebla, Puebla, México.

ABSTRACT

An effective n-type doping of ZnO using Cl was demonstrated in thin films electrochemically synthetized by adding different amounts of chlorine ions in the starting electrolyte. The ratio between chlorine and zinc cations was varied between 0 and 2 while the zinc concentration in the solution was kept constant. When the concentration of chloride in the bath increases an effective n-type doping of ZnO films takes place. n-type doping is evidenced by the rise of donors concentration, obtained from Mott-Schottky measurements, as well as from the blue shift observed in the optical gap owing to the Burstein-Moss effect.

INTRODUCTION

ZnO is an intrinsic, n-type semiconductor with a broad range of applications in optoelectronics. As a semiconducting oxide material ZnO has low resistivity and high transmittance down to the UV spectral range making it thus a well-suited material to be used as transparent electrode for photovoltaic solar cells and electrodes. However, some challenges have yet to be reached or improved for a broad usage of ZnO as TCO film, for example, the environmental stability of ZnO [1]. Although unintentionally doped ZnO is always n-type, due to the unavoidable presence of native defects that act as donors, a higher level of n-type doping can be attained by using group III metal elements such as Al, Ga, In, which substitute Zn. Substituting O by VII group elements, such as F or Cl, also results in an effective n-type doping [2].

Cu(In,Ga)Se$_2$-based thin-film solar cells use ZnO as window layer. Best results are obtained by sputtering a ZnO bilayer made of an unintentionally doped ZnO layer followed by a n-type ZnO:Ga layer. However, ZnO thin films can also be prepared by wet routes such electrodeposition (ED). ED presents some advantages compared to other techniques because deposition of semiconductors is performed at low temperature, atmospheric pressure and over large areas. Moreover it allows good control of film thickness through the control of deposited charge. Depending on the electrolyte, ZnO thin films can be deposited under several morphologies ranging from nanostructured and discontinuous films to extremely flat, compact and smooth films [3]. For PV window applications films with moderate surface textures are desired because of their effective light trapping in a wide wavelength range [4].

This work reports on the synthesis and characterization of ZnO thin films prepared by ED from an organic electrolyte like DMSO to be used as optical window in PV devices. Effective n-type doping of ZnO was achieved by varying the chlorine ion concentration in the electrolyte. The electrodeposited layers were characterized by XRD to study their structural properties, SEM for morphology details, Optical Spectroscopy to estimate the bandgap energy and electrochemical impedance spectroscopy (EIS) to calculate the doping donor concentration. Results showed that good quality ZnO polycrystalline films are obtained by ED. XRD results showed that as-deposited ZnO thin films exhibit narrow peaks, an indication of good crystallinity. SEM images revealed very compact and uniform layers, which is at the origin of the high transmittance. Further, when the chloride concentration in the bath increases an effective n-type doping of ZnO films takes place. n-type doping is evidenced by the drop of resistivity and the rise of donors concentration, obtained from Mott-Schottky measurements, as well as from the blue shift observed in the optical gap due to the Burstein-Moss effect.

EXPERIMENT

ZnO films were electrodeposited from DMSO solutions always containing a constant concentration 25 mM of Zn^{2+}. Zn^{2+} cations were obtained from two different precursors, chlorides ($ZnCl_2$) and perchlorates $Zn(ClO_4)_2$. The ratio between chloride and zinc ions was varied while keeping constant the zinc concentration as shown in **Table I**. The bath solution was saturated by oxygen gas bubbling through the electrolyte. The electrodeposition process of ZnO was carried out in a three-electrode electrochemical cell on glass slides coated with FTO by applying a constant potential, V= -0.9V. Thin films with thicknesses of about 700 nm were obtained for deposited charges of -1C/cm².

Table I. Content of dissolved salts and $[Cl^-]/[Zn^{2+}]$ ratio in the starting electrolyte (a, b, c). Cl content (at%) in ZnO films measured by EDX (d). Donor concentration (N_d) and Flatband Potential obtained from Mott-Schottly plots (e, f). Bandgap of ZnO:Cl films (g).

(a) $[ZnCl_2]$ (mM)	(b) $[Zn(ClO_4)_2]$ (mM)	(c) $[Cl^-]/[Zn^{2+}]$	(d) Cl (at%) Measured by EDX	(e) N_d (cm^{-3})	(f) Flatband Potential (V)	(g) Eg (eV)
25.0	0.0	2.00	5.0	8.06×10^{19}	-0.47	3.442
19.0	6.0	1.52	4.5	6.21×10^{19}	-0.42	3.423
12.5	12.5	1.00	3.6	5.16×10^{19}	-0.38	3.409
6.0	19.0	0.48	3.2	3.34×10^{19}	-0.38	3.386
0.0	25.0	0.00	0.7	1.89×10^{19}	-0.31	3.372

DISCUSSION

EDX analysis reveals that the content of Cl atomic concentration ranges from 0.7 to 5 at% for a pure perchlorate electrolyte to pure chloride electrolyte (**Table I (a, b)**). Chlorine is detected in the film even when no $ZnCl_2$ is present into the electrolyte. This means that ClO_4 ions also can act as a source of chlorine in the films. But this effect is much weaker than that of adding chloride ions in the bath. The atomic concentration of chlorine in the film is proportional to chloride concentration in the bath. So, EDX data indicate the incorporation of chlorine atoms into ZnO films (**Table I (c)**) and it is found to be proportional to $[Cl^-]/[Zn^{2+}]$ ratio in electrolyte.

Rousset et al. [5] have demonstrated that when Cl substitutes O in ZnO an extrinsic donor level is incorporated. Cl_O behaves as a shallow donor. According to that, a higher n-type level of doping of ZnO films can be achieved by increasing the content of Cl into the ZnO lattice.

Mott-Schottky measurements have been carried out in order to evaluate the impact of electrolyte nature on the electrical properties of the deposited material. This method is based on the Schottky barrier formation between the semiconductor material and the electrolyte. If the electrolyte is concentrated enough, the voltage drop due to the inverse polarization and the Schottky barrier are completely distributed in the semiconductor material. It causes the creation of a depletion zone that can be characterized by a capacitance measurement. Then, when the system obeys Mott-Schottky behavior, the evolution of $1/C^2$ is a linear function of the applied potential. The sign and value of the slope indicate type of doping and carrier concentration, respectively. A positive slope means that the semiconductor is n-type and the shallow donor concentration. The latter can be determined from the slope of this curve, according to the following equation:

$$\frac{1}{C^2} = \left(\frac{2}{e\varepsilon_0\varepsilon_r N_d A^2}\right)\left(V - V_{FB} - \frac{kT}{e}\right) \tag{1}$$

where C is the capacitance of the space charge region of the film at potential V, V_{FB} is the flatband potential, N_d is the donor concentration, A is the delimited area in contact with the electrolyte, the assumption of a perfectly smooth surface is made, ε_0 is the permittivity of the free space, and ε_r the relative dielectric constant taken as the typical value for bulk ZnO ($\varepsilon_r=8.0$). The donor concentration and the flatband potential obtained from Mott-Schottky measurements are displayed in **Table I (d, e)**.

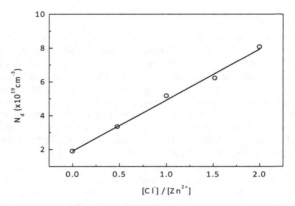

Figure 1. Donor concentration versus $[Cl^-]/[Zn^{2+}]$ ratio obtained from Mott-Schottky plots.

In order to better show the linearity of the doping with Cl^- concentration in the electrolyte, N_d values calculated from Mott-Schottky plots are displayed as a function of the $[Cl^-]/[Zn^{2+}]$ ratio present in the bath **Figure 1**. The calculated donor concentration exhibits a linear dependence with the amount of Cl^- and the trend is the higher the concentration of isolated Cl^- ions in the electrolyte the higher the donor concentration (N_d). This is a clear evidence of the n-type doping character of Cl atoms in ZnO. The carrier concentration is minimum when the material is synthesized in the pure perchlorate electrolyte (1.89×10^{19} cm^{-3}) and increases more than four times (8.06×10^{19} cm^{-3}) when the films are grown in pure chloride electrolyte.

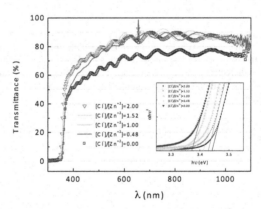

Figure 2. Transmittance for ZnO:Cl thin films with different $[Cl^-]/[Zn^{2+}]$ ratios. **Inset:** Variation of $(\alpha h\upsilon)^2$ versus $h\upsilon$.

To assess the increase in the conductivity of ZnO:Cl layers the sheet resistance was measured with a four points probe. It was found that the sheet resistance varies from 9.7 Ω/sq for FTO substrates to 5.4 Ω/sq. for the most doped ZnO:Cl sample. Even if this is an approximated method for obtaining the sheet resistance because there are two conductive layers and it is not possible to know accurately the contribution of each layer, the measures can always be compared among them.

High transparency is the second important property, besides high conductivity, for the use of ZnO:Cl films as PV window layers. The visible transmission spectra for ZnO:Cl films with different amount of chloride doping are presented in **Figure 2**. The transmission is nearly 70% for the films grown in electrolytes without Cl⁻ ions and nearly 90 % transmission is achieved for the films deposited in presence of chloride ions. Higher doping of Cl⁻ in film leads to higher transmittance. Further, a shift to shorter wavelengths is observed when raising the content of Cl⁻ ions in ZnO:Cl films. The inset of **Figure 2** shows the extrapolation of $(\alpha h\upsilon)^2$ versus $h\upsilon$ for obtaining the bandgap values in ZnO:Cl films. Calculated E_g values are displayed in **Table I (g)**.

The shift towards higher energies of the bandgap in ZnO:Cl films is a result of Cl doping and is explained by the Burstein-Moss (BM) effect. BM effect is attributed to the occupation of the conduction band (CB) from the electrons coming from the valence band (VB). BM effect results in higher effective bandgap as consequence of the longer distance between unoccupied states of the CB and the top of the VB. The effective optical gap (E_g) can be calculated as the sum of the optical gap for the intrinsic material (E_{g0}) and the increment due to the BM effect (ΔE_{BM}).

$$E_g = E_{g0} + \Delta E_{BM} \; ; \; \Delta E_{BM} = \frac{h^2}{8\pi^2 m^*} \left(3\pi^2 n\right)^{2/3} \qquad (2)$$

where h is Planck's constant, m^* is the effective mass of the electron and n is the electron free carrier concentration.

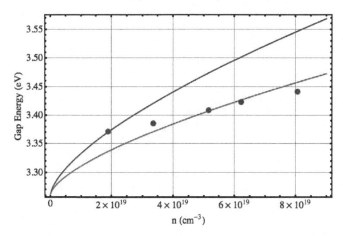

Figure 3. Theoretical calculations of E_g as a function of electron concentration (n) for two values of effective electron mass in ZnO found in the literature. Up: m*=0.24·me. Down: m*=0.35·me.

Figure 3 shows the experimental values obtained by extrapolation of $(\alpha h\upsilon)^2$ versus $h\upsilon$ (dots) as well as the theoretical values obtained for the bangap of ZnO as a function of the free electron concentration. Up and down lines are the theoretical bandgap energies for two different effective electron mass found in the literature ($m^*=0.24 \cdot m_e$ and $m^*=0.35 \cdot m_e$) and for a value of the optical bandgap for the intrinsic material, $E_{g0}=3.263$ eV. The experimental points fall inside the theoretical curves for both values of effective electron masses. In this plot the abscissa is the free electron concentration, n, while the experimental points calculated from Mott-Schottky plots provide the donor concentration, N_d. It is likely to get a better fit of theoretical and experimental values if compensation involving donor and acceptor levels is taken into account.

CONCLUSIONS

Electrochemical deposition in electrolytes containing chlorides and DMSO appears to be an attractive way for synthetizing ZnO layers with very high transparency and good control of the conductivity. Due to their low cost synthesis route and good performance, electrodeposited ZnO:Cl layers are well suited to be used in photovoltaic window applications .

ACKNOWLEDGMENTS

This work was supported by European Commission through NanoCIS project FP7-PEOPLE-2010-IRSES (ref. 269279).

REFERENCES

1. A. Asvarov, A. Abduev, A. Akhmedov, A. Abdullaev; *Phys. Status Solidi C* **7**, 1553 (2010).
2. E. Chikoidze, M. Nolan, M. Modreanu, V. Sallet, P. Galtier; *Thin Solid Films* **516**, 8146 (2008).
3. H. Cui, M. Mollar, B. Marí; *Optical Materials* **33**, 327 (2011).
4. Y. Wang, X. Zhang, L. Bai, Q. Huang, C. Wei and Y. Zhao, *Appl. Phys. Lett.* **100**, 263508 (2012).
5. J. Rousset, E. Saucedo and D. Lincot; *Chem. Mater.* **21**(3), 534 (2009).

Mater. Res. Soc. Symp. Proc. Vol. 1538 © 2013 Materials Research Society
DOI: 10.1557/opl.2013.985

Study of optical losses in mechanically stacked dye-sensitized/CdTe tandem solar cells

Vincent Barrioz[1]*, Simon Hodgson[1], Peter Holliman[2], Arthur Connell[2], Giray Kartopu[1], Andrew J. Clayton[1], Stuart J.C. Irvine[1], Shafiul Monir[1] and Matthew L. Davies[2]

[1]Centre for Solar Energy Research, Glyndŵr University, OpTIC, St Asaph, LL17 0JD, UK.
[2]School of Chemistry, Bangor University, Bangor, Gwynedd, LL57 2UW, UK.
* v.barrioz@glyndwr.ac.uk (+44 1745 535 159)

ABSTRACT

In a constant effort to capture effectively more of the spectral range from the sun, multi-junction cells are being investigated. In this context, the marriage of thin film and dye-sensitized solar cells (DSC) PV technologies may be able to offer greater efficiency whilst maintaining the benefits of each individual technology. DSC devices offer advantages in the nature of both the metal oxide photo-electrode and dye absorption bands, which can be tuned to vary the optical performance of this part of a tandem device, while CdTe cells absorb the majority of light above their band-gap in only a few microns of thickness. The key challenge is to assess the optical losses with the goal of reaching a net gain in photocurrent and consequently increased conversion efficiency. This study reports on the influence of optical losses from various parts of the stacked tandem structure using UV-VIS spectrometry and EQE measurements. A net gain in photocurrent was achieved from a model developed for the DSC/CdTe mechanically stacked tandem structure.

INTRODUCTION

In order to increase the photovoltaic (PV) competitiveness and achieve grid parity globally, the research community constantly strives to increase conversion efficiency and, where possible, go beyond the physical limits of single junction cells [1]. In the case of a near optimum direct band gap material (i.e. 1.5 eV), the conversion efficiency limit would be 32 % with a short-circuit current density (J_{sc}) of 28.9 mA/cm^2. A NREL-certified record efficiency of 18.7 %, has recently been reported by First Solar for thin film CdTe solar cells [2]. This follows GE Global Research previous record [3] at 18.3 % which had an impressive 26.95 mA/cm^2 over the spectral range of 350 – 850 nm. The exact source from this enhanced blue response is not known, however, it is generally achieved either by thinning the CdS front window layer or by increasing its band gap (E_g) by alloying the CdS (e.g. with Zn [4]). For dye-sensitized solar cells (DSC) conversion efficiencies have reached 12.3 % ($J_{sc} \sim$ 17.3 mA/cm^2 within spectral range of 400 – 700 nm) using Zn-porphyrin dye co-sensitized with a triphenylamine and a cobalt complex redox couple [5].
Exploring alternative routes to higher cell efficiencies is essential. One path to enhance performance is to use multi-junction solar cells in order to minimize thermalization losses. The preference is to use monolithically integrated sub-cells in series as widely reported for III-V concentrated multi-junction solar cells [6] and of increasing attraction for organic/inorganic cells [7,8]. This arrangement requires current matching between the sub-cells and the solar irradiance conditions need to be optimal to avoid losses. In the simplest case of using a series connected two-junction cells, the top cell should be 1.7 eV, while the bottom cell should be 1.1 eV [9]. In

the case of an absorbing material such as CdTe, this means that it could be used as a top cell by widening its band gap. However, use of CdTe as an optimum bottom cell in a two-junction series-connected cell is not straightforward. Consider that the multi-junction device could contain up to four-terminals, connected in parallel. Here, the constraint associated with current matching is relieved, and the sub-cells would only need to be voltage matched. This alternative design would also enable the retrofitting of a top cell in front of existing PV modules. The key challenge is then to identify and minimize the optical losses to ensure an overall increase in photocurrent.

In this context, CdTe single junction cells operate on a fixed absorber band gap (E_g) which effectively sets the V_{oc}, whilst consistent spectral response at photon energies > E_g set the photo generated current. By comparison, DSC devices operate differently because photons are captured by dyes and adsorbed onto a metal oxide surface. During operation, excited electrons from the dye are injected into the conduction band of the metal oxide (usually anatase TiO_2). Here, the theoretical V_{oc} limit for a singly dyed DSC device is set by the TiO_2 Fermi level and the energy level of the redox couple, which means the spectral response can be tuned using different dyes without affecting the voltage. Both CdTe and DSC devices have reported open-circuit voltage (V_{oc}) with similar value of ~700 mV.

DSC/CIGS tandem cells reported in the literature [10] tend to focus on demonstrating the feasibility of the tandem approach and place less emphasis on the origin of optical losses. Preliminary results using a mechanically stacked tandem structure [11] using a DSC with D149 dye and a CdTe cell had excessive losses. This paper reports on the optical losses and associated photocurrent generated using the case of a mechanically stacked top cell based on a yellow dye (YD) DSC [12] and a bottom CdS/CdTe solar cell [13].

THEORY AND EXPERIMENT

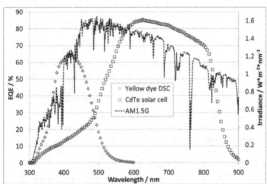

Figure 1. EQE of individual cells for a YD DSC using a TEC8/TEC15 float glass structure with DSL18NRT TiO_2 [12] and a CdS/CdTe device in superstrate configuration on TEC10 substrate.

A Varian Cary 5000 spectrophotometer was used with an integrating sphere setup for measuring the total (i.e. specular and diffuse) transmittance and reflectance. From these measurements, the haze and absorptance can be extracted from a given optical arrangement. In terms of spectral response from each individual cell, the EQE, shown in Fig. 1, for the YD DSC was measured as described in [12], while the EQE for the CdTe cell was carried out using a Bentham PVE300 system. Individually, the short-circuit current densities (J_{sc}) have been calculated, from EQE measurements, to be 3.3 and 18.4 mA/cm^2, respectively. If the cells were side-by-side, the total J_{sc} would be their sum, while for a stacked arrangement; the goal is to improve transparency towards the bottom cell without incurring excessive losses on the top cell. Light interacting with a material obeys the following relation as a function of wavelength (λ):

$$R_t(\lambda) + T_t(\lambda) + A_t(\lambda) = 1 \qquad (1)$$

where R_t is the total reflectance of the system, T_t is the total transmittance, and A_t is the absorptance from the optical arrangement under investigation. Therefore, for an incident photon flux at a given wavelength, there will be reflection, transmission and absorption portions. There should be sufficient difference in the spectral absorption range between the DSC (350 – 500 nm) and the CdTe cell (500 – 900 nm), as seen in Fig. 1, in order to minimize absorption losses in a 4-terminal tandem cell stack. As a first assessment, the analysis in terms of optical losses can therefore be separated into average R_t, T_t and A_t over 2 spectral ranges, for each sub-cell to be evaluated individually.

The losses caused by the different parts of the chosen optical arrangement can be quantified against the incoming photon flux, which can then be normalized as photon fraction as shown by Demastu and Sites [14]. The contribution from different sections of the sub-cells can be identified and an equivalent model be made where the EQE is built back up, inter-changing optical parts, such as glass composition, thickness, etc. From this, the potential short-circuit current density (gains and losses) can be estimated in each case.

RESULT & DISCUSSION

In this study, the optical losses are measured by decoupling the different parts of the mechanically stacked structure [11], situated above the CdS window layer of the CdTe cell. Different glass and TCO types, photo-electrodes and DSC shells (i.e. by DSC shell, it is meant a DSC structure without electrolyte) were investigated. Corning boro-aluminosilicate (700μm thick) and NSG TEC^{TM} float glass standard (3.2 mm thick) and low iron (300 μm & 700 μm thick) were selected for this investigation. In some cases, the Corning glass, supplied by Delta-technologies was coated with 135 nm indium tin oxide (ITO) with a sheet resistance of 5 – 15 $\Omega/\square$. 3.2 mm NSG TEC^{TM} float glass were also investigated with fluoride doped tin oxide (FTO) branded as TEC15, TEC10 and TEC8 for sheet resistance of 15, 10 and 8 $\Omega/\square$, respectively. No anti-reflection (A/R) coatings were intentionally used in this study.

Optical measurements from glass and other 'substrates'

Firstly, loss of transparency through the glass stack, and whether it can be improved upon shall be discussed. Comparison was made between the different glass types. Over the 350 – 500 nm spectral range, transmittance is ~ 90 % for all types of glass and any fluctuation appears to be mainly caused by composition rather than thickness, the worst case being for the 3.2 mm thick float glass due to the Fe^{3+} absorption peaks. The influence of the Fe^{3+} is more pronounced for the 500 – 900 nm spectral range where the thickest float glass drops to an average T_t of 88 % compared to the boro-aluminosilicate being at 92 %. Over the 500 – 900 nm range, this

223

transparency difference corresponds to a maximum J_{sc} of 24.1 mA/cm^2 compared with 25.4 mA/cm^2 respectively. In all cases, R_t is as expected with ~8 % for all glass caused by both air/glass interfaces. In the proposed mechanical stack, up to 3 glass substrates may be used and therefore boro-aluminosilicate or glass with low iron content would be preferable.

Table I. Average T_t, R_t and Haze of different TCO/glass investigated for 2 spectral ranges.

	ITO*			TEC15**			TEC10**			TEC8**		
Wavelength range	T_t (%)	R_t (%)	Haze (%)	T_t (%)	R_t (%)	Haze (%)	T_t (%)	R_t (%)	Haze (%)	T_t (%)	R_t (%)	Haze (%)
350 – 500 nm	78.6	15	0.3	78.3	10.8	2.7	78.6	9.5	6.6	75	10	24.8
500 – 900 nm	85	11.4	0.2	79.3	10.1	0.6	82	9.1	1.8	79	7.7	7.5

* Deposited on 700μm Boro-aluminosilicate glass.
** Deposited on 3.2 mm Float glass.

In a cell, the glass will be coated with a transparent conductive oxide (TCO) layer. Optically, the primary effect of the TCO is to increase the reflectivity and haze (scattering) of the glass. The optical measurement results for TCO coated glass are summarized in table I. It should be noted that high conductivity and high haze are preferred for DSC cells, but too much (such as in TEC8) could be detrimental to the bottom cell. ITO and TEC10 offer the highest average T_t. However, overall TEC10 appears to offer the best compromise between low haze, high transmittance and low reflectance and so this was selected in this study as the substrate for the CdTe bottom cell.

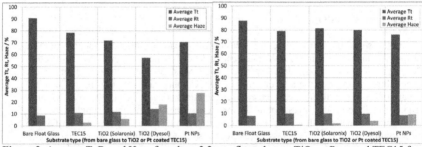

Figure 2. Average T_t, R_t and Haze from bare 3.2 mm float glass to TiO$_2$ or Pt coated TEC15 for the spectral range (left) 350 – 500 nm and (right) 500 – 900 nm

The DSC requires electrodes to be deposited on top of the TCOs, TiO$_2$ on one side and platinum on the other. A trade-off needs to be made between the performance of the DSC and long wavelength transparency. TEC15 was used as a reference substrate for the optical evaluation of these electrodes due to its wider availability. One type of Pt nanoparticles (from Dyesol Pt1 paste) and two types of TiO$_2$ were assessed on TEC15, namely DSL18NRT from Dyesol and Solaronix-HT from Solaronix. The T_t, R_t and haze results for all electrodes can be found in Fig. 2. The data show that Solaronix-HT presents reduced haze along with superior transparency. The average R_t for both types of TiO$_2$ coating is constant over the 500 – 900 nm spectral range measured. Therefore, these data show that Solaronix-HT should be the preferred option for a tandem structure where the aim is to transmit as much light as possible to the lower cell in the stack.

Optical and J_{sc} losses from individual and stacked cells

For the individual solar cells, Table II summarizes the potential, actual and losses in J_{sc} from the optical and EQE measurements. For the DSC [12], the absorption losses are from the front TEC15 with DSL18NRT TiO_2 layer. For the CdTe cell, the absorber is thick enough (> $2\mu m$) not to contain significant transmission losses in the spectral range studied. The data indicate that the highest losses are from the Fe content within the glass and TCO, resulting in a loss of 2.4 mA/cm^2.

Table II. Calculated photocurrent based on optical measurements of T_t, R_t, A_t and EQE for individual cells: (left) YD DSC with TEC8/TEC15 [12] and (right) CdTe on TEC10. Calculated J_{sc} are over the 350 – 500 nm and 500 – 900 nm spectral ranges for DSC and CdTe respectively.

	Calculated J_{sc} for YD DSC (mA/cm^2)	Calculated J_{sc} for CdS/CdTe (mA/cm^2)
Potential J_{sc} with current structure	5.2	22.3
Actual J_{sc} from EQE	3^a	17.5
Losses caused by reflection	0.5	2.2
Losses caused by absorption	1.7	2.4
Losses caused by transmission	1.1	n/a

^{a} J_{sc} from the EQE includes absorption from the TiO_2 layer and the sum does not equate to the potential J_{sc} shown.

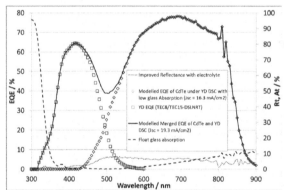

Figure 3. Modeled EQE of tandem CdTe and YD solar cells with electrolyte in place and use of low absorption glass.

First attempts at constructing an un-optimized tandem DSC/CdTe device showed net losses as detailed by Barrioz et al. [11]. By first modeling a more idealized device structure, where optical losses, related to reflectance from the electrolyte and absorption from the iron content in the float glass, are subtracted from the optical characteristics of the overall stack. Then combining with current CdTe/DSC cell performance (i.e. EQE), introduced earlier, it is possible to predict a scenario with a possible net current gain of 0.9 mA/cm^2 compared with the bare CdTe device (18.4 mA/cm^2), as detailed in Fig. 3. The spikes seen at 810nm are an artifact from the optical measurements.

CONCLUSIONS

A study in the source of optical losses has been carried out in 4-terminal mechanically stacked tandem cells. It was shown that improvements can be made to reduce the absorption losses in a tandem structure. These predominantly relate to the glass/TCO composition. A model has been demonstrated, resulting in a net gain of photocurrent generated by ~1 mA/cm^2. These results provide the opportunity to explore alternative designs for tandem cells without excessive losses.

ACKNOWLEDGMENTS

The authors would like to thank Paul Warren from NSG for supplying the float glass and for interesting discussions. The low carbon research institute (LCRI) is also gratefully thanked for funding the SPARC Cymru project under the European Convergence Region Programme.

REFERENCES

1. W. Shockley, H.J. Queisser, *J. Appl. Phys.* 32 (1961) 510.
2. First solar, http://www.firstsolar.com/en/Innovation/CdTe-Technology/CdTe-Resources (08/03/2013).
3. M.A. Green, K. Emery, Y. Hishikawa, W. Warta and E.D. Dunlop, *Progr. Photovolt.: Res. Appl.* 21 (2013) 1.
4. G. Kartopu, A.J. Clayton, W.S.M. Brooks, S.D. Hodgson, V. Barrioz, A. Maertens, D.A. Lamb and S.J.C. Irvine, *Prog. Photovolt: Res. Appl.* (2012) (DOI: 10.1002/pip).
5. A. Yella, H.-W. Lee, H.N. Tsao, C. Yi, A.K. Chandiran, M.K. Nazeeruddin, E.W-G. Diau, C.-Y. Yeh, S.M. Zakeeruddin, M. Grätzel, *Science* 334 (2011) 629.
6. J.M. Olson, D.J. Friedman, S. Kurtz, Handbook of Photovoltaic Science and Engineering (1st Ed.), A. Luque, and S. Hegedus; Wiley: New York, 2003, Chapter 9, pp. 359–411.
7. J. Yang, W. Chen, B. Yu, H. Wang and D. Yan, *Organic Electronics* 13 (2012) 1018.
8. W.-S. Jeong, J.-W. Lee, S. Jung, J.H.Yun and N.-G. Park, *Sol. Energ. Mat. Sol. Cells* 95 (2011) 3419.
9. T.J. Coutts, K.A. Emery and J.S. Ward, *Prog. Photovolt: Res. Appl.* 10 (2002) 195.
10. P. Liska, K.R. Thampi, M. Grätzel, D. Bremaud, D. Rudmann, H.M. Ipadhyaya, A.N. Tiwari, *Appl. Phys. Lett.* 88 (2006) 203103.
11. V. Barrioz, P. Holliman, A.J. Clayton, A. Connell, S.J.C. Irvine, M.L. Davies, S. Hodgson, *PVSAT-8 Conference Proceedings*, Newcastle (2012).
12. P.J. Holliman, M. Mohsen, A. Connell, M.L. Davies, K. Al-Salihi, M.B. Pitak, G.J. Tizzard, S.J. Coles, R.W. Harrington, W. Clegg, C. Serpa, O.H. Fontes, C. Charbonneau and M.J. Carnie, *J. Mater. Chem.* 22 (2012) 13318.
13. S.J.C. Irvine, V. Barrioz, D. Lamb, E.W. Jones and R.L. Rowlands-Jones, *J. Cryst. Growth* 310 (2008) 5198.
14. S.H. Dematsu and J.R. Sites, *Proc. IEEE PV Spec. Conf.* 31 (2005) 347.

Mater. Res. Soc. Symp. Proc. Vol. 1538 © 2013 Materials Research Society
DOI: 10.1557/opl.2013.1012

Point Contact Admittance Spectroscopy of Thin Film Solar Cells

Anthony Vasko, Kristopher Wieland and Victor Karpov
University of Toledo, 2801 W. Bancroft, Toledo, OH 43606, U.S.A.

ABSTRACT

We present the new characterization technique of multi-dimensional admittance measurements. In standard admittance measurements, a semiconductor device is probed in the transverse dimension, between flat plate contacts. We extend such measurements to distributed, possibly non-uniform solar cells where one of the two contacts has very small (point-like) dimensions. As a result, both the real and displacement currents spread into lateral directions while flowing between the electrodes. Correspondingly, the probing electric field may result in contact voltages that are laterally not equipotential. The spatial voltage distribution will depend on the probing DC bias and AC frequency. The resulting measurement will give information about the system's lump parameters, such as open circuit voltage, sheet and shunt resistances, as well as the presence and location of shunts. Understanding of the measurement is developed through intuitive and analytic models. Numerical models, utilizing finite element circuits, are used to verify the analytic results, and also may be directly compared to or used to fit experimental data. While our focus is on introducing the physical theory, early experimental results demonstrating spatial scaling are shown.

INTRODUCTION

Large area solar cells may have non-uniformities or inhomogeneities that can impair device performance or influence degradation mechanisms. Techniques for studying such non-uniformities include laser beam induced current (LBIC) [1], electron beam induced current (EBIC) [2], lock-in thermography [3], and local surface photovoltage [4]. Numerical modeling has also been used to simulate devices with non-uniformities[5]; in this work, we extend such modeling to simulate admittance measurements on devices with non-uniformities, with the potential for using these measurements on real devices to reverse engineer their parameters.

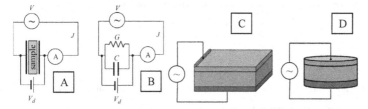

Figure 1. The experimental connections for an admittance measurement (A) and equivalent circuit (B). Contacting for (C) 1-dimensional and (D) 2-dimensional measurements. In C and D, the lower layer is a low resistance contact, the middle layer is the active semiconductor, and the upper layer is relatively resistive. In C, a low resistance 1-dimensional contact causes the voltage variation in the resistive contact to be in one dimension (perpendicular to the contact) only.

Admittance measurements have long been used as a characterization method for thin-film solar cells, giving information such as doping profiles and density of states[6]. The typical admittance measurement is done on a semi-conductor structure sandwiched between two parallel, flat, equipotential electrodes (such as transparent conducting oxide and metal). The resulting complex current I response to an applied voltage V is interpreted in terms of a leaky capacitor, as shown in figure 1, giving the admittance Y in terms of conductance G and capacitance C:

$$Y \equiv \frac{dI}{dV} = G + i\omega C \tag{1}$$

In general terms, G and C may depend on applied DC bias and applied AC frequency. For a uniform device, a DC bias will change the depletion width of a diode, so that capacitance as a function of bias gives information on carrier density as a function of depth in the device.

By *point contact admittance*, we specially refer to a situation for which one of the electrical contacts is relatively resistive, so that it cannot be assumed to be at equipotential. We use this term whether the variation in voltage is 1-dimensional (Figure 1c) or 2-dimensional (Figure 1d). Realistic situations for which this might apply include **(1)** A normal CdTe superstrate device for which the low resistance contact is the metal back contact while the high resistance contact is the front transparent conducting oxide (TCO). **(2)** A partially unfinished CdTe superstrate device for which the low resistance contact is the TCO while the high resistance contact may be, for example, a doped ZnTe back layer. **(3)** A CIGS device for which the low resistance contact is the back metal electrode and the front electrode could either be a finished TCO layer or a high-resistance transparent (HRT) buffer layer before TCO deposition.

For a device that is not uniform or that has very resistive electrode, a DC bias will induce a spatially varying voltage over the electrode that controls the area of the device that is probed along the dimensions of the electrode. Similarly, the frequency of the applied AC signal can also control the probed area.

THEORY

Single Node Admittance

A typical solar cell is shown in figure 2 next to a greatly simplified AC equivalent circuit, in which the diode has been replaced by its dynamic resistance and combined with parasitic shunt resistance R_{sh} and series resistance R_s to form the equivalent resistance R'. The cell capacitance is C, and ρ represents (to the accuracy of a numerical multiplier) the cell sheet resistance. As a further simplification, we neglect possible parallel contribution to the admittance from dielectric relaxation [7], assumed to be relatively small here. As the solar cell is modeled with a single diode and no distributed spatial effects are considered, we denote the resulting admittance as single node.

Figure 2. Left, single node model of solar cell and, right, simplified AC equivalent circuit

The resulting admittance can easily be found

$$Y = \frac{\left(\rho\left(1+(R'\omega C)^2\right)+R'\right)+i\left(CR'^2\omega\right)}{(\rho+R')^2+(\rho R'\omega C)^2} \tag{2}$$

which has a low frequency limiting case

228

$$G \approx \frac{1}{\rho+R'} + \frac{C^2 R'^3 \rho}{(\rho+R')^3} \omega^2 \approx \frac{1}{R'} + C^2 \rho \omega^2 \tag{3}$$

$$C_{effective} \approx C \frac{R'^2}{(\rho+R')^2} - \frac{C^3 R'^4 \rho^2}{(\rho+R')^4} \omega^2 \approx C - C^3 \rho^2 \omega^2 \tag{4}$$

The second approximations assume R' >>ρ, which is usually the case, particularly when the single node interpretation is valid.

The expression also has a high frequency limiting case

$$G \approx \frac{1}{\rho} - \frac{\rho+R'}{\rho^3 R' C^2} \frac{1}{\omega^2} \tag{5}$$

$$C_{eff} \approx \frac{1}{C\rho^2 \omega^2} \tag{6}$$

However, for many typical solar cells we do not expect the above high frequency limiting cases to be observed; rather, at high frequency, the single node admittance model will fail and admittance will be controlled by capacitive screening lengths, as discussed below.

Screening Length

Consider a 1-dimensional system with shunt conductance per length $1/R_{sh}$ and finite series resistance per length ρ. If the system is divided into nodes of length Δx, some of which are shown in Figure 3, then the resistances can be approximated by finite element resistors.

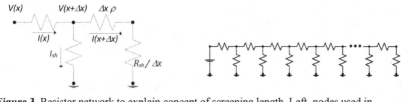

Figure 3. Resistor network to explain concept of screening length. Left, nodes used in derivation. Right, entire network.

The value of the resistors is as shown in Figure 3. The circuit can be analyzed with Kirchoff's laws, giving

$$I(x + \Delta x) = I(x) - I_{sh} = I(x) - V(x + \Delta x)\Delta x/R_{sh} \quad \text{or} \quad \frac{\Delta I}{\Delta x} = -\frac{V}{R_{sh}} \tag{7}$$

$$V(x + \Delta x) - V(x) = -\Delta x \, \rho \, I(x) \quad \text{or} \quad \frac{\Delta V}{\Delta x} = -\rho I \tag{8}$$

As Δx tends to zero, the system becomes continuous, and equations (7) and (8) combine to become

$$\frac{d^2 V}{dx^2} = \frac{\rho}{R_{sh}} V \tag{9}$$

An infinitely long resistor network with applied voltage V_0 at x=0 and with the boundary condition that voltage must decay to zero at x= ∞ will have a voltage distribution of

$$V(x) = V_0 e^{-x\sqrt{\frac{\rho}{R_{sh}}}} \tag{10}$$

We consequently define the characteristic length

$$L_{sh} = \sqrt{\frac{R_{sh}}{\rho}} \tag{11}$$

to be the (shunt conductance) *screening length*. This is the scale over which an applied voltage influences the voltage elsewhere on the device.

Qualitatively similar relations apply for finite-sized and 2-dimensional circuits (for the latter, ρ must be replaced with the sheet resistance and $1/R_H$ with the shunt conductance per area). In all cases, L_{sh} is the length scale over which the total shunt and series resistances are equal[8], a condition which also allows general derivation of screening lengths.

The above describes a shunt conductance, but equally applies to a shunt capacitance per length (or area, for the 2-dimensional case) c'. For a capacitance per length c', the absolute value of the complex impedance per length is $1/(\omega c')$. Replacing R_{sh} in Equation 11 with this impedance results in the screening length for capacitance

$$L_C = \sqrt{\frac{1}{\omega \rho c'}} \qquad (12)$$

As the probing frequency tends to infinity, the capacitive screening length tends to zero. When the linear size of the cell is much larger than the capacitive (or shunt conductance) screening length, the entire cell will contribute to the admittance, and the single node admittance interpretation will be valid. However, at frequencies for which the capacitive screening length is much less than the cell size, then the applied AC voltage only probes an area of the cell over the length scale L_C, leading to effective capacitances

$$C_{eff}^{1D} \propto L_C c' \propto \omega^{-1/2} \qquad (13)$$
$$C_{eff}^{2D} \propto L_C^2 c' \propto \omega^{-1} \qquad (14)$$

The 2-dimensional effective capacitance, for example, is the capacitance per area times the area of a circle, centered at the probe, of radius L_C. The factor of π in the area of the circle is dropped in Equation 14 as an insignificant numerical multiplier, and because the screening length argument is by nature qualitative. The precise behavior of the admittance with frequency is obtained numerically, discussed below.

Note that the high frequency power dependence of the capacitance for a distributed cell is quantitatively different than expected for a single node, for which it is expected to be -2. We consider this high-frequency dependence to be evidence of *scaling*, due to its dependence on the dimensionality of the cell and its contact.

An additional important length in nonuniform cells is the distance from the contact to a point of high shunt conductance; this length, compared to the capacitive screening length, determines the frequencies for which the presence of the shunt affects the admittance.

Finally, there is a screening length for a cell in large forward DC bias, $q(V-V_{OC}){>}{>}kT$. This screening length is derived elsewhere[9], but is

$$L_{V+} = \sqrt{\frac{V-V_{OC}}{j_L \rho}}\, e^{-\frac{q(V-V_{OC})}{2nkT}} \qquad (15)$$

V is the applied DC bias, V_{OC} is the cell open circuit voltage, j_L is the light generated current per area, ρ is the sheet resistance, n is the ideality factor, k is Boltzmann's constant, and T is temperature. When the forward bias screening length is smaller than the cell size and the capacitive screening length, this screening length will control the effective capacitance, analogous to L_C in Equations 13 and 14.

Numerical Modeling

Screening length arguments give intuitive explanations for the power dependencies of capacitance and conductance but are by their nature merely qualitative. To obtain more quantitative results, we have employed finite element numerical modeling. To do so, we divide a 2-dimensional cell into many nodes, one of which is shown in figure 4, consisting of resistive electrode sheet resistance (ρ), a diode with non-ideal shunt and series resistances (R_{sh} and R_s, respectively), light generated current, and local capacitance. Converting between continuous 1-dimensional, 2-dimensional, and discrete numeric models is described in the reference[10]. The resolution must be such that the size of each node is less than any screening length.

Figure 4. An elemental node used in numerical modeling.

As the model is numerical, it is possible to include non-uniformity effects by, for example, varying the local diode properties (local open circuit voltage, shunt resistance, etc.)

To model the admittance, the following procedure is used: The DC bias is used as a boundary condition at the contacts, while the cell edges define the other boundary conditions; the local DC voltage is found at each point, using Kirchhoff's Laws, so the operating point and dynamic resistance of the local diodes may be found; then an AC voltage of 1 (in arbitrary units) is applied as the contact boundary condition, again using Kirchhoff's Laws, and the resulting complex current flowing to the contact is the admittance.

RESULTS

Simulation Results

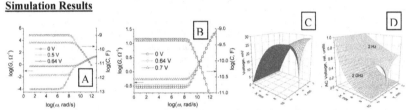

Figure 5. Simulated admittance of an unshunted cell (A) and shunted cell (B). (C) Voltage distribution for shunted cell at short circuit. (D) AC voltage distribution for shunted cell at two frequencies, demonstrating changing screening length.

Figure 5 shows the admittance of an unshunted (left) and shunted (right) cell. The cell simulated is 1 cm x 1cm. The parameters used in this simulation were j_L=30mA/cm^2, Voc = 0.64V, $\rho = 3\Omega/\square$, Rsh = 10^4 Ωcm^2, Rs = 0.1 Ωcm^2, c' = 1nF/cm^2. Examples of DC (C) and AC (D) voltage distributions calculated in the course of calculating B are shown; in this case (for both A and B), the contact is 1-dimensional along an edge of the cell; for B, a shunt is on an edge of the cell opposite the contact. Note the shunted cell has, at low frequency, comparatively lower capacitance and higher conductance. At high frequency, when the capacitive screening length becomes much less than the distance from the contact to the shunt, the admittances become indistinguishable. The resistive contact of the modeled cell has properties similar to TCO, which is why the point admittance features do not begin to appear until over 10^7Hz, in comparison to typical admittance measurements done at less than 10^5Hz.

231

At low frequency, increasing forward DC bias decreases the effective capacitance due to the bias screening length discussed above. This effect is less pronounced in the shunted cell, as there, to be observed, the bias screening must decrease the effective capacitance more than the existence of the shunt already has. Consequently, admittance curves for larger bias values are shown for the shunted cell so an effect from DC bias may be seen.

Admittance simulating software capable of generating similar data is available at http://photon.panet.utoledo.edu/~vkarpov/download.html

Experiment and Results

Point contact admittance would be most interesting on unfinished devices, one with an electrode with sheet resistance much higher than $1k\Omega/\square$, while still being less than the shunt resistance of the device over a reasonably large area. We believe that a CIGS cell, finished to the point of the deposition of the highly resistive and transparent layer, but before the deposition of the transparent conducting oxide layer, would naturally fulfill this requirement.

Our laboratory fabricates CdTe cells, most often in the typical superstrate configuration, which has greater challenges for contacting the high resistance CdTe material. Although various options exist for this, such as a more conducing layer of ZnTe, or a temporary conductive liquid contact, our initial tests used a thin, highly resistive metal contact.

Figure 6 shows admittance data using such a contact. The device was on commercial TEC15 glass and TCO, and used sputtered CdS/CdTe, using processing typical for our lab. The contact was a 2 inch by 2 inch square of thin metal ($>10k\Omega/\square$), on top of which, in the middle, was deposited a 2mm diameter contact of thicker ($<1\ \Omega/\square$) metal. The purpose of this thicker contact was so that the conductance would be dominated by the shunt resistance of the CdTe and not a $R_{Sheet}Log(L_{cell}/L_{contact})$ term, with L_{cell} on the order of the size of the entire cell and $L_{contact}$ on the order of the contact, due to compressing the current over the entire cell down to the contact. As the logarithm grows slowly, under many circumstances this term can be neglected, but may become significant if R_{sheet} is large and there are many orders of magnitude difference between L_{cell} and $L_{contact}$. Such a condition may not be limiting in the future so long as contact size is chosen appropriately, or if conductance is measured at high enough precision for this effect to be subtracted.

Equipment used in the measurement is given in the reference[11].

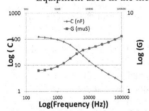

Figure 6. Admittance data of device described in text

The resulting data shows the qualitative features of the simulated admittance, such as near constant admittance at low frequency, and power-dependent roll-off of capacitance at high frequency. The numerical model used did not include the frequency-dependence capacitance of real equipotential cells, which may be included in later modeling and may explain some non-ideal features in the data.

CONCLUSIONS

We have modeled the admittance of extended area photovoltaic devices, including effects such as local device parameter variation. Early experimental measurements demonstrate dimensional scaling, establishing the potential of the technique and for future work on different contacting schemes, more direct extraction of parameters from the data, and a more statistically significant set of samples.

ACKNOWLEDGEMENTS

This work was performed under the auspice of the NSF Award No. 1066749.

REFERENCES

1. Stavros Busenberg, Weifu Fang and Kazufumi Ito, "Modeling and Analysis of Laser-Beam-Induced Current Images in Semiconductors", *SIAM J Appl Math,* **53**, 187-204, (1993)
2. R. Harju, V. G. Karpov, D. Grecu, and G. Dorer, "Electron-beam induced degradation in CdTe photovoltaics", *JAP*, **88**, 1794-1801 (2000)
3. Datong Wu, Gerd Busse, "Lock-in thermography for nondestructive evaluation of materials", *REV GEN THERM,* **37**, 693-703 (1998)
4. D. Eich, U. Hereber, U. Groh, U. Stahl, C. Heske, M. Marsi, M. Kiskinova, W. Reidl, R. Fink, and E. Umbach, "Lateral inhomogeneities of Cu(In,Ga)Se2 absorber films", Thin Solid Films **361–362**, 258-262 (2000).
5. Diana Shvydka and V. G. Karpov , "Power Generation in Random Diode Arrays", Phys. Rev. **B 71**, 115314 (2005)
6. D. K. Schroder, Semiconductor Material and Device Characterization (John Wiley, 1998).
7. A. R. Long, "Frequency-dependent loss in amorphous semiconductors", Adv. Phys. 31, 553 (1982)
8. C. Christopolous, *The Transmission Line Modeling Methods* (IEEE, New York, 1995)
9. V. G. Karpov, G. Rich, A. V. Shubashiev, G. Dorer, "Shunt screening, size effects, and I/V analysis in thin-film photovoltaics", J. Appl. Phys. **89**, 4975 (2001)
10. K. Wieland, A. Vasko, and V. G. Karpov, "Multi-dimensional admittance spectroscopy", J. Appl. Phys. **113**, 024510 (2013)
11. V. G. Karpov, Diana Shvydka, U. Jayamaha, A. D. Compaan, "Admittance spectroscopy revisited: Single defect admittance and displacement current", J. Appl. Phys. **94**, 5809 (2003)

Mater. Res. Soc. Symp. Proc. Vol. 1538 © 2013 Materials Research Society
DOI: 10.1557/opl.2013.1014

Ternary Cu$_3$BiY_3 (Y = S, Se, and Te) for Thin-Film Solar Cells

Mukesh Kumar[1] and Clas Persson[1,2]
[1]Department of Materials Science and Engineering, Royal Institute of Technology, SE-100 44 Stockholm, Sweden.
[2]Department of Physics, University of Oslo, P.O. Box 1048 Blindern, NO-0316 Oslo, Norway.

ABSTRACT

Very recently, Cu$_3$BiS$_3$ has been suggested as an alternative material for photovoltaic (PV) thin-film technologies. In this work, we analyze the electronic and optical properties of Cu$_3$BiY_3 with the anion elements Y = S, Se, and Te, employing a first-principles approach within the density function theory. We find that the three Cu$_3$BiY_3 compounds have indirect band gaps and the gap energies are in the region of 1.2–1.7 eV. The energy dispersions of the lowest conduction bands are small, and therefore the direct gap energies are only ~0.1 eV larger than the fundamental gap energies. The flat conduction bands are explained by the presence of localized Bi p-states in the band gap region. Flat energy dispersion implies a large optical absorption, and the calculations reveal that the absorption coefficient of Cu$_3$BiY_3 is larger than 10^5 cm^{-1} for photon energies of ~2.5 eV. The absorption is stronger than other Cu-S based materials like CuInS$_2$ and Cu$_2$ZnSnS$_4$. Thereby, Cu$_3$BiY_3 has the potential to be a suitable material in thin-film PV technologies.

INTRODUCTION

Development of photovoltaic materials (PV) that involves only inexpensive, non-toxic, and earth abundant elements is considered to be a major criterion to meet the increasing demand for electricity production [1]. For that reason, Cu(In,Ga)Se$_2$ [2,3], CdTe [4,5], and Cu$_2$ZnSn(S,Se)$_4$ [6,7] compounds are utilized in thin-film PV technologies. However, due to cost issues (In, Ga), supply issues (In, Te), and environmental issues (Cd), also these solar cell materials are discussed to be replaced in future clean-electricity technologies. That is, thin-film PV devices based on these ternary copper-based chalcogenides may not has reached its optimum device level, and therefore research work are dedicate to improve the quality of the thin-films as well as to find alternative solutions. For example, Cu$_3$BiS$_3$ has been suggested as an alternative material for thin-film PV cells [8-13]. It has been reported that Cu$_3$BiS$_3$ prepared by sulfurization of Bi-Cu metal precursors can achieve an external quantum efficiency of 10% [9]. Even though Bi is not an optimal element, its content ratio is only 1/7 in this compound. In order to explore these alternative PV materials, it is important to carefully study and analyze the fundamental physical properties of the compounds.

In this work, we study the electronic structure and the optical activity of Cu$_3$BiY_3 (with Y = S, Se, and Te). For comparison, we present calculations also of Cu$_3$SbS$_3$. We used the projector augmented wave method within the density functional theory (DFT) and in conjunction with a screened hybrid functional [8,14-17]. We find that all three compounds of Cu$_3$BiY_3 as well as Cu$_3$SbS$_3$ are indirect band gap semiconductors. The fundamental band gap energy is estimated to be $E_g \approx$ 1.5–1.7 eV in Cu$_3$BiS$_3$, $E_g \approx$ 1.3–1.5 eV in Cu$_3$BiSe$_3$, $E_g \approx$ 1.2–1.4 eV in Cu$_3$BiTe$_3$, and $E_g \approx$ 1.8–2.0 eV in Cu$_3$SbS$_3$. The corresponding direct band gap energy is only ~0.1 eV larger than the corresponding fundamental gap energy. Furthermore, the calculations reveal that the absorption coefficient of Cu$_3$BiY_3 is about twice as large as in CuInS$_2$ and Cu$_2$ZnSnS$_4$.

THEORY

Cu_3BiY_3 crystallizes in an orthorhombic structure (space group $P2_12_12_1$). The Bi atoms are five-coordinated and the Cu atoms are three-coordinated with respect to the neighboring anion Y atoms [8,13]. All atoms occupy $4a$ Wyckoff sites. The calculations of the electronic, and optical properties of Cu_3BiY_3 (with Y = S, Se, and Te) and Cu_3SbS_3 are based on the DFT, employing the projector augmented wave method with the Heyd-Scuseria-Ernzerhof (HSE06) screened hybrid functional [14-17]. This hybrid functional combines the Perdew-Burke-Ernzerhof (PBE) exchange-correlation and the Hartree-Fock exchange interaction. We use the standard range-separation parameter ω = 0.2 to decompose the Coulomb kernel, which is known to fairly accurately describe the crystal volume and the band gap energies. The calculations are performed with a cut-off energy of 400 eV. The density and the density-of-states (DOS) is generated from a tetrahedron Brillouin zone integration with a 6×6×6 Monkhorst-Pack k-mesh. The cell volume and the alloy configurations are fully relaxed within the HSE06 method, using a quasi-Newton algorithm. The lattice parameters are relaxed to a convergence of the total energy of 0.1 meV, and the Wyckoff $4a$ atoms positions are relaxed to a convergence of 10 meV/Å for the forces of each atom. The optical properties of the copper-based chalcogenides are analyzed by the means of the complex dielectric function $\varepsilon(\omega) = \varepsilon_1(\omega) + i\varepsilon_2(\omega)$ and the optical absorption coefficient $\alpha(\omega) = (\sqrt{2\omega^2}/c)\ [|\varepsilon(\omega)| - \varepsilon_1(\omega)\ ]^{1/2}$ where c is the speed of light. Here, the dielectric function is calculated from the joint DOS and the optical matrix elements, which directly determines the absorption coefficient [18,19].

DISCUSSION

Relaxation of the crystalline lattice parameters of Cu_3BiY_3 (Y = S, Se, and Te) in the orthorhombic $P2_12_12_1$ structure reveals that increase of cell volume directly follows the covalent radii for the anion atoms Y = S, Se, and Te. The lattice parameters are $\{a, b, c\}$ = {7.704 Å, 10.415 Å, 6.743 Å} for Cu_3BiS_3, {8.073 Å, 10.944 Å, 7.092 Å} for Cu_3BiSe_3, and {8.568 Å, 11.615 Å, 7.526 Å} for Cu_3BiTe_3. Thus, the increase of lattice parameters is ~5% going from Cu_3BiS_3 to Cu_3BiSe_3, and also ~5% going from Cu_3BiSe_3 to Cu_3BiTe_3. This is directly connected to the Cu–Y and Bi–Y bond lengths: δ(Cu–S) = 2.28 Å, δ(Cu–Se) = 2.42 Å, δ(Cu–Te) = 2.60 Å, δ(Bi–S) = 2.58 Å, δ(Bi–Se) = 2.71 Å, and δ(Bi–Te) = 2.90 Å. This is in agreement with measured crystal structure. For instance, measured data for Cu_3BiS_3 are $\{a, b, c\}$ = {7.723 Å, 10.395 Å, 6.715 Å} with the bond lengths δ(Cu–S) = 2.296 Å, and δ(Bi–S) = 2.587 Å [9,20].

Electronic properties

We analyze the electronic properties of Cu_3BiY_3 from both the atomic and angular resolved DOS (figure 1) and the electronic band structure (figure 2). DOS clearly shows that the valance band of these compounds is mainly composed of Cu-d and S-p states, while the conduction band contains Bi/Sb-p states with a small contribution from S-p. The band structure along the main symmetry directions (Fig. 2) demonstrates that the four compounds have their valence-band maximum (VBM) at the Γ-point. The conduction band minimum (CBM) is located along the line between the T-point (0,1/2,1/2) and the R-point (1/2,1/2,1/2) in both Cu_3SbS_3 and Cu_3SbSe_3, whereas at the Z-point (0,0,1/2) in Cu_3SbTe_3 [21]. It is a tendency that the conduction band states in the Γ-Z-U region are lowered when the anion atom number is increased. The

236

valence bands are very similar for the three Cu_3BiY_3 compounds, and these bands are relatively flat in the U-R-T-Z region; the topmost states in this region are ~0.5 eV below the VBM. Cu_3SbS_3 has similar band structure as Cu_3BiS_3, although ~0.3 eV larger energy gap.

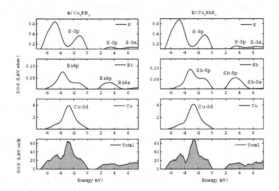

Figure 1. Atomic and angular-resolved DOS of (a) Cu_3BiS_3 and (b) Cu_3SbS_3 obtained by using the HSE06 potential and presented with a 0.05 eV broadening.

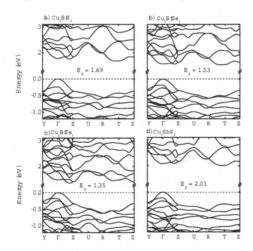

Figure 2. The electronic band structures of Cu_3BiY_3 (Y = S, Se, and Te) and Cu_3SbS_3 demonstrate that these compounds have flat CBs and indirect band gaps. The energies refer to the VBM, located at the Γ-point. The fundamental band-gap energies are presented in the figure.

The four compounds have thus indirect gap, and the fundamental band gap energies are 1.69, 1.53, and 1.35 eV for Y = S, Se, and Te, respectively, and 2.01 eV for Cu_3SbS_3. However, the lowest conduction band state at the Γ-point is only ~0.1 eV higher than the CBM, thus the direct gap is $E_g^d \approx E_g + 0.1$ eV. More specifically, $E_g^d \approx 1.79$, 1.64, and 1.45 eV for $Y = $ S, Se, and Te, respectively, and 2.14 eV for Cu$_3$SbS$_3$. We believe that the calculated gap energies with HSE06 potential are overestimated by 0.1–0.2 eV because such behavior has been reported earlier for similar compounds [8, 22-24]. The flat valence bands in the U-R-T-Z region are due to localized anion p-states at the VBM, and the flat conductions bands are explained by the presence of localized Bi $6p$ or Sb $5p$ states at the CBM.

Optical properties

The optical properties are analyzed in terms of the complex dielectric function $\varepsilon(\omega)$; see figure 3. Overall, the three Cu$_3$BiY$_3$ compounds and Cu$_3$SbS$_3$ have comparable response functions. The average high-frequency dielectric constant is determined as $\varepsilon_\infty \equiv \varepsilon_1(0) \approx \varepsilon_1(E_g/2)$ when the electron-optical phonon interaction is neglected. For Cu$_3$BiS$_3$, Cu$_3$BiSe$_3$, Cu$_3$BiTe$_3$, and Cu$_3$SbS$_3$ the estimated data are $\varepsilon_\infty \approx 7.87$, 8.60, 10.05, and 11.21 respectively. These values are similar to values for CuInS$_2$ (6.17) and Cu$_2$ZnSnS$_4$ (6.31). Thus, for Cu$_3$BiY$_3$ the size of ε_∞ follows the inverse of the band gap energy. The dielectric constant ε_∞ is related to the bands near the VBM and CBM which thus involves energies close to the band gap energies.

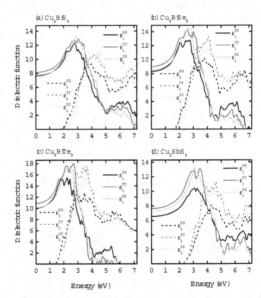

Figure 3. The complex dielectric function $\varepsilon(\omega) = \varepsilon_1(\omega) + i\varepsilon_2(\omega)$ of Cu$_3$BiY$_3$ and Cu$_3$SbS$_3$. Solid lines show the real part while dotted lines represent the imaginary part of dielectric functions. The compounds show only weak anisotropy.

The dielectric function for larger photon energies (*i.e.*, > E_g + 2 eV) involves contributions from several conduction bands and energetically low-lying valence bands [25], and therefore $\varepsilon(\omega)$ in this energy region is more dependent of the full band structure of the materials. The optical activity is represented by the absorption coefficient (figure 4). Since the Bi-6p and the Sb-5p states form flat energy dispersions of the lowest conduction band, the optical activity is large in Cu$_3$BiY_3 and Cu$_3$SbS$_3$ compounds. The absorption coefficients for $\hbar\omega$ = 2.5 eV are $\alpha(\omega)$ = 0.7, 1.4, 2.7, and 0.3 ×10^5 cm^{-1} for Cu$_3$BiS$_3$, Cu$_3$BiSe$_3$, Cu$_3$BiTe$_3$, and Cu$_3$SbS$_3$, respectively. The larger value for Cu$_3$BiTe$_3$ is due to a smaller band gap, and in order to better compare optical activity we present the absorption with an energy scale referenced to the gap energy. The absorption coefficient for $\hbar\omega$ – E_g = 1.0 eV is $\alpha(\omega)$ = 1.1, 1.5, 2.1, and 1.3 ×10^5 cm^{-1} for Cu$_3$BiS$_3$, Cu$_3$BiSe$_3$, Cu$_3$BiTe$_3$, and Cu$_3$SbS$_3$, respectively, demonstrating that the compounds have comparable $\alpha(\omega)$. The corresponding values for CuInS$_2$ and Cu$_2$ZnSnS$_4$ are 0.6 and 0.8 ×10^5 cm^{-1}. Thus, the absorption coefficient of Cu$_3$BiY_3 is about twice as large as in the conventional Cu-based chalcogenides. Furthermore, the HSE06 calculation in this work of the absorption in Cu$_2$ZnSnS$_4$ agrees with GW Green's function calculations [26].

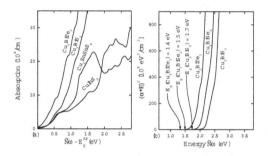

Figure 4. (a) The absorption coefficients $\alpha(\omega)$ of four Cu-based compounds, where the energy scale is referred to the band gap energy. (b) The absorption $(\alpha \cdot \hbar\omega)^2$ of Cu$_3$BiY_3 demonstrating that the absorption edges are similar although shifted according to the different band-gap energies.

CONCLUSIONS

To conclude, Cu$_3$BiY_3 is an orthorhombic type semiconductor composed of fairly inexpensive and earth abundant elements. From first-principles DFT/HSE06 calculations, we find that the compounds have an indirect band gaps of 1.69, 1.53, and 1.35 eV for Y = S, Se, and Te, respectively. The direct Γ-point gap energy is only ~0.1 eV larger than the fundamental gap energy. We estimate that the calculated gap energies are overestimated by 0.1–0.2 eV. Thereby, the indirect gap energies of the Cu$_3$BiY_3 compounds are in the energy region $E_g \approx$ 1.2–1.7 eV and the direct gap energies are thus $E_g^d \approx$ 1.3–1.8 eV. The reason why the direct and indirect gaps are so similar is explained by the flat conduction band generated by localized Bi 6p states near the band edges. Flat energy dispersion implies large optical absorption, and the calculations reveal that the optical activity in Cu$_3$BiY_3 is considerably stronger than in other Cu-S based materials.

Cu_3BiY_3 have absorption coefficients that are $\alpha(\omega) > 10^5$ cm^{-1} for photon energies $\hbar\omega > 2.5$ eV. Hence, Cu_3BiY_3 can be regarded as a potential absorber material in thin-film PV technologies.

ACKNOWLEDGMENTS

This work is supported by the Swedish Energy Agency (contract 34138-1) and the Swedish Research Council (contract C0485101). We acknowledge access to high-performance computer resources at NSC and HPC2N through SNIC/SNAC and Matter network. The authors acknowledge financial support from the Swedish Institute and the Erasmus Mundus program India4EU.

REFERENCES

1. C. Wadia, A. P. Alivisatos, and D. M. Kammen, Environ. Sci. Technol. **43**, 2072 (2009).
2. I. Repins, M. A. Contreras, B. Egaas, C. DeHart, J. Scharf, C. L. Perkins, B. To, and R. Noufi, Prog. Photovolt: Res. Appl. **16**, 235 (2008).
3. S. Siebentritt, M. Igalson, C. Persson, and S. Lany, Prog. Photovolt: Res. Appl. **18**, 390 (2010).
4. J. Britt and C. Ferekides, Appl. Phys. Lett. **62**, 2851 (1993).
5. X. Wu, Sol. Energy **77**, 803 (2004).
6. H. Katagiri, Thin Solid Films **480**, 426 (2005).
7. T. K. Todorov, K. B. Reuter, and D. B. Mitzi, Adv. Mater. **11**, E156 (2010).
8. M. Kumar and C. Persson, Appl. Phys. Lett. 102, 062109 (2013).
9. D. Colombara, L. M. Peter, K. Hutchings, K. D. Rogers, S. Schäfer, J. T. R. Dufton, and M. S. Islam, Thin Solid Films **520**, 5165 (2012).
10. V. Estrella, M. T. S. Nair, and P. K. Nair, Semicond. Sci. Technol. **18**, 190 (2003).
11. P. K. Nair, L. Huang, M. T. S. Nair, H. Hailin, E. A. Meyers, and R. A. Zingaro, J. Mater. Res. **12**, 651 (1997).'
12. N. J. Gerein and J. A. Haber, Chem. Mater. **18**, 6297 (2006).
13. F. Mesa, A. Dussan, and G. Gordillo, Phys. Status Solidi C **7**, 917 (2010).
14. G. Kresse and J. Furthmüller, Phys. Rev. B **54**, 11169 (1996).
15. G. Kresse and D. Joubert, Phys. Rev. B **59**, 1758 (1999).
16. P. E. Blöchl, Phys. Rev. B **50**, 17953 (1994).
17. J. Heyd, G. E. Scuseria, and M. Ernzerhof, J. Chem. Phys. **118**, 8207 (2003).
18. C. Persson and A. Ferreira da Silva, Appl. Phys. Lett. **86**, 231912 (2005).
19. M. Gajdoš, K. Hummer, G. Kresse, J. Furthmüller, and F. Bechstedt, Phys. Rev. B **73**, 045112 (2006).
20. V. Kocman and E. W. Nuffield, Acta Crystallogr. **B29**, 2528 (1973).
21. In Ref. 8, the T- and R-points are incorrectly described.
22. L. Yu, R. S. Kokenyesi, D. A. Keszler, and A. Zunger, Adv. Energy Mater. **3**, 43 (2013).
23. D. J. Temple, A. B. Kehoe, J. P. Allen, G. W. Watson, and D. O. Scanlon, J. Phys. Chem. C **116**, 7334 (2012).
24. M. Kumar and C. Persson, J. Renewable Sustainable Energy **5**, 0311616 (2013).
25. S.G. Choi, R. Chen, C. Persson, T.J. Kim, S.Y. Hwang, Y. D. Kim, and L. M. Mansfield, Appl. Phys. Lett. **101**, 261903 (2012).
26. H. Zhao and C. Persson, Thin Solid Films **519**, 7508 (2011).

CdTe Solar Cell and Electronic Structure

Mater. Res. Soc. Symp. Proc. Vol. 1538 © 2013 Materials Research Society
DOI: 10.1557/opl.2013.1016

Development of CdTe on Si Heteroepilayers for Controlled PV Material and Device Studies

T.A. Gessert, R. Dhere, D. Kuciauskas, J. Moseley, H. Moutinho, M.J. Romero,
M. Al-Jassim, E. Colegrove*, R. Kodama[†], and S. Sivananthan*

National Renewable Energy Laboratory, Golden, Colorado 80401
*University of Illinois at Chicago, Physics Department, Chicago, Illinois 60612
[†]EPIR Technologies, Bolingbrook, Illinois 60440

ABSTRACT

The objective of the National Renewable Energy Laboratory's (NREL) current three-year CdTe plan under the U.S. Department of Energy's SunShot Initiative is to identify primary mechanisms that limit the open-circuit voltage and fill factor of polycrystalline CdTe photovoltaic (PV) devices, and develop CdTe synthesis processes and/or device designs that avoid these limitations. Part of this project relies on analysis of crystalline materials and pseudo-crystalline CdTe layers where point and extended defects can be introduced sequentially without the complications of extensive impurities and grain boundaries that are typical of present polycrystalline films. The ultimate goals of the project include producing CdTe PV devices that demonstrate $\geq$20% conversion efficiency, while significantly improving our understanding of processes and materials capable of attaining cost goals of <$0.50 per watt.

While NREL is investigating several options for the routine fabrication of high-quality CdTe layers, one pathway involves CdTe molecular beam heteroepitaxy (MBE) on Si in collaboration with the University of Illinois at Chicago. Although CdTe/Si heteroepitaxy is relatively unfamiliar to researchers in the PV community, it has been used successfully for more than 20 years to produce high-quality CdTe surfaces required for commercial production of large-area single-crystal HgCdTe infrared detectors and focal-plane arrays. The process involves chemical and thermal preparation of Si (211) wafers, followed by deposition of As-passivation and ZnTe-accommodation layers. MBE-grown CdTe layers deposited on top of this "template" have been shown to demonstrate low etch-pit density (EPD, preferably $\leq \sim 5 \times 10^5$ cm^{-2}) and high structural quality (full width at half maximum $\sim$ 60 arcs). These initial studies indicate that 10-μm-thick CdTe layers on Si are indeed epitaxial with cathodoluminescence-determined dislocation density consistent with historic EPD measurements, and that recombination rates are distinct from either as-deposited polycrystalline or crystalline materials.

INTRODUCTION

Photovoltaic (PV) devices based on polycrystalline CdTe absorbers have attained a level of performance sufficient for large-scale commercial deployment. 2012 global production of CdTe-based thin-film PV is expected to surpass 1.5 GW$_p$, with average commercial modules attaining 12.5% efficiency, and champion modules and cells confirmed at 15.3% and 18.7%, respectively [1]. Although these benchmarks continue to attract commercial interest, many believe achieving future cost goals (in $ watt^{-1}) will require CdTe laboratory devices to demonstrate at least 20% conversion efficiency, and this will require improvement in open circuit voltage (V_{oc}) above the $\sim$860 mV that is typical of present record-efficiency cells. The search for performance limitations in *polycrystalline* CdTe PV devices encompasses some insightful guidance from *crystalline* CdTe PV devices. In 1987, Nakazawa described the fabrication of single-crystal devices that are likely CdTe homojunctions. In this study, In was diffused into a high-quality bulk crystalline CdTe substrate from an evaporated In_2O_3 layer to form an n-CdTe layer [2]. The study reports devices with V_{oc} values up to 910 mV; this is ~50 mV higher than the best record-efficiency polycrystalline device reported to date. Another noteworthy study by Carmody (2010) reports devices in which In is diffused from a thin metal overlayer into p-CdZnTe (CZT) that was grown heteroepitaxially by molecular beam epitaxy (MBE) on Si [3]. Although this study does not report devices with *binary* CdTe absorber layers, extrapolation of the performance of the CZT devices suggests the plausibility of using similar processes to produce CdTe devices that could demonstrate V_{oc} approaching or exceeding the 1000 mV threshold.

In this study, we report on the collaborative production of heteroepitaxial CdTe layers by MBE on Si substrates for use in the development of high-performance CdTe PV devices. Although the use of CdTe layers on Si is common for infrared (IR) detectors based on HgCdTe (MCT), it represents a relatively new research pathway for the PV technology community. For MCT IR detectors, the use of Si substrate encompasses many benefits over lattice-matched CdZnTe (CZT) or CdTe substrates, including: wide availability in various sizes, orientations, and electrical properties; wafer reproducibility; mechanical integrity; high purity; and low cost. Although all these benefits are important, the fact that Cu diffusion from Si is much lower than that from high-purity CZT or CdTe substrates is often considered paramount. Cu is detrimental to the function of the MCT IR detectors, and requires either insightful substrate fabrication (e.g., incorporation of Te inclusions into the CZT during growth to getter Cu impurities), or inclusion of layers that similarly limit Cu diffusion before it can enter the MCT. Presently, although MCT IR detectors are still produced on CdZnTe and CdTe substrates, and research continues on other alternative substrates (primarily GaAs), the availability of MCT IR detectors on Si substrates is rapidly expanding.

At this time, the use of heteroepitaxial CdTe on Si for the development of high-performance CdTe PV devices has not been widely investigated. Processes developed for the MCT IR detectors are designed to suppress dislocations in the CdTe epilayer to the level required for low dark current in the MCT applications (e.g., a dislocation density $< 5x10^5$ cm^{-2} for long-wave IR detectors [LWIR, 9-12 µm operation] is preferred) [4]. However, at this time, relatively little is known about the maximum dislocation density that that may be acceptable for a CdTe PV device. In this study, we report the results of initial collaborations between the National Renewable Energy Laboratory (NREL) and the University of Illinois at Chicago (UIC) involving both synthesis and analysis of CdTe layers with a particular view toward identifying the parameters that will likely be critical to ultimate PV device functionality.

EXPERIMENTAL DETAILS

The CdTe heteroepitaxial layers analyzed in this study were produced at the UIC Microphysics Laboratory. Further, several different MBE systems were used during this study to investigate to what extent system differences could be observed in the structural or optoelectronic data. CdTe/Si heteroepitaxy techniques have been developed over the past 25 years at UIC [5,6] and include the use of a double-side polished (211) 3"-diameter Si wafer that is boron doped to 30–70 Ohm-cm. The wafer is cleaned ex situ using a modified RCA cleaning process, after which the substrates are loaded into one of three RIBER MBE systems: a 3-inch Compact 21 (sample labeled D), a 5-inch Opus (samples labeled W), or a 9-inch V100 (samples labeled E). Systems were pumped to pressures $<1 \times 10^{-9}$ Torr where the substrates undergo a dehydration bake at ~400°C followed by a thermal oxide strip at >1000°C (note that all MBE temperatures quoted are uncorrected thermocouple values). Deposition of CdTe layers on Si begins with the fabrication of a multi-layer, lattice-accommodating "template." Although the template layer can vary depending on intended application, the template produced for this study included a sub-monolayer of As followed by ~15 nm of ZnTe. In this template design, the As layer passivates dangling bonds at the Si, improving structural and adhesion integrity, while the ZnTe pseudomorthic layer (i.e., thinner than the critical thickness at which misfit dislocations occur) maintains the (211) orientation. The ZnTe layer is deposited using migration-enhanced epitaxy where beam flux from elemental sources of Zn and Te (99.99999%-purity) are alternated with a short time between atomic layers to allow surface migration to improve two-dimensional growth. Following ZnTe deposition, the substrate temperature is reduced to the CdTe growth temperature of 325°C. Although the CdTe layer used for MCT devices is generally ≥ 10 μm, this initial study included CdTe layers that were thinner to allow correlation of the optoelectronic properties of the CdTe layer with structural properties and dislocation densities (nominal CdTe thicknesses for this study were 0.1, 1.0, 5.0, and 10.0 μm, see Table 1). All the CdTe films produced for this study were undoped and deposited from a nominally stoichiometric CdTe source (99.99995%-purity CdTe source) at a rate of ~1 μm hr⁻¹. The 0.1- and 1-μm-thick films were grown without any in-situ anneal cycles, and the 5-μm film underwent two anneal cycles, while the 10-μm film underwent four anneal cycles.

Films were analyzed using a combination of techniques at UIC or NREL. Analyses at UIC included film thickness uniformity using mapping IR spectroscopy [(~35-point maps, Fourier transform infrared spectroscopy (FTIR)] in the wavenumber range of 400–6000 cm⁻¹ [Thermo Nicolet Nexus model 870 with KBr beamsplitter and DTGS-TEC detector], surface-quality uniformity using a combination of multi-point Nomarski optical imaging (5-point), whole-wafer optical image capture of low-angle scattered white light, and crystalline uniformity using mapping X-ray diffraction double-crystal rocking curve (DCRC, ~35-point maps) of the full width at half maximum (FWHM) of the CdTe <422> peak (Bruker AXS Diffraktometer D8 with a high-resolution X-ray diffraction system employing a Ge(200) four-bounce beam filter for the Cu Kα source). For the 10-μm-thick film, the etch-pit density (EPD) was measured at UIC using a solution of 1HF:4HNO₃:25 lactic acid for 30 sec (standard Everson Etch) and etch pit density was assessed using Nomarski optical microscopy.

Table 1. As-grown parameters for CdTe films used in this study.

System ID	Nominal Thickness (μm)	Actual Thickness (μm)		DCRC† (arcsec)		EPD (cm⁻²)
		Min	Max	Min	Max	
D-12008	10	10.6		63		~7e6*
E-13003	5	7.76	7.8	112	120	Unknown
W-13004	5	6.95	7.44	100	120	at this
E-13004	1	1.23	1.24	360	400	time
W-13003	1	1.65	1.73	390	430	
E-13005	0.1	Unk	Unk	Unk	Unk	

*Measured at UIC with Everson Etch. † Minimum resolution < 8 arcsec

Analysis at NREL included electron backscatter diffraction (EBSD) to assess if the films demonstrated any polycrystalline artifacts [HKL Technology EBSD system with Nordlys detectors and electronics mounted in a FEI NanoSEM 200 scanning electron microscope (SEM)]

and low-temperature cathodoluminescence (LTCL) to assess dislocation density at the film surface and subsurface (JEOL JSM-7600F FE-SEM equipped with an Oxford MonoCL system and a germanium photodiode [North Coast EO-187R]). Room-temperature time-resolved photoluminescence (TRPL) was performed on the 10-μm-thick CdTe films at an analysis wavelength of 820 nm, using an excitation wavelength of 630 nm, a laser repetition rate of 1.1 MHz, and 0.3-ps laser pulses. Three excitation powers were used (2, 1, and 0.2 mW) to investigate the possible influence of space-charge separation of minority carriers competing with recombination [7].

RESULTS AND DISCUSSION

Because the present generation of heteroepitaxial CdTe template layers used for MCT IR detectors are typically at least 10 μm thick, the thinner CdTe layers produced for this initial study (i.e., 0.1, 1, and 5 μm thick) were expected to be of lower structural quality. Although Nomarski and optical scattering analysis indicated that all CdTe films were continuous and without surface features, as shown in Table 1, DCRC data indicates the structural quality of the CdTe layers decreases with thickness. The thickness and crystallinity of the 0.1-μm-thick layer could not be assessed at this time with available equipment. Figure 1 illustrates the reduction in crystal quality with thickness, and confirms that the different MBE systems produce CdTe films with structural quality that follows very similar trends. The FWHM value of the ~10-μm-thick CdTe film (63 arcsec) is consistent with the best structural quality required for CdTe template layers used for commercial heteroepitaxial MCT on Si focal-plane arrays (FPAs) [4]. For comparison, high-quality crystalline CdTe and CdZnTe substrates used for IR detectors are typically specified with a FWHM not to exceed 30 arcsec.

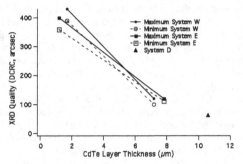

Figure 1. DCRC crystal quality vs. FTIR-measured CdTe layer thickness. For the nominal 1-μm and 5-μm-thick films, the dotted and solid lines show minimum and maximum FWHM, respectively. A minimum and maximum range is not provided for the nominal 10-μm-thick sample because only single-point FTIR was performed.

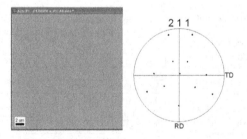

Figure 2. Left – EBSD image showing lack of multiple crystal orientations on surface of ~10-μm-thick CdTe layer on Si. Right – Pole figure confirming crystalline film with (211) orientation slightly off surface normal (spots enlarged for clarity).

EBSD analysis of this same ~10-μm-thick film is shown in Figure 2. The uniform gray color indicates, within the resolution of the instrument, that the entire film is oriented in the same direction, and there are no features observed suggesting a polycrystalline nature to the film. Pole figure analysis further shows the orientation of the surface to be (211), consistent with the orientation of the Si substrate. A slight misorientation (0.28°) was also detected uniformly across the sample, but at this time it is not known if this is due to wafer misorientation or sample mounting. Taken together, the analysis presented thus far indicates the CdTe is epitaxial, monocrystalline, and (211) oriented.

Figure 3 shows LTCL analysis of the 10-μm-thick heteroepitaxial CdTe layer. The analysis indicates that the dislocation density, averaged between five different locations on the sample, is $1.1x10^7$ cm^{-2}. The observation that LTCL yields a higher dislocation density compared to etch-pit density (i.e., EPD = ~$7.6x10^6$ cm^{-2}, see Table 1) has been observed by others [4] and is believed to be due to the historically used chemical etches (e.g., Everson Etch) producing nano-ridges that obscure the visibility of pits. Because dislocation density of these heteroepitaxial films may correlate with minority carrier lifetime, studies are underway to more fully understand the indications of defect density assessed by EPD vs. LTCL. These studies include the use of new etches (e.g., Benson Etch) that have been found to delineate etch-pit clusters that may correlate better with MCT IR device performance, and the use of SEM in addition to optical microscopy of resulting etch pits [8]. Taken together with the DCRC analysis, LTCL analysis indicates the structural quality of the 10-μm-thick layers produced for this study are consistent with templates used for present commercial MCT IR detectors.

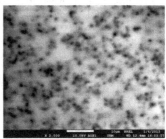

Figure 3. LTCL image on 10-μm-thick CdTe on Si sample.

TRPL analysis of the 10-μm-thick epitaxial CdTe layer (Figure 4) revealed similar trends to that observed on polycrystalline CdTe samples measured previously at NREL, and a two-exponential function is needed to provide a good fit to the luminescent decay [9]. For this sample, the fast decay constant (τ_1) is ~200 ps and represents 95%–97% of the total decay signal. The longer decay constant (τ_2) represents 5%–7% of the total decay signal and indicates a decay constant of ~700–1000 ps. Very little effect is seen with varying laser power. In comparison, *as-deposited* polycrystalline CdTe films demonstrate a shorter τ_1 decay constant ranging from 10 to 100 ps [9]. Although the epitaxial and polycrystalline decay constants are similar when measured with the single-photon excitation TRPL technique, efforts are underway to determine the extent that surface recombination and or near-surface fields may be influencing the measured result [7,10].

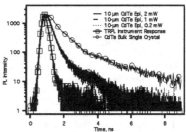

Figure 4. Single-photon TRPL of heteroepitaxial 10.6-μm CdTe on Si showing τ_1 decay rate of ~200 ps for as-deposited films. Note that three sets of indistinguishable data exist between data indicated by squares and circles that represent various laser powers.

CONCLUSIONS

The fabrication of crystalline CdTe layers formed onto Si substrates represents an important option for the development of test structures to probe the limitations of related PV devices. This study reports initial analysis of CdTe/Si samples produced through an NREL/UIC collaboration where the thickness of the CdTe layer has been systematically varied from ~0.1 to ~10 µm. Correlations between structural quality and surface dislocation density confirm initial speculation that future material and device studies should use CdTe heteroepilayers that are ≥ 10 µm thick. These 10-µm-thick CdTe heteroepilayers demonstrate DCRC crystal quality of ~65 arcsec and LTCL-measured dislocation density of ~$1x10^7$ cm^{-2}.

The initial fast decay constant (τ_1) for the 10-µm-thick CdTe measured by single-photon TRPL (~200 ps maximum) is only slightly longer than that for as-deposited polycrystalline CdTe layers on CdS/glass produced at NREL (~100 ps for polycrystalline CdTe deposited in oxygen, but without CdCl$_2$ or Cu treatments). At this time we believe the reason for the relatively short TRPL decay constant is that surface recombination and/or that near-surface effects may be influencing the measured result.

ACKNOWLEDGEMENTS
This research was supported under DOE Contract No. DE-AC36-08-GO28308 to NREL, and under NREL Agreement No. XEU-2-2208-01 with the University of Illinois at Chicago.

REFERENCES
1. M.A. Green, K. Emery, Y. Hishikawa, W. Warta, and E.D. Dunlop, "Solar Cell Efficiency Tables (Version 42)," To be published in *Prog. Photovolt: Res. Appl.* (2013).
2. T. Nakazawa, K. Takamizawa, and K. Ito, "High Efficiency Indium Oxide/Cadmium Telluride Solar Cells," *Appl. Phys. Lett.* **50** (5), 279 (1987).
3. M. Carmody, S. Mallick, J. Margetis, R. Kodama, T. Biegala, D. Xu, P. Bechmann, J.W. Garland, and S. Sivanathan, "Single-Crystal II-VI on Si Single-Junction and Tandem Solar Cells," *Appl. Phys. Lett.* **96**, 153502 (2010).
4. J.D. Benson, P.J. Smith, R.N. Jacobs, J.K. Markunas, M. Jamie-Vasquez, L.A. Almeida, A. Stoltz, L.O. Bubulac, M. Groenert, P.A. Wijewarnasuriya, G. Brill., Y. Chen, and U. Lee, "Topography and Dislocations in (112)B HgCdTe/CdTe/Si," *J. Elect. Mater.* **38** (8) 1771–1775 (2009).
5. S. Rujirawat, L.A. Almeida, Y.P. Chen, S. Sivananthan, and D.J. Smith, "High quality large-area CdTe(211)B on Si(211) grown by molecular beam epitaxy," *Appl. Phys. Lett.* **71**, 1810–1812 (1997).
6. D.J. Smith, S.C.Y. Tsen, D. Chandrasekhar, P.A. Crozier, S. Rujirawat, G. Brill, Y.P. Chen, R. Sporken, and S. Sivananthan, "Growth and characterization of CdTe/Si heterostructures - effect of substrate orientation," *Materials Science and Engineering: B* **77**, 93–100 (2000).
7. D. Kuciauskas, J.N. Duenow, P. Dippo, A. Kanevce, M. Young, J.V. Li, D.H. Levi, and T.A. Gessert, "Spectrally and Time Resolved Photoluminescence Analysis of the CdS/CdTe Interface in Thin-Film Photovoltaic Solar Cells," Submitted to *Appl. Phys. Lett.* (2013).
8. J.D. Benson, L.O. Bubulac, P.J. Smith, R.N. Jacobs, J.K. Markunas, M. Jamie-Vasquez, L.A. Almeida, A.J. Stoltz, P.S. Wijewarnasuriya, G. Brill, Y. Chen, U. Lee, M.F. Vilela, J. Peterson, S.M. Johnson, D.D. Lofgreen, D. Rhiger, E.A. Patten, and P.M. Goetz, "Characterization of Dislocations in (112)B HgCdTe/CdTe/Si," *J. Elect. Mater.* **39** (7) 1080–1086 (2010).
9. W. K. Metzger, D. Albin, M. J. Romero, P. Dippo, and M. Young, "CdCl$_2$ Treatment, S Diffusion, and Recombination in Polycrystalline CdTe," Journal of Applied Physics **99**, 103703 (2006).
10. A. Kanevce, D. Kuciauskas, T. Gessert, D.H. Levi, and D. Albin, "Impact of Interface Recombination on Time Resolved Photoluminescence (TRPL) Decays in CdTe Solar cells (Numerical Simulation Analysis)," Proc. 38th IEEE Photovoltaic Specialists Conf., Austin, TX (2012).

Mater. Res. Soc. Symp. Proc. Vol. 1538 © 2013 Materials Research Society
DOI: 10.1557/opl.2013.1017

CdTe Solar Cells: Processing Limits and Defect Chemistry Effects on Open Circuit Voltage

Brian E. McCandless[1]
[1]Institute of Energy Conversion, 451 Wyoming Road,
University of Delaware, Newark, DE 19716, U.S.A.

ABSTRACT

The role of CdTe solar cell processing on the defect chemistry that limits open circuit voltage (V_{OC}) is addressed in the thermochemical processing regimes commonly encountered in present-generation CdTe devices. The highest V_{OC} is 0.91 V for a bulk CdTe crystal with ITO which is only marginally higher than V_{OC} = 0.86 V obtained for polycrystalline CdTe films with CdS. Both fall ~0.4 V short of the V_{OC} expected for CdTe, having band gap E_G = 1.5 eV. The present >16% efficient superstrate CdTe cell uses a process based on high-temperature, T > 500°C, CdTe growth on CdS, coupled with optimized methods for incorporating oxygen, sulfur, copper, and chloride species in the CdTe film. Pushing cell conversion efficiencies beyond 20% will require increasing V_{OC} beyond 1V. However the present pathway of processing optimization will likely yield V_{OC} and efficiency converging on 0.9 V and <20%, respectively.

INTRODUCTION

Development of polycrystalline thin film solar cells based on cadmium telluride (CdTe) continues to advance, with AM1.5 terrestrial conversion efficiency η = 19%[1]. This efficiency represents 59% of the Schockley-Queisser (SQ) efficiency upper limit, η_{SQ} ~ 32%, expected for a material having an optical band gap E_G = 1.5 eV[2]. The recent CdTe cell efficiency advancements by General Electric and First Solar are attributed to optimization of short circuit current (J_{SC}) and fill factor (FF), a point which highlights cell open circuit voltage (V_{OC}) as the critical challenge for developing cells with efficiency approaching the SQ efficiency limit.

Based on SQ limit estimates, an ultimate V_{OC} = 1.2 V is expected for E_G = 1.5 eV assuming: entropy of loss in directivity of spontaneous emission; no light trapping losses; and no quantum efficiency losses. The maximum measured room temperature V_{OC} in polycrystalline thin film CdTe/CdS cells is 0.865 V, while that for a bulk single crystal CdTe/ITO cell is 0.91 V, with an acceptor concentration of 10^{15} cm^{-3} [3]. The V_{OC} deficit (= SQ V_{OC} minus measured V_{OC}) in CdTe thin film cells is thus ~0.34 V, corresponding to 72% of the ultimate V_{OC}. The highest V_{OC} single crystal CdTe cell has a V_{OC} deficit of 0.29 V, corresponding to 76% of the ultimate V_{OC}. In contrast, single crystal GaAs (E_G = 1.43 eV) and Si (E_G = 1.15 eV) solar cells, have reach 99% and 83% of their respective SQ V_{OC} limits.

Resolving the CdTe V_{OC} problem, where both single crystal and thin film cells fall far short of expectation, requires discovery of the predominant limiting mechanism(s) and development of processing routes capable of mitigating those mechanisms. Given the wide array of effort already expended, a new approach is required to overcome what appear to be limiting characteristics associated with CdTe doping and electron lifetime in the Cd-Te-S-O-Cl-Cu melange. To develop a new approach, it is necessary to understand where other approaches have succeeded and failed. This paper presents a phenomenological interpretation of the 30+ years of CdTe cell development, separating thermodynamics from kinetic and dynamic effects of film growth and post-deposition processing on the CdTe electronic properties affecting V_{OC}.

HIGH EFFICIENCY THIN FILM CDTE BACKGROUND

Cell Processing Overview

The highest efficiency CdTe cells, with η >16%, are obtained in a superstrate stack configuration wherein a weakly p-type CdTe film is deposited onto an n-type CdS film[4]. The CdTe deposition is carried out at a relatively high temperature, >500°C to a thickness of 2-4 μm. The CdS film thickness is typically <50 nm to maximize transmittance of short wavelength photons. Following the CdTe deposition, the semiconductor stack is subjected to a thermal treatment in the presence of cadmium chloride (CdCl$_2$) and oxygen. This treatment passivates CdTe grain surfaces (improves surface luminescence emission) and increases p-type conductivity in the CdTe film (van der Pauw and Seebeck analysis of author's films, unpublished). At the CdTe-CdS interface, high temperature deposition, reactive post-deposition thermal treatment, and enhanced grain boundary diffusion promote interdiffusion which converts the semiconductor interface to an abrupt junction between CdTe$_{1-x}$S$_x$ and CdS$_{1-y}$Te$_y$ alloys, where the values of x and y obtained depend on the temperature reached and are typically x~0.05 and y~~0.03[5]. Alloying reduces lattice mismatch ($\Delta a_0/a_0$) from 11% to about 9% and decreases E$_G$ on each side of the junction due to the optical bowing parameter induced by non-ideal mixing[6].

Contacts to either side of the CdTe/CdS couple consist of a thin interfacial layer and a current-carrying conductor. On the CdS side, the interfacial layer acts as a buffer layer to maintain high cell V$_{OC}$ by reducing forward diode current between the TCO and CdTe in regions where CdS is discontinuous. Many high resistivity oxides, such as Zn$_2$SnO$_4$, Ga$_2$O$_3$, Al$_2$O$_3$, or SnO$_2$, deposited 10-30 nm onto a transparent conductive oxide (TCO), such as Cd$_2$SnO$_4$, SnO$_2$:F, or In$_2$O$_3$:Sn are effective. If no buffer is used, then V$_{OC}$ decreases as CdS thickness is reduced below 100 nm. The buffer layer actually has no beneficial effect when CdS is thicker than 100 nm. On the CdTe side, the interfacial layer is a transition region between CdTe and the conducting material designed to minimize the electrical barrier arising from the high CdTe hole affinity (5.8 eV). Low resistance contacts are achieved by manipulating the interface composition with respect to Te and Cu, either by formation of a Cu-doped Te layer, deposition of a Cu-doped ZnTe coating, or formation of Cu$_2$Te.

Device Operation

High efficiency cells, with η > 16%, have typical parameters: 0.875 V < V$_{OC}$ < 0.825 V; J$_{SC}$ > 26 mA/cm^2; and FF > 75%. Current-voltage-temperature analysis of high efficiency CdTe/CdS cells reveals V$_{OC}$-T intercept equal to the band gap, and diode ideality factor ~1.5 ±0.2, consistent with diode current arising from space-charge recombination through mid-gap states in CdTe, i.e., Shockley-Read-Hall (SRH) recombination[7]. The CdS free carrier concentration films (~10^{16} cm^{-3}) is 10-100X more conductive than CdTe films (~5x10^{14} cm^3), with the space-charge region in the CdTe film. Although the CdS heteropartner yields the highest V$_{OC}$ and FF with CdTe, it contributes insignificant photocurrent to the cell, scavenging photons with higher energy that greater than E$_G^{CdS}$ = 2.4 eV. Therefore the CdS, necessary to form the electrical junction, is thin as possible to maximize CdTe photocurrent. High efficiency cells with vanishingly thin CdS, < 10 nm, have been fabricated by numerous groups, indicating that its primary role is junction formation by inversion of the near-surface region in CdTe and that its hole transport properties are inconsequential. As a parasitic optical layer, the CdS controls light generated current density, J$_L$, for photons with wavelength less than 2.4 eV and therefore has a small influence on V$_{OC}$, since:

$$V_{OC} \propto \ln\left(\frac{J_L}{J_0}\right) \qquad\qquad (1)$$

The CdTe photocurrent density can be calculated from CdS film absorptivity, allowing the V_{OC}-CdS thickness response to be determined over a range of diode current densities (Figure 1a). For a given diode current density, J_0, reducing CdS thickness to near zero contributes only 20 mV to V_{OC} by the slight increase in J_L. The oval on the lower curve in Figure 1a, for $J_0 = 10^{-8}$ mA/cm^2 at final CdS thickness from 25 to 50 nm, corresponds to present-generation cells on commercial glass superstrates. By this facile analysis, arbitrarily reducing J_0 by 1-2 orders of magnitude seems a viable route to reaching $V_{OC} = 1V$ for a given CdS final thickness.

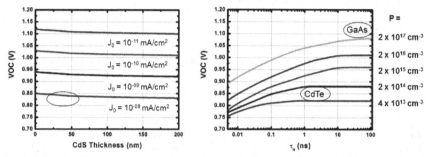

Figure 1. Calculated V_{OC} versus: a) final CdS thickness for different diode current densities, using equation (1), assuming diode ideality factor A = 1.5 (left); b) electron lifetime for different hole concentrations assuming a Gaussian distribution of mid-gap states at 2 x 10^{13} cm^{-3} and capture cross sections $\sigma_n = 10^{-11}$ cm^2 and $\sigma_p = 10^{-14}$ cm^2 (right). Simulations carried out uing wxAMPS[8] (2011) using device input parameters from J. Pan[9] (2007).

Electron Lifetime

The problem is that in a polycrystalline device, the simple diode model does not provide guidance in cell design or CdTe layer manipulation. A better guide is to more fully model the sensitivity of device current-voltage behavior from a benchmark baseline cell to semiconductor properties such as acceptor doping concentration and minority carrier lifetime. Figure 2 shows simulated V_{OC} results for SnO$_2$/25 nm CdS/4 μm CdTe versus electron lifetime at different acceptor concentrations, assuming a Gaussian distribution of mid-gap states at 2x10^{13} cm^{-3} and capture cross sections $\sigma_n = 10^{-11}$ cm^2 and $\sigma_p = 10^{-14}$ cm^2. This calculation accurately conveys the V_{OC} realm of present-generation CdTe, and coincidentally, GaAs solar cells as indicated in the graph, with the implication that increasing CdTe acceptor concentration is a more important pathway for V_{OC} than increasing minority carrier lifetime. Specifically, it suggests that present generation CdTe cells, with electron lifetime ~3 ns, measured by time-resolved photoluminescence (TRPL), could reach $V_{OC} = 1V$ by increasing hole concentration from the present level of 10^{14} cm^{-3} to 2x10^{16} cm^{-3}. However, the calculations assume that the defect composition controlling recombination does not change under the conditions needed to increase acceptor concentration. Furthermore, as p-doping increases, the decreasing space charge width shifts recombination to the CdTe-CdS interface and increases reliance on diffusion of deeply generated photocarriers. Present generation cells have CdTe with electron lifetime ~ 3 ns, already on the edge, so that a drop in crystal quality or change in surface chemistry might offset gains due to doping increases.

All of this assumes that the CdTe film is a uniform medium and does not consider the presence or specific action of grain boundaries or interfaces. CdTe film electron lifetime measurements obtained by time-resolved photoluminescence decay (TRPL) typically exhibit two exponential decay functions whose interpretation has been quantitatively interpreted as a combination of recombination from bulk (grain interior) and surfaces[10,11]. Apart from this, if individual grain dimensions are within a carrier diffusion length, then the decay time estimates from TRPL measurements also combine effects from the grain interior (τ_{Bulk}) and surfaces/interfaces (τ_{Surf}). The effective lifetime is:

$$\frac{1}{\tau_{eff}} = \frac{1}{\tau_{Bulk}} + \frac{1}{\tau_{Surf}} \qquad (2)$$

Bulk lifetime is comprised of radiative, Auger, and Shockley-Read-Hall (SRH) recombination mechanisms:

$$\frac{1}{\tau_{Bulk}} = \frac{1}{\tau_{Rad}} + \frac{1}{\tau_{Auger}} + \frac{1}{\tau_{SRH}} \qquad (3)$$

The Auger lifetime is not limiting in CdTe: $1/\tau_{Auger}=C(N_A)^2$ where C = the Auger coefficient = $2x10^{-29}$ cm^6/s and $N_A = 1x10^{14}$ cm^{-3}, giving $\tau_{Auger} = 5$ sec. Electron lifetime at surfaces depends on the surface recombination velocity (S) and geometry. Approximating grains as spheres, for films with height/width aspect ratio ~ 1, with the minimum dimension being the film thickness, d, the surface recombination lifetime can be approximated by[12]:

$$\frac{1}{\tau_{Surf}} = \frac{2S}{d} \qquad (4)$$

The sensitivity of effective electron lifetime in films to a range of surface recombination velocities for different bulk lifetimes is calculated in Figure 2. At effective lifetimes less than 100 ps, the measured effective lifetime is completely dominated by surface recombination. Bulk recombination only becomes apparent when S drops below 10^5 cm/s.

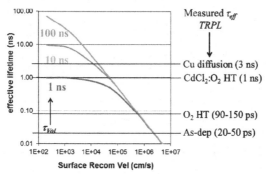

Figure 2. Effective electron lifetime versus surface recombination velocity for bulk lifetimes of 1, 10 and 100 ns. Measured effective lifetimes of CdTe/CdS films are indicated to right of figure for different processing.

In Figure 2, measured values of effective CdTe lifetime of IEC-deposited vapor transport CdTe films on CdS/Zn$_2$SnO$_4$/Cd$_2$SnO$_4$/glass window stacks at different stages of processing are indicated. As-deposited films (T$_{SS}$ = 590°C) show τ_{eff} < 50 ps, consistent with S > 10^6 cm/s suggesting poor as-deposited surface passivation. Heat treatment for 10 min at 420°C in O$_2$ increases τ_{eff} 3-5X, suggesting the onset of surface passivation by native oxidation alone. Performing a CdCl$_2$ treatment at 420°C for 30 min in air further increases τ_{eff} to 1 ns. Finally, diffusion of Cu into the cell at 185°C for 30 min in air yields maximum τ_{eff} = 3 ns. Assuming bulk lifetime ~50 ns, then the surface recombination velocity is estimated to be 10^4 cm/s. The recombination velocity is:

$$S = \sigma_{Recom} v_{th} N_{Surf} \tag{5}$$

Where σ_{Recom} is the recombination cross-section, v_{th} is the carrier thermal velocity, and N$_{Surf}$ is the density of surface states. If we take σ_{Recom} = 10^{-18} cm^2, much lower than the geometric cross-section (~10^{-15} cm^2), and v_{th} = 4x10^7 cm/s, then the density of surface states is 2.5x10^{14} cm^{-2} in completed solar cells. This is comparable to CdTe dangling bond density based on termination of different crystallographic planes (δ = 2d$_{hkl}$/a$_0^3$): $\delta_{(100)}$ = 4.8x10^{14} cm^{-2} and $\delta_{(111)}$ = 5.5x10^{14} cm^{-3}. Assuming each bond carries a charge of 1, the dangling bonds and CdTe$_{1-x}$S$_x$-CdS$_{1-y}$Te$_y$ alloy interface mismatch could account for the residual 10^4 cm/s surface recombination velocity and correspondingly low effective lifetimes.

p-type Doping

The binary equilibrium between Cd and Te exhibits one compound, CdTe, at Cd:Te = 50:50% and a eutectic near the Te liquidus (Figure 3). Detailed examination of the CdTe phase regime at elevated temperatures shows asymmetric broadening of the single phase region on the Te-rich side at T > 600°C. At 750°C, the Te stoichiometric deviation is +0.0002 at%, corresponding to a Cd deficiency of 10^{18} cm^{-3}.

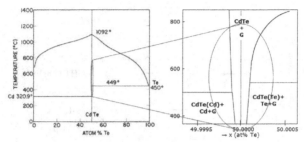

Figure 3. Cd-Te binary equilibrium phase diagram, T-x projections over entire range (left) and near Cd/Te = 1 (right).

At elevated temperatures, such as those used for cell processing, CdTe sublimes:

$$CdTe \rightarrow Cd + \tfrac{1}{2}Te_2 \tag{6}$$

With equilibrium constant given by:

$$K_{CdTe} = p_{Cd}\,p_{Te_2}^{\frac{1}{2}} \tag{7}$$

The solid state stoichiometry is fixed by the disproportionate Cd and Te partial pressures over CdTe, with Cd having higher partial pressure than Te over CdTe:

$$\log p_{Cd}^{CdTe} = -\frac{6949.5}{T} + 5.393 \text{ (atm)} \tag{8}$$

$$\log p_{Te}^{CdTe} = -\frac{13019}{T} + 6.811 \text{ (atm)} \tag{9}$$

Within the CdTe lattice, stoichiometric deviations are accommodated by native defects: vacancies, interstitials, and substitutions. Electronically, cation vacancies form acceptor states, anion vacancies form donor states, and substitutions tend to compensate. At CdTe deposition temperature of 600°C, 2×10^{16} cm^3 cadmium vacancies are expected in equilibrium, and if these are electrically active, ionized, with charge = 1e⁻, and in the absence of compensating defects, then the acceptor concentration would be $p = N_A = 2\times10^{16}$ cm^3. However, films deposited at 600°C typically exhibit $p < 10^{14}$ cm^3, suggesting a high degree of compensation.

Upper limits to intrinsic doping are thus limited by solubility limit, ionization state of the defect, and the degree of compensation. Foreign impurities introduced to the lattice add another control, since they can form interstitial or substitutional sites and can bond with native defects to form additional states. Table 1 lists pertinent examples of intrinsic and selected extrinsic point defects of CdTe. To these, we must add defect complexes formed between native defects and substitutional impurities such as the A-center V_{Cd}-Cl_{Te} and with interstitial impurities such as V_{Cd}-Cl_i, and V_{Cd}-Cu_i.

Table 1 – Selected CdTe point defects and complexes.

	Vacancy	Interstitial	Antisite	Extrinsic Acceptor	Extrinsic Donor	Cation Anion Subst
CdTe	V_{Cd}	Cd_i	Te_{Cd}	Cu_{Cd}	Cl_i, Cl_{Te}	Zn_{Cd}
	V_{Te}	Te_i	Cd_{Te}	N_{Te}	In_i	O_{Te}, S_{Te}

Assignment of defect energy levels is abundant in literature, with each state covering a range of energies, depending on whether it has been calculated from first principles, measured in bulk or epitaxial single crystals, or detected in polycrystalline films. Table 2 is a compilation of energies for selected point defects. The defects V_{Cd}, Cu_{Cd}, N_{Te}, and Na_{Cd} form shallow acceptors. During processing, Cu is introduced from the back contact, resulting in slightly higher doping levels but improved lifetime, hence passivation. N may be incorporated as a by-product of N_2 decomposition in the deposition ambient. Na may diffuse from commercial glasses, depending on how effective the window layers serve as diffusion barriers at high processing temperatures. Deep levels can act as traps, and of particular concern are F_{Te} from SnO_2:F used in some cells, and the native defects V_{Te}, Te_i, and the antisite defect Te_{Cd}. The latter exists as both a deep state and donor state depending on the ionization state. Finally, the antisite defect Cd_{Te} forms a shallow donor state.

Table 2 – Selected CdTe point defect and complex energy level ranges.

Defect	Energy Above VBM (eV)
Na_{Cd}	0.02-0.05
N_{Te}	0.05-0.10
V_{Cd}	0.11-0.16
$V_{Cd}-Cl_{Te}$	0.10-0.14
Cu_{Cd}	0.20-0.35
F_{Te}	0.8-0.9
Te_i	0.45-0.60
V_{Te}	0.60-0.75
$Te_{Cd}(++/+)$	0.90-0.95
Cl_{Te}	1.10-1.25
Cd_i	1.10-1.25
$Te_{Cd}(+/0)$	1.10-1.25
Cd_{Te}	1.35-1.40

During crystal or film growth of CdTe at elevated temperatures, the disproportionate vapor-solid kinetics produces cadmium vacancies (V_{Cd}) which lowers the Fermi energy, or chemical potential, towards the valence band maximum (VBM). This in-turn lowers the formation energy of compensating defects such as Te on Cd (Te_{Cd})[13].

CELL PROCESSING EFFECTS

Cadmium Chloride Treatment Chemistry and Dynamics

Although the chemical kinetics and dynamics of $CdCl_2$ treatment with respect to CdTe thickness, grain size, and ambient have been previously addressed[14,15], a few new points are worth noting. First is that $ZnCl_2$ vapor treatments yield similar results as $CdCl_2$, highlighting the role of the halide component. Second is that residual water vapor can significantly alter the surface chemistry. Post-deposition heat treatments in dry O_2-containing ambient or dry $CdCl_2:O_2$ promotes oxidation of CdTe at the surface, whereupon preferential oxidation of Cd from the lattice creates V_{Cd} in the near-surface region[16]. The primary surface chemical products are CdO, TeO_2 and $CdTeO_3$ - CdTe film p-type conductivity is enhanced. Treatment in humid O_2-containing ambient enhances oxidation of Te from the lattice, resulting in formation of $CdTe_2O_5$ and conversion of the film to n-type[17]. $CdTeO_3$ and $CdTe_2O_5$ crystal phases and differing Te-O bond states obtained on the CdTe surface by dry-treated and humid-treatment were detected by x-ray diffraction and x-ray photoemission spectroscopy, respectively.

From a defect formation perspective, surface chemistry predominates, and the thermochemical balance between CdTe, $CdCl_2$, O_2, and the catalytic effects of H_2O, determine the surface and sub-surface defect composition. The role of subsequent treatment time is to: 1) uniformly redistribute defects via diffusion into each grain; 2) modify surface passivation; and 3) anneal dislocations and other crystallographic defects out of each grain. In extreme examples, such as low-temperature deposited CdTe having small grain size and high twin and $\Sigma3$ boundaries, the $CdCl_2$ treatment also promotes significant grain re-growth, but this is not observed in state-of-the-art CdTe cells.

To highlight the onset of electronic changes incurred at the CdTe surface and grain boundaries at the beginning of treatments, CdTe device structures with 40 nm CdS were treated in 30 sec intervals in a rapid thermal treatment system. The films were exposed to 150 mTorr O_2 for 30 sec intervals with 9 mTorr $CdCl_2$ at 420°C and 480°C. Diagnostic V_{OC} was measured using a quinhydrone:Pt (QHPt) liquid junction contact, which is non-reactive with CdTe and yields V_{OC} linearly proportional to those obtained in completed cells. In each case, the QHPt V_{OC} decreased 250 mV after the first 30 sec treatment but recovered after an additional 30 sec treatment. For the 420°C sample, the V_{OC} progressively increased after three subsequent treatments and reached saturation after a total of 150 sec. For the 480°C treated sample, the QHPt V_{OC} also recovered but thereafter progressively decreased, due to CdS consumption. A third sample was treated in 150 Torr O_2 with no $CdCl_2$ present and exhibited no drop in QHPt V_{OC}; in this case, the QHPt V_{OC} progressively reached nearly the same value as the V_{OC} for the sample treated with $CdCl_2$ at 420C, as though the only change was progressive but poor passivation. A final experiment was conducted by exposing the CdTe to $CdCl_2$ and O_2 vapors at 420°C for 30 sec and then continuing the treatment in vacuum for another 20 min, after removing the $CdCl_2$ source boat. This cell exhibited the same V_{OC} as cells treated in the presence of $CdCl_2$ and O_2 for the entire 20 min, demonstrating that the surface chemistry in the first 30 sec is sufficient to provide the necessary point defects for doping, but optimal V_{OC}, requires time to distribute those defects, with the time depending on the grain size according to: time = r^2/D, where r is the grain radius and D the limiting diffusion coefficient.

Defect Properties of Vapor Transport CdTe Cells

Cells made using vapor transport deposited CdTe[18] were analyzed by admittance spectroscopy, transient photocapacitance, and transient photocurrent methods. In one study, cells on $CdS/Ga_2O_3/SnO_2$:F were processed with and without $CdCl_2$ treatment and Cu treatment[19]. The CdTe was 5 microns thick and was deposited at 550°C at 8 microns per minute. Admittance analysis of the baseline cell, having both $CdCl_2$ and Cu treatments yielded free acceptor concentration $N_A = 2x10^{14}$ cm^{-3} with an activation energy $E_A = 0.14$ eV, consistent with the A-center V_{Cd}-Cl_{Te}. Admittance and transient photocapacitance analysis found states at 0.35 eV above VBM (Cu_{Cd}), 0.64 eV (V_{Te} or Te_i) and 1.24 eV (Te_{Cd}, Cl_{Te}). Cells receiving $CdCl_2$ treatment but no Cu treatment exhibited states at 0.16 eV (A-center) and 0.45 eV (Te_i?). Cells with no $CdCl_2$ treatment exhibited states at 0.35 eV (with Cu = Cu_{Cd}) and 0.45 eV (no Cu = Te_i?). If the 0.45 eV state is indeed due to Te_i, this may be due to excess Te precipitation from the vapor transport deposition and subsequent cool-down. No deep states associated with F_{Te} were identified.

In a second study, a 13% efficient cell deposited on $CdS/Zn_2SnO_4/Cd_2SnO_4$ was analyzed by transient photocurrent (TPI) and transient photocapacitance (TPC) at 160K[20]. The CdTe was 6 microns thick and was deposited at 590°C. As with cells on Ga_2O_3/SnO_2:F, the acceptor concentration was determined to be $2x10^{14}$ cm^{-3}. Both TPI and TPC exhibit Urbach tails with similar energy, $E_U = 22$ meV, suggesting some degree of structural disorder in CdTe. A comparative study with respect to CdTe deposition temperature or $CdCl_2$ anneal would be needed to confirm this. The TPI/TPC ratio indicates electron capture, since the ratio of electron to hole emission rates into their respective bands is 90%. Finally, a broad Gaussian-like defect band centered at 1.24 eV was detected, having a total defect density of $\sim10^{15}$ cm^{-3}. This defect band, also detected on the completed cell in the first study, may be associated with Te_{Cd} antisite and/or Cl_{Te} substitutional defects. If it is the antisite defect, then the source of high CdTe compensation in completed polycrystalline devices is suggested. To rule out artifacts arising from low doping and interference from the back contact, a second cell with 10-11 micron thick CdTe was analyzed and yielded similar results.

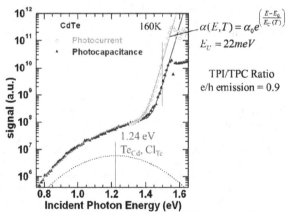

Figure 4. Transient photocurrent and transient photocapacitance of IEC CdTe cell at 160K.

257

Other IEC cells fabricated and optimized with this same cell stack and substrate temperature have attained conversion efficiencies = 16.5% through a change in diode properties resulting in an increase in FF, not V_{OC}. Cells with FF = 82.5% were fabricated and exhibited low diode ideality factor $1.0 < A < 1.4$, indicative of a shift away from SRH recombination, and low diode current $J_0 \sim 10^{-12}$ mA/cm^2. Extrapolation of V_{OC}-T to 0K yielded activation energies less than the band gap, consistent with a shift in recombination mechanism. Post fabrication diffraction analysis of the junction region on these cell revealed that the CdS films consistently exhibited no detectable c-axis grains parallel to the substrate (1% detection limit).

The role of Cu on the CdTe-CdS junction was investigated by fabrication of CdTe cells on single crystal CdS platelets (Figure 5). For this, three samples were prepared consisting of 2 micron thick CdTe deposited onto 2 mm thick crystals in three conditions: 1) Etched in HCl to expose a fresh surface; 2) Diffusion of 10 nm Cu at 500°C for 24 hr in argon followed by HCl etching to remove excess Cu and prepare a fresh surface; 3) Diffusion of Cu as above but with no etch. The cells were completed with $CdCl_2:O_2$ treatment at 425°C for 25 min and contacted with a bilayer of 5 nm Cu/50 nm Ni to make a low resistance contact.

Figure 5. Ni/CdTe/CdS device fabricated on 2 mm thick CdS platelet.

The etched CdS crystal with Cu diffusion yielded V_{OC} = 0.79 V, while the HCl etched sample had V_{OC} = 0.69 V and the Cu diffused sample with no etch had V_{OC} = 0.71 V. Noting that the solubility limit of Cu in bulk CdS is $\sim 5\times10^{15}$ cm^{-3} at 500C[21], and that the surface was etched of excess Cu, the V_{OC} result for the diffused/etched sample suggests that relatively low Cu concentrations can be delivered from the CdS side of the device and passivate the CdTe-CdS interface. Elevated Cu levels reportedly found in CdS films of high efficiency CdTe devices, based on SIMS depth profiles, can probably be attributed to accumulation in CdS grain boundaries, which constitute about 0.1 vol% of the CdS film.

PROSPECTS AND CHALLENGES

Attaining CdTe cell efficiencies beyond 20% will require increasing the V_{OC} beyond 1V. The other cell parameters, J_{SC} and FF are approaching the practical limits for superstrate devices, with $J_{SC} \sim 29$ mA/cm^2 (First Solar) and FF > 82% (IEC). In the p/n cell with a CdS heteropartner, increasing V_{OC} will require additional control over CdTe doping and lifetime, which are coupled through a high degree of self-compensation. As doping increases, compensating defect density follows, limiting free hole concentration, and the overall defect density will ultimately limit bulk lifetime. High temperature CdTe growth controls to some degree the crystalline quality of the CdTe film and establishes weak p-type conduction due to equilibrium between Cd and Te$_2$ vapors over solid CdTe yielding V_{Cd}. Oxygen during CdTe growth and in CdS was not addressed here specifically but affects CdTe nucleation and, although often incorporated into as-deposited

CdS films, may exit the CdS film once CdTe growth temperatures are reached. CdTe-CdS interface alloying only partially relieves lattice mismatch, and a wide range of bulk intermixing, up to the solubility limit, allows fabrication of cells with $V_{OC} > 0.8V$. The various post-deposition treatments and constituents added to the p/n CdTe-CdS system affect CdTe surface chemistry at the free surface and along grain walls which in-turn: 1) promote additional bulk doping and 2) improve surface and junction passivation, leading to effective electron lifetime ~3 ns in devices with $\eta > 16\%$. However, in spite of many alternative cell fabrication recipes V_{OC} remains pinned at 0.85-0.9V suggesting a fundamental limitation associated with the equilibrium-driven processing presently employed to fabricate cells. It is reasonable to conclude therefore that means to offset equilibrium formation of compensating defects, such as the Te_{Cd} antisite, by non-equilibrium processing methods is required to increase acceptor concentration. Such processing methods may include quenching-in specific defects generated during CdTe growth and manipulating the microstoichiometry to allow a wider range of substitutional dopants to be explored.

An alternative, p/i/n, device structure that leverages the low conductivity of CdTe may offer a path to $V_{OC} > 1V$. In this case, emphasis is shifted towards perfectly stoichiometric CdTe, with lower conductivity and therefore lower compensating defect density. In a cell incorporating intrinsic CdTe, improved control of contact junctions will be critical to maximizing the built-in potential. Here, analogous with amorphous silicon cells, the CdTe absorber thickness will be reduced to optimize field collection, to about one absorption length, ~0.8 micron. Light trapping techniques will be required to maximize photocurrent. In either device, p/n or p/i/n, the apparent "requirement" that CdTe be deposited onto CdS still warrants investigation, especially in view of single crystal results, developments in alternative window and buffer layer materials, and increasing methods for manipulating surface chemistry such as atomic layer deposition.

ACKNOWLEDGMENTS

The author wishes to thank his colleagues at the University of Delaware, especially Robert W. Birkmire, Karl W. Boer, Wayne Buchanan, Shannon Fields, Brett Guralnik, Steve Hegedus, Bill Shafarman, Chris Thompson, and Curtis Walkons, and his colleagues and collaborators outside of the University of Delaware, especially David Cohen and Jason Boucher at the University of Oregon, Alan Fahrenbruch at Stanford University, Fred Seymour, formerly at the Colorado School of Mines, and Jim Sites at the Colorado State University. This work would not have been possible without the collaboration over many years with members of the thin film photovoltaic community and CdTe National R&D Team (1996-2006).

REFERENCES

[1] M. Gloeckler, I. Zankin, Z. Zhao, Presented at 39th IEEE PVSC (May, 2013).
[2] W. Shockley and H. J. Queisser, *J. Appl. Phys.* 32 (1961) 510-519.
[3] T. Nakazawa, K. Takamizawa, K. Ito, *Appl. Phys. Lett.* 50 (5) (1987) 279-280.
[4] For a description of CdTe and CdS cell fabrication methods, see: B. E. McCandless and J. R. Sites, "Cadmium Telluride Solar Cells", Chapter 14 in Handbook of Photovoltaic Science and Engineering, Wiley (2011).
[5] B. E. McCandless, M. Engelmann, R. W. Birkmire, *J. Appl. Phys.* 89(2) (2001) 988-994.

[6] B. E. McCandless and J. R. Sites, "Cadmium Telluride Solar Cells", Chapter 14 in Handbook of Photovoltaic Science and Engineering, Wiley (2011) p.619.

[7] S. S. Hegedus and W. N. Shafarman, Progress in Photovoltaics: Research and Applications, B. E. McCandless, Editor, 12 (2-3) (2004) 155-176.

[8] A. Rockett, University of Illinois Urbana-Champaign (2011).

[9] J. Pan, Doctoral Dissertation, Colorado State University (2007).

[10] A. Kanevce, D. Kuciauskas, T. A. Gessert, D. H. Levi, D. S. Albin, *Conf. Rec. 40th IEEE PVSC* (2012).

[11] E. S. Barnard, et. al., Scientific Reports, 3:2098 (28 June 2013).

[12] R. Ahrenkiel, D. Levi, J. Arch, Solar Energy Materials and Solar Cells, 41/42 (1996) 171-181.

[13] S.-H. Wei, Mtg. Record, National CdTe R&D Team Meeting (2001), also this conference, paper C13.01.

[14] B. E. McCandless and W. A. Buchanan, *Conf. Rec. 33rd IEEE PVSC* (2008).

[15] V. Plotnikov, D. Kwon, K. Wieland, A. Compaan, *Conf. Rec. 34th IEEE PVSC* (2009) 1435-1438.

[16] Y. Marfaing, Thin Solid films, 387 (2001) 123-128.

[17] B. E. McCandless, unpublished analysis of Seebeck voltage on treated CdTe films on glass.

[18] G.M. Hanket, B.E. McCandless, W.A. Buchanan, S. Fields, R.W. Birkmire, J. Vac. Sci. and Technology A., **24**(5), (2006) 1695-1702.

[19] F. Seymour and J. Beach, National CdTe Team Collaboration (2006) unpublished.

[20] D. Cohen, University of Oregon, private communication, 22 September 2012.

[21] Aven and Prener (Editors), Physics and Chemistry of II-VI Compounds, Wiley (1967) p.234.

Mater. Res. Soc. Symp. Proc. Vol. 1538 © 2013 Materials Research Society
DOI: 10.1557/opl.2013.983

Low-temperature Photoluminescence Studies of CdTe Thin Films Deposited on CdS/ZnO/Glass Substrates

Corneliu Rotaru[1], Sergiu Vatavu[1,2,3], Christoph Merschjann[2], Chris Ferekides[3], Vladimir Fedorov[1], Tobias Tyborski[2], Mihail Caraman[1], Petru Gaşin[1], Martha Ch. Lux-Steiner[2] and Marin Rusu[1,2]

[1]Faculty of Physics and Engineering, Moldova State University, 60 A. Mateevici str., MD-2009 Chisinau, Republic of Moldova

[2]Institut für Heterogene Materialsysteme, Helmholtz-Zentrum Berlin für Materialien und Energie, Lise-Meitner-Campus, Hahn-Meitner-Platz 1, 14109 Berlin, Germany

[3]Department of Electrical Engineering, University of South Florida, 4202 East Fowler Ave, Tampa, FL 33647, USA

ABSTRACT

The CdTe photoluminescence spectra of CdTe/CdS/ZnO heterojunctions annealed in the presence of $CdCl_2$ have been analyzed in the 4.7-100K temperature range. The analysis has been performed for laser excitation power between 0.01 mW and 30 mW. The analysis showed that the photoluminescence spectrum in the 1.1-1.6 eV region consists of a defect band (1.437 eV) having complex structure and revealing well contoured LO phonon replicas and bound exciton annihilation in the 1.587-1.593 eV region. The band analysis has been carried out by deconvoluting the spectra. It has been shown that the defect band consists of two elementary bands and their phonon replica. An "unusual" temperature dependence of the defect band has been found.

INTRODUCTION

CdTe-based thin film heterojunctions used for PV device fabrication continue to be a promising means of direct solar to electricity conversion. Defects and dopants are important to further enhancement of PV parameters. One of the possibilities to explore the direct consequences of technology influence on the development of CdTe solar cells is thorough study of the low-T photoluminescence (PL). This technique can reveal photo-active levels and can be used to relate their influence to solar cell performance.

In particular, PL analysis is suitable for investigating the influence of the technology variations during the fabrication of CdS/CdTe heterojunctions [1] for photovoltaic use. The analysis publications related to the topic of investigation shows that the results and the interpretation of the physical processes in CdTe related to luminescence differ and are sometimes ambiguous. Such situations often occur when authors cross-reference to results measured in different experimental set-ups (i.e. different excitation lasers, etc.) or to different samples (annealed/unannealed, thin film or single crystal etc.).

An attempt to add more information to the analysis of the radiative transitions in CdTe thin film is made. This work is focused on the investigation of the evolution of the PL of CdTe thin films deposited onto CdS/ZnO/Glass substrate in device configuration.

EXPERIMENT

CdS/CdTe heterojunctions have been manufactured by a modified CSS system as described in a previous work [2]. The analysis has been carried out for heterojunctions deposited onto i-ZnO/ZnO:Al/Glass substrates. The deposition temperatures for CSS CdS onto *i*-ZnO/ZnO:Al/Glass substrates are: T_{sub}=320°C, T_{sour}=535-540°C and deposition temperatures for CSS CdTe thin films: T_{sub}=465°C, T_{sour}=555°C. Annealing of the Glass/TCO/CdS/CdTe structure followed the CSS deposition; it was carried out in presence of $CdCl_2$ at 385-410°C for 25 min. Bromine-methanol solution was used for etching the samples prior to contact deposition.

In this work, a comparative analysis of the localized levels has been investigated by PL in the 4.5-100K temperature range at different excitation intensities (0.01-30 mW, beam spot size of about 100 µm) of the Ar-Ion laser (514.5 nm). A measurement system consisting of a Czerny-Turner imaging spectrograph having 1200 l/mm grating (with a focal length of 303 mm (Andor Technologies, Shamrock 303), approx. 0.7 meV spectral resolution), a Peltier-cooled Si-CCD camera (Andor iDus DU 401-BR-DD), covering a spectral range of 200-1100 nm, and a He flow cryostat (Oxford Instruments, Variox) equipped with spectrosil B windows was used. x-y-z translation stages enhanced measurements reproducibility of the PL collecting setup.

All measurements have been carried out for Glass/TCO/CdS/CdTe heterojunctions annealed in the presence of $CdCl_2$ and etched in bromine-methanol solution. The samples were excited on the CdTe surface (i.e. not from CdS/CdTe interface).

RESULTS AND DISCUSSION

The changes in the photoluminescence spectra of the CdTe films at different temperatures, in device configuration (i.e components of the Glass/ZnO:Al/i-ZnO/CdS/CdTe heterojunctions) are given for comparison in figure 1. At low excitation intensities (less than 10^{15} photon/cm^2s) one can clearly distinguish peaks at 1.458 eV, 1.437 eV, 1.416 eV, 1.395 eV, 1.374 eV, 1.353 eV due to D-A and e-A transitions and a distinct radiative annihilation of bound excitons in the 1.587-1.593 eV region (4.7K). A peak at 1.566 eV is clearly seen at 4.7K only. The spectra recorded at higher excitation rates show a redistribution of the radiative transitions, the spectral composition being determined by a broad band at 1.423 eV and a band, which is shifting to lower energy as the temperature increases, having peak intensity at 1.575 eV (4.7K).

Temperature increase causes the quenching of the PL intensity for both bands (figure 1 B) for high excitation rates, while at low excitation rates (figure 1 A) the temperature increase results in the disappearance of the distinct features of the 1.3-1.47 eV band, the intensity vs temperature dependence revealing an inflection point at about 35K.

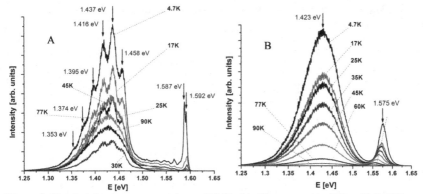

Figure 1. The photoluminescence spectrum of CdTe thin films as a part of Glass/ZnO:Al/i-ZnO/CdS/CdTe heterojunctions annealed in presence of CdCl₂ (laser excitation power A – 0.1 mW, B – 30 mW)

<u>**Photoluminescence spectra of thin film CdTe – 1.4 eV band**</u>

A typical deconvolution analysis of the 1.4 eV band is given in figure 2. The spectra can be well fitted by Lorentzian shaped elementary bands (FWHM varying between 12 and 36 meV), the convolution result (red line) corresponds to the experimental curve. Analytically the band's contour can be described by expression 1 [3].

$$I(E) = A \sum_i \exp(-S) \frac{S^i}{i!} \frac{1}{1 + \frac{(E - E_0 + iE_{LO})^2}{\gamma^2}}$$

(1)

It has been observed that the band contains well defined equidistant peculiarities (peaks) located at approx. 21 meV apart from each other. This value (21.2 eV is the energy of LO phonons in CdTe [4]) and the shape of the spectra suggest that the peaks at 1.416 eV 1.395 eV, 1.374 eV, 1.353 eV are LO phonon replica of the 1.437 eV band (4.7K) which can be considered as the zero-phonon line. The Huang-Rhys factor has a value of 1.24. Another well defined peak is positioned at 1.458 eV. One might consider this peak as the zero-phonon line as it was previously shown in several papers [5,6]. Additional analysis of the dependence of the PL intensity for these two bands vs excitation power at different temperatures shows that both of them have different origin (figure 3). A similar situation of PL band behavior has been observed in [7]. The 1.458 eV band (4.7K) manifests almost no saturation tendency, a fact which can be assigned to a band to defect level radiative transition. A power function describes PL intensity vs. excitation intensity with an exponent of 0.94-0.96 (depending on temperature). Considering the band gap at this temperature 1.606 eV [8], radiative recombination through a 148 meV recombination level takes place. Considering the technology used for sample preparations, one can assign it to an acceptor level - Cd vacancy or anti-site Cd$_{Te}$ defect [9]. On the contrary, the

1.437 eV band has a saturation tendency (figure 3). One can conclude that in this case D-A radiative transitions occur.

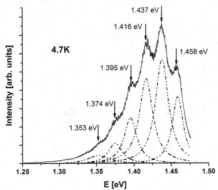

Figure 2. Example of deconvolution of the 1.4 eV band (T=4.7K, excitation power 0.1 mW).

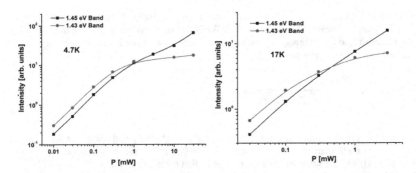

Figure 3. Dependence of the intensity of 1.45 eV and 1.43 eV deconvoluted bands on PL excitation power at different temperatures.

An unusual behavior of both bands mentioned above has been noticed (figure 4 – A). Thermal quenching of both bands can be described by the well-known formula (2).

$$I(T) = \frac{I_0}{1 + A \exp\left(\frac{\Delta E}{kT}\right)} \tag{2}$$

Activation energies calculated using the above formula, give values in the range of few meV. A plausible explanation can be assumed if the presence of a shallow donor is considered. It is interesting to follow a theoretical line shape of the 1.43 eV band [9] (often called A-band) modeled considering an activation parameter of 5 meV (and V_{Cd}-Cl_{Te} complex), being close (at least as a order of magnitude) to activation energy determined in experiment. Another spectral peculiarity is the behavior of the 1.45 eV and 1.43 eV bands at low excitation conditions. A well-defined minimum at approx. 30K can be noticed. A physical model, which might explain such an unusual behavior, lies in the fact that another recombination channel is opened (a very low concentration defect level, probably another shallow donor, considering the kT at this temperature is 2.6 meV).

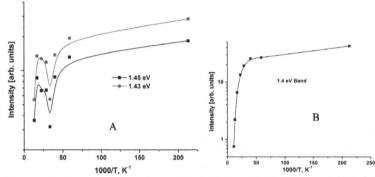

Figure 4. A - temperature dependence of the 1.45 eV and 1.43 eV deconvoluted band (excitation power 0.1 mW); B - temperature dependence of the 1.423 eV band.

The intensity of the PL vs. inverse T (figure 4 – B) shows a typical dependence of the PL for this material and determines activation energy of 34 meV (possible assignment – Te vacancy [10]).

Radiative annihilation of excitons in CdTe polycristalline thin films

The rate of radiative annihilation of excitons in CdTe is comparable to the D-A and e-A radiative mechanisms intensity in heterojunctions. The PL with peak intensities at 1.592 eV (and 1.593 eV) are attributed to annihilation of the (D^0, X) center [11,12]. The donor nature might be related to chlorine presence, resulted from annealing procedure [13]. The activation energy for this peak is 5.3 meV, determined by formula 3.

$$I(T) = \frac{I_0}{1 + AT^{3/2} exp\left(\frac{\Delta E}{kT}\right)}$$

(3)

The dependence of the 1.592 eV peak on laser excitation intensity can be described by a power function with an exponent of 1.23. Super-linear dependence of the peak on laser excitation power is indicative of bound exciton radiative transitions. The calculation of the neutral donor energy is using formula 3 in [14] and FE position [15] gives a value of 72 meV. The 1.587 peak is determined by A^0X transitions (V_{Cd}-Cl_{Te} acceptor) [13].

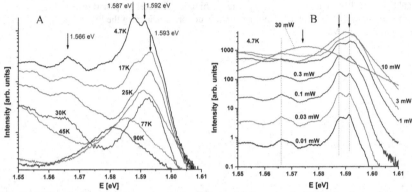

Figure 5. The photoluminescence spectrum of the excitonic region in CdTe: A - thermal quenching and redistribution of the radiative probabilities; B - photuluminescence, T=4.7K, variation of the incident beam power for PL excitation.

CONCLUSIONS

The high quality of the films obtained has been proven by the presence of a distinct excitonic line at 1.592 eV. The situation with respect to assignment of the band becomes more complex if one analyzes the work of Panosyan [16] where the 1.45 and 1.43 eV bands' intensities varied with respect to impurity composition of the samples. So, the problem consists of finding the zero phonon line. In this work we suggest a possible interpretation of the experimental results. 1.45 eV band is caused by e-A transitions, while 1.43 eV band is determined by a shallow chlorine related donor to acceptor transitions. A possible mode of PL band deconvolution is given. 1.566 eV band (4.7K) and its further evolution is not completely understood at this time. Energetically, it coincides with A^0X – 1LO transitions, but additional data and analysis is needed.

ACKNOWLEDGMENTS

This work was supported by Bilateral Moldova-Germany ASM/BMBF grant MDA 09/023 and 10.820.05.03GA and institutional grant 11.817.05.12A. S.V. would like to acknowledge, the Fulbright Scholar Program.

266

REFERENCES

1. S. Vatavu, H. Zhao, V. Padma, R. Rudaraju, D.L. Morel, P. Gaşin, Iu. Caraman and C.S. Ferekides, Thin Solid Films, 515, 6107 (2007).
2. S. Vatavu, C. Rotaru, V. Fedorov, T. A. Stein, M. Caraman, Ig. Evtodiev, C. Kelch, M. Kirsch, P. Chetruş, P. Gaşin, M. Ch. Lux-Steiner and M. Rusu, Thin Solid Films, http://dx.doi.org/10.1016/j.tsf.2012.11.105 (2013).
3. C. Onodera, T. Taguchi, J. Cryst. Growth, 101, 502 (1990).
4 *Physics and Chemistry of II-VI compounds*, edited by M. Aven and J.S. Prener (North Holland Publishing Company, Amsterdam, 1967).
Physics and Chemistry of AIIBVI compounds, S.A. Medevedev (Ed.), MIR, Moskow, 1970.
5. J. Prochazka, P. Hlidek, J. Franc, R. Grill, E. Belas, M. Bugar, V. Babentsov and R. B. James, J. Appl. Phys., 110, 093103 (2011).
6. J. Krustok, J. Madasson, K. Hjelt, H. Collan, J. Mater. Sci. Lett., 14, 1490 (1995).
7. H.L. Cotal, A.C. Lewandowski, B.G. Markey, S.W.S. McKeever, E. Cantwell, J. Aldridge, J. Appl. Phys., 67 (2), 975 (1990).
8. J. Aguilar-Hernandez, G. Contreras-Puente, H. Flores-Llamas, H. Yee-Madeira and O. Zelaya-Angel, J. Phys. D: Appl. Phys., 28, 1517 (1995).
9. S. Biernacki, U. Scherz, B.K. Meyer, Phys. Rev. B, 48, 11726 (1993).
10. X. Mathew, J.R. Arizmendi, J. Campos, P.J. Sebastian, N.R. Mathews, C.R. Jimenez, M.G. Jimenez, R. Silva-Gonzalez, M.E. Hernandez-Torres, R. Dhere, Sol. Energy Mater. & Solar Cells, 70, 379 (2001).
11. N. C. Giles, R. N. Bicknell, and J. F. Schetzina, J. Vac. Sci. Technol., A 5, 3064 (1987).
12. P. Hiesinger, S. Suga, F. Willmann, W. Dreybrod, Phys. Stat. Sol. (b), 67, 641 (1975).
13. H.-Y. Shin, C.-Y. Sun, J. Cryst. Growth, Sun, J. Cryst. Growth, 186, 354 (1998).
14. S. Vatavu, H. Zhao, Iu. Caraman, P. Gasin, C. Ferekides, Thin Solid Films, 517, 2195 (2009).
15. Z.C. Feng, M.J. Bevan, S.V. Krishnaswamy, W.J. Choyke, J. Appl. Phys., 64, 2595 (1988).
16. Zh. R. Panosyan in *Radiative Recombination in Semiconductor Crystals*, (D. V. Skobeltsyn, Proceedings of the P.N. Lebedev Physics Institute, Consultant Bureau, New York, 1975), Vol. 68, p. 145.

Mater. Res. Soc. Symp. Proc. Vol. 1538 © 2013 Materials Research Society
DOI: 10.1557/opl.2013.984

Metal chloride passivation treatments for CdTe solar cells

Jennifer Drayton, Russell Geisthardt, John Raguse, James R. Sites
Colorado State University, Physics Department, Fort Collins, CO 80523

ABSTRACT

The traditional $CdCl_2$ passivation of CdTe is expanded by adding other chlorides such as $MgCl_2$, NaCl, and $MnCl_2$ into the process through a two-step passivation procedure that combines closed space sublimation step with a vapor process. This allows the possibility of forming a highly doped field at the back of the device that could act as an electron reflector that could boost device performance by directing electrons back into the absorber layer and increasing the voltage while limiting recombination at the back of the device. The effects the two-step passivation process on device performance are characterized by current-voltage measurements, and by electroluminescence and laser-beam induced current images to show the degree of device uniformity. Additionally, capacitance voltage measurements are used to study doping density, depletion width, and possible formation of a field at the back of the device.

INTRODUCTION

The use of $CdCl_2$ to passivate CdTe-based photovoltaic devices is well known. The passivation process is multipurpose; it enhances the electronic properties of the p-n junction between the n-type CdS and p-type CdTe, it promotes recrystallization of the CdTe from small to larger grain polycrystalline film, and it modifies the grain boundaries of the polycrystalline film [1].

The best CdTe devices exhibit voltages of ~850 mV [2]. This is much less than the theoretical limit of 1.5 V for this type of device. The challenge is to increase the voltage of the device without decreasing other parameters. One way to do this is to limit or significantly decrease carrier recombination at the back contact by directing the carriers back into the absorber layer. Theoretically, an effective electron reflector (ER) to a CdTe device need only have a conduction band offset to the CdTe of 0.2 eV. One approach to this is to include an additional thin layer of material at the back of the device that would be an ER. Another method is to build a field in the CdTe layer that could act like an ER without adding a separate layer [3].

In our experiments we utilize a two-step passivation process in an effort to build a field at the back of the device that would act as an ER. The first passivation step is a CSS-type $CdCl_2$ process immediately after the CdTe without any break in vacuum. We then move to a second system and utilize a vapor $CdCl_2$ passivation step that incorporates other solid chloride materials with the $CdCl_2$. In this way, we attempt to build a field at the back of the CdTe by doping with Mg, Mn, or Na. The results of our experiments are reflected in the Voc, measured with current – voltage (J-V), and changes in the carrier density of the CdTe demonstrated with capacitance voltage (CV) data. We will also show physical evidence of the changes to the semiconductor layers using the luminescence techniques of electroluminescence (EL) and laser-beam induced current (LBIC) imaging.

EXPERIMENT

All CdS/CdTe devices used in this study were made in the PV Manufacturing Lab at Colorado State University. The Advanced Research Deposition System (ARDS) is an inline tool that utilizes closed space sublimation (CSS) to deposit thin film CdS and CdTe on 3 inch glass plates with a commercial TCO. The ARDS also has the capability to do inline processing of $CdCl_2$ passivation and Cu doping. For each experiment, a plate was made using the standard ARDS process to compensate for drift in the system. The back contacts are composed of a thin layer of carbon paint and a layer of nickel paint that is roughly twice the thickness of the carbon layer.

Early bell jar passivation experiments with a mixture of $CdCl_2$ and $MgCl_2$ indicated that the use of He was not essential to obtaining complete passivation. Comparable devices were made using Ar as the ambient gas. The use of O_2 caused discoloration of the $MgCl_2$ source material which produced inferior devices and was not pursued farther. All subsequent passivation was completed with Ar as the ambient gas.

The two-step passivation process entails a shortened CSS $CdCl_2$ process in the ARDS immediately following the CdTe deposition. The plates are then removed to a second system where for vapor passivation process in a Pyrex bell jar at 400 Torr in Ar atmosphere. Times of passivation in both the ARDS and bell jar were varied as specified in the remainder of the manuscript: ARDS time in seconds/ bell jar time in minutes.

The best $CdCl_2$ passivation processes in the bell jar occurred at 400 to 405°C. Between passivation steps, samples were held in an evacuated desiccator with desiccant present. After bell jar passivation, samples were returned to the ARDS for Cu doping. The back contact was then applied and small area devices (SADs) were delineated. SAD area is ~0.7 cm^2.

In order to determine that the bell jar passivation was indeed improving the devices with the shortened passivation times in the ARDS, the experiment was performed in which the devices received 60, 90, 120, and 180 seconds of passivation in the ARDS, with 180 s being the standard process. Devices with less than 180 s of passivation measured lower in all J-V parameters. Plates were also completed with only the bell jar passivation. The devices on these plates showed similar distributions in measured J-V parameters compared to plates made with the standard ARDS process.

For vapor passivation in the bell jar, the sample is placed film side down on 1.5 mm quartz spacers above a graphite boat loaded with the chloride source material. A solid graphite plate is placed on top of the glass to insure uniform heating of the glass and minimize heat loss. The graphite top and bottom are heated by IR lamps mounted in an enclosure that is placed around the bell jar. Passivation temperature is ramped and maintained by Watlow EZ-Zone controllers that communicate with thermocouples embedded in the graphite top and bottom. Process gas is introduced to the system through stainless steel lines near the base of the bell jar. Pressure is monitored using a 275 series convection gauge and readout from Kurt J. Lesker Co.

C-V measurements were done with a Hewlett Packard HP 4192A impedance analyzer at room temperature in the dark. C-V measurements were acquired at 100 kHz, voltage was swept from -2 V to 1 V. J-V data were measured using a solar simulator calibrated with a known GaAs standard. EL images were acquired with an Apogee Alta U8300 camera using a cooled Kodak KAF-8300 silicon CCD detector using the setup described by [4]. The measurements were performed in a light-tight enclosure using a Hoya 720 nm long-pass filter. LBIC maps were created by shining light from an optically chopped laser diode on the cell and measuring current with a lock-in amplifier. The cell is rastered under the light spot using stepper motors.

DISCUSSION

All solid chlorides used for bell jar passivation were Sigma Aldrich anhydrous beads, 10 mesh, with 99.9% or higher purity. The source materials were stored in a N_2 atmosphere glove box and were removed only when loaded into the bell jar. The process gas used was ultra-high purity Ar in order to minimize water being introduced to the system. Mixtures of roughly 50/50 ratios of $MgCl_2$, NaCl, and $MnCl_2$ to $CdCl_2$ were determined by the formula weights of the chlorides and were baked at 200°C in flowing Ar for 20 minutes prior to any passivation processing.

J-V data were measured for all devices. C-V data were acquired for the SAD on each plate with the highest Voc. EL and LBIC imaging were completed on devices which received identical passivation treatments.

The devices that underwent a second passivation step in the bell jar showed a notably tighter distribution in Voc, shown in Figure 1. Note that the devices made with standard ARDS $CdCl_2$ are not identical but are comparable as the average Voc for the 3 plates is nearly the same. For the devices passivated with the mixture of $MgCl_2$ and $CdCl_2$, current, fill factor, and efficiency were comparable to the ARDS standard processed device. Devices passivated with the $NaCl/CdCl_2$ mixed source had comparable Voc to the ARDS standard but had lower Jsc and fill factor. Devices passivated with $MnCl_2/CdCl_2$ correlated with the ARDS standard process for the 120/6 and 90/8 passivations but the devices with 60/10 passivation had degraded fill factors indicating that the Cu doping of these devices was not optimal.

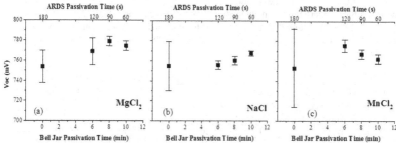

Figure 1. Average Voc (vertical axis) with the standard deviation across the plate as a function of passivation time for (a) $MgCl_2$ (b) NaCl and (c) $MnCl_2$. All data is averaged over 9 devices.

C-V data of the device with the highest Voc from each plate for the various passivation processes are shown in Figure 2. C-V measurements show changes to the hole density in the CdTe and changes occurring at the backs of the devices. For devices passivated with $MgCl_2$ (a), the hole density increases and there appears to be a field forming at the back of the device but these changes do not directly trend with passivation time. The devices passivated with NaCl (b) appear to be nominally the same regardless of passivation time. In (c) it appears that for devices passivated with $MnCl_2$ there is a trend with passivation time but it is very slight.

271

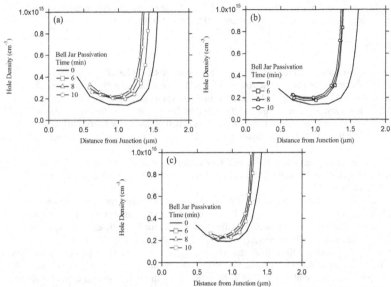

Figure 2. C-V data for devices processed with (a) MgCl$_2$ (b) NaCl and (c) MnCl$_2$.

EL and LBIC images for devices which received 120/6 passivation treatments were acquired for comparison. EL imaging shows the radiative recombination in the CdTe and LBIC imaging gives indication of high and low performing areas as a function of laser wavelength. Table I lists the measured J-V parameters of each device presented in the EL and LBIC images.

Table I. Measured parameters of devices imaged with EL and LBIC

Device ID	Bell jar source material*	Voc (mV)	Jsc (mA/cm^2)	Fill Factor (%)	Efficiency (%)
303-6-8	MgCl$_2$	781	20.6	76.2	12.2
324-4-9	NaCl	761	20.5	70.4	11.0
333-4-1	MnCl$_2$	787	21.1	72.1	11.9

*anhydrous beads, mixed with CdCl$_2$ anhydrous beads

EL on the devices was measured at room temperature with a current density of 40 mA/cm^2 and an exposure time of 100 seconds. Each EL image is of an entire SAD. LBIC measurements were made with a 100 um laser spot size. Two scans were performed on each cell; one with 638 nm laser light at 0.6 V forward bias and the other with 850 nm light at short circuit conditions. LBIC images are of the upper right quadrant of the EL image presented.

Figure 3 contains EL and LBIC images for the device passivated with MgCl$_2$. The device looks fairly uniform in the EL image (a). However, the LBIC images (b and c) show there are areas with high and low current contributions. These could be from variations in thickness or they could be a result of non-uniform doping with Mg. The dark areas in the image obtained at forward bias with 638 nm laser light (b) do not correlate with the bright areas obtained at 850 nm laser light (c).

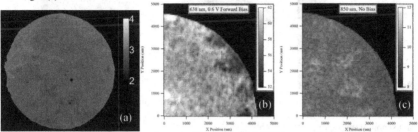

Figure 3 (a) EL image of SAD passivated with MgCl₂ (b) LBIC image of upper right quadrant of SAD scanned with 638 nm laser light at 0.6 V forward bias (c) and LBIC image of same quadrant scanned with 850 nm laser light.

Figure 4 shows the same roster and orientation of images as above but for the device passivated with NaCl. This EL image indicates that this device is less uniform than the device in Figure 3. The brighter areas of the device have higher current than the darker areas. LBIC images (b and c) demonstrate this also with dark and bright areas. For this device some of the dark areas in the 638 nm forward bias image (b) coincide with the bright areas in the 850 nm image (c).

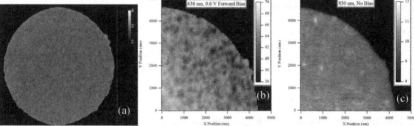

Figure 4 (a) EL image of SAD passivated with NaCl (b) LBIC image of upper right quadrant of SAD scanned with 638 nm laser light at 0.6 V forward bias (c) and LBIC image of same quadrant scanned with 850 nm laser light.

Figure 5 shows EL and LBIC images for the device passivated with MnCl₂. The EL image (a) indicates that the device is very non-uniform. The non-uniformity accentuated in the LBIC images (b and c). The dark areas in in the image acquired with 638 nm laser light in forward bias (b) directly correspond to bright areas in the image acquired with 850 nm laser light (c). The features shown in the images are spatially large. If these bright and dark areas were due to the formation of weak diodes or variations in film thickness, the device would perform poorly when compared to the devices in Figures 3 and 4. The J-V data presented in Table II does not support that assumption.

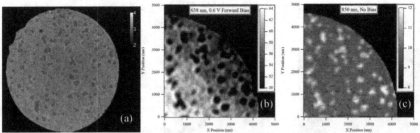

Figure 5 (a) EL image of SAD passivated with MnCl$_2$ (b) LBIC image of upper right quadrant of SAD scanned with 638 nm laser light at 0.6 V forward bias (c) and LBIC image of same quadrant scanned with 850 nm laser light.

CONCLUSIONS

Plates that received the two-step passivation showed a much tighter distribution in Voc than plates passivated either with the CSS or vapor process only. This could be due to more uniform heating of the plate in the bell jar system. C-V data indicate that there is an effect at the back of the devices driven by the two-step passivation. EL and LBIC images show non-uniformities in the devices which do not correspond to poor performance. This could an indication that the backs of the devices are being doped with Mg, Mn, or Na resulting in an electron reflector as a consequence of a built in field.

ACKNOWLEDGMENTS

This work is funded by the U.S Department of Energy, F-PACE program. The authors are particularly grateful for Kevan Cameron for assistance in fabricating the solar cells.

REFERENCES

1. Brian E. McCandless and James R. Sites in *Handbook of Photovoltaic Science and Engineering 2nd Edition* edited by Luque, Antonio and Hegedus, Steven (John Wiley & Sons, New York, 2011) pp. 615 – 619.
2. M.A. Green, K. Emery, Y. Hishikawa, W. Warta, E.D. Dunlop, Progress in Photovoltaics **20** (2012) pp. 606-614.
3. Kuo-Sui Hsaio and James R. Sites, in *Proceedings of the 34th IEEE Photovoltaic Specialists Conference* (Philadelphia, PA, 2009) pp. 1352-1356.
4. Raguse, John, J. Tyler McGoffin, and James R. Sites. "Electroluminescence system for analysis of defects in CdTe cells and modules." *Photovoltaic Specialists Conference (PVSC), 2012 38th IEEE.* IEEE, 2012.

Mater. Res. Soc. Symp. Proc. Vol. 1538 © 2013 Materials Research Society
DOI: 10.1557/opl.2013.1009

Developing Monolithically Integrated CdTe Devices Deposited by AP-MOCVD

S.L. Rugen-Hankey[1*], V. Barrioz[1], A. J. Clayton[1], G. Kartopu[1], S.J.C. Irvine[1], C. White[2], G. Rutterford[2], G. Foster-Turner[2]

[1]Centre for Solar Energy Research, Glyndwr University, OpTIC Technium, St Asaph, North Wales, UK
[2] OpTek Systems, Unit 14 Blacklands Way, Abingdon Business Park, Abingdon, Oxford, OX14 1DY
*s.rugenhankey@glyndwr.ac.uk (01745 535 213)

ABSTRACT

Thin film deposition process and integrated scribing technologies are key to forming large area Cadmium Telluride (CdTe) modules. In this paper, baseline $Cd_{1-x}Zn_xS/CdTe$ solar cells were deposited by atmospheric-pressure metal organic chemical vapor deposition (AP-MOCVD) onto commercially available ITO coated boro-aluminosilicate glass substrates. Thermally evaporated gold contacts were compared with a screen printed stack of carbon/silver back contacts in order to move towards large area modules. P2 laser scribing parameters have been reported along with a comparison of mechanical and laser scribing process for the scribe lines, using a UV Nd:YAG laser at 355 nm and 532 nm fiber laser.

INTRODUCTION

A key advantage in the large scale production of thin film photovoltaics (PV) is that an inline process does not require the assembly of smaller cells into modules, as in the case of crystalline or polycrystalline silicon wafer based systems, but instead uses monolithically integrated cells. This well-known approach reduces cost and allows for continuous inline processes to be used, such as roll-to-roll production lines [1]. CdTe solar cells, deposited by atmospheric pressure metal organic chemical vapour deposition (AP-MOCVD), have achieved > 15 % [2] using evaporated gold back contacts. This back contacting process is convenient at the research scale but when moving to large scale, evaporated gold is expensive compared to alternative contacting materials and is a fast diffuser in CdTe. Moving towards thin film PV modules means that alternative back contacts need to be assessed without resulting in excessive losses in device performance. Furthermore, the alternating scribing and deposition of different layers to form monolithic integration of cells, connected in series with each other, can be challenging as electrical characteristics of thin film devices are influenced by the scribing parameters [3,4]. The resolution and repeatability of the scribed lines, precision in substrate-pattern alignment, cell interconnects having low series resistance and high shunt resistance, are key to module performance [5]. It is therefore essential to optimise the laser parameters to create scribes with minimal heat affected zone (HAZ), smooth edges and no recast debris. This is further complicated as the area between the transparent conducting oxide (TCO) and back contact scribes (P1 and P3, respectively) is not active, as seen in Figure 1. With a sub-cell width generally limited to 10 mm to reduce the lateral conduction losses through the TCO and scribe lines, in the order of several tens of microns in width, with the separation between P1 and P3 can be from tens to hundreds of microns. This non-active area must be minimised to increase the photocurrent generated within the entire PV module aperture.

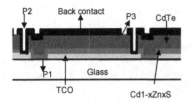

Figure 1. Schematic of a monolithically integrated thin film device in superstrate configuration, including scribe lines P1, P2 and P3.

This paper investigates the challenges of scaling up back contacting and structural characteristics of thin film devices with respect to the scribing and contacting parameters used to form monolithic cell interconnects, as seen in Figure 1. A comparative study will be given between laser and mechanical scribing techniques with several metal back contacts. The aim will be to develop a recipe to form monolithic interconnections of cells in CdTe thin film PV deposited by AP-MOCVD.

EXPERIMENTAL

$Cd_{1-x}Zn_xS$/CdTe solar cells were deposited by AP-MOCVD onto commercial ITO coated boro-aluminosilicate glass substrates, 25×50 mm^2, from Delta Technologies Ltd. and further experimental details can be found elsewhere [2,6,7]. Laser scribing was completed at OpTek Systems using a UV Nd:YAG laser at 355 nm and a 532 nm fiber laser. Line and 3D mapping profilometry were obtained using a DekTak 150 stylus profilometer. Scanning electron microscope (SEM) images and energy dispersive X-ray (EDX) spectra were collected using a Hitachi TM3000 operated at 15 kV acceleration voltage. The current-voltage (I-V) measurements were carried out using an ABET Technologies Sun 2000 solar simulator under AM1.5 irradiation, calibrated using a Fraunhofer ISE a-Si reference cell. Screen printing pastes were supplied by DuPont and cured for 30 minutes at 170 °C.

DISCUSSION

Contacting Process

An alternative to thermal metal evaporation is the use of conducting metal pastes which can be screen printed directly onto the PV devices. In this study, pastes were first printed onto glass substrates to assess the resistivity and curing parameters before being printed on CdTe devices. It was found that the silver paste did not form a suitable ohmic contact when printed directly onto the CdTe surface. A carbon/silver stack was necessary to successfully form an ohmic contact as reported in the literature [8,9] to form large area CdTe modules. Comparable efficiencies using the screen printed pastes compared to the standard gold contacts were obtained when using laboratory scale contacts of 0.25 cm^2 as seen in Table I.

Table I. Device parameters for CdTe cells comparing carbon/silver and gold as a back contact.

Back Contact	η (%)	Jsc (mA/cm²)	Voc (mV)	FF (%)	Rs (Ω cm²)	Rsh (Ω cm²)
Gold	14.7	25.4	767.8	75.5	2.3	4111
C/Ag	13.3	24.5	788	69.1	5.7	979

P1 Process

The P1 process was carried out using a 355 nm laser with a frequency of 2.5 khz. It was found that a scan speed of 15 mm/s produced scribe lines with considerable cracking along the edges, while successful removal of the ITO was achieved using lower frequency of 100 Hz and a scan speed of 7 mm/s. Minimal cracking of the ITO in the HAZ was observed, Figure 2.

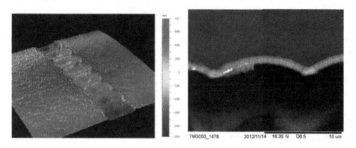

Figure 2. 3D map profilometry of P1 scribe (left) and SEM image of minimal cracking in the HAZ around the P1 scribe (right)

P2 Process

A 50 μm diameter laser beam spot was used and verified from analysing a typical ~2 μm thick $Cd_{1-x}Zn_xS/CdTe$ p-n junction. In order to assess the effect of structure thickness on the scribe lines width, samples with different p-n junction thicknesses were laser scribed. It was found that the scribe width is dependent on the thickness of the p-n junction, as expected for a focused laser beam. A scribe width of 26 μm was observed on a sample with absorber thickness 3.3 μm while an absorber 0.8 μm thick produced scribed widths of 56 μm, instead of the desired 50 μm.

The P2 laser scribes through the p-n junction were analysed using SEM and EDX. It was found that for film-side laser processing a single pass process could not remove the entire p-n junction. Even though a continuous scribe was produced, EDX line scans revealed that the CdTe absorber layer was being preferentially removed and some of the 240 nm $Cd_{1-x}Zn_xS$ window layer remained (Figure 3). In order to reduce the contact resistance at the back contact/ITO interface, the P2 scribe line must be free from any residue and avoid damage to the ITO itself. Without full removal of the p-n junction in the scribe region, an ohmic contact between the back metal contact and the TCO could not be made effectively and a double diode current-voltage response was obtained upon testing.

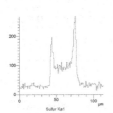

Figure 3. EDX sulfur line scan across the scribed p-n junction (left, y-axis shown in counts per second), SEM image of P2 laser scribe (right).

To demonstrate that the removal of the window layer was not a limitation of the 355 nm laser, a window layer was grown. Removal of the window layer alone can result in cracking of the underlying ITO as shown in Figure 4. With optimisation of the laser parameters the successful removal of the window layer was achieved without damaging the ITO and hence concluded that the P2 scribe could be achieved.

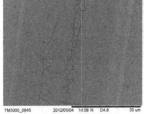

Figure 4. SEM image of P2 scribe of a CdZnS window layer with damage to the underlying TCO

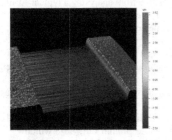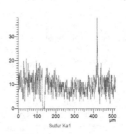

Figure 5. 3D line map of a P2 laser scribe through a p-n junction (left), sulfur EDX line scan (right).

Full removal of the p-n junction was thereafter achieved by modifying the laser scribing parameters (1 kHz, 0.05 mm/s) to leave a clean scribe of 284 µm with no recast debris, delamination or cracking of the material and good electrical contact was achieved, as seen in Figure 5.

A p-n junction was mechanically scribed and analysed for comparison to the laser scribing process. SEM pictures, shown in Figure 6, showed that both the window and the absorber layer were removed in the process but there were torn edges and re-deposited material around the mechanically scribed regions along with an inconsistent scribe width.

 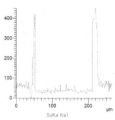

Figure 6. SEM image of manual scribe (left) and EDX sulfur line scan of the manual scribe (right).

Due to the increased resistance associated with the carbon contacting the TCO, (carbon 76 Ω/▨, 4 μm compared to 15.76 mΩ/▨, 12.7 μm for the silver paste) an investigation was undertaken into the removal of the p-n junction including a screen printed carbon contact. This would allow the highly conductive silver paste to be in direct contact with the TCO and reduce the contact resistance. It was found that the use of the 355 nm laser could not remove the carbon layer and hence the use of a 532 nm laser was adopted due to the strong adsorption of the carbon. SEM (Figure 7) and EDX confirmed that the clean removal of the carbon layer could be achieved. However, due to the accessibility of the 532nm laser, which was based at OpTek Systems in South Carolina, America, a mechanical scribing method was investigated.

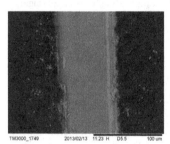

Figure 7, SEM image of a p-n junction with screen printed carbon mechanically scribed

Re-deposited material was apparent around the scribe edge but the material did not tear as in the case of the p-n junction alone. The mechanical scribe edges were not as clean as the laser scribed process and some delamination was observed at the edges. This method did however produce desired scribe widths of around 100 μm, Figure 7.

CONCLUSIONS

A preliminary investigation into the technological issues that occur when developing large scale CdTe devices has been presented. The use of commercially available pastes for screen printing which allow low cost and fast throughput have been found to be compatible with CdTe devices grown by AP-MOCVD. Full optimization of the laser parameters has been achieved in order to remove the complete $Cd_{1-x}Zn_xS/CdTe$ p-n junction without damaging the underlying TCO. Mechanical scribing was also tested and found not to be a practical method of forming a scribe through the p-n junction due to the torn edges and re-deposited material. However mechanical scribing through the p-n junction and the carbon contact showed that this method could be a possible alternative for forming the P2 scribe. Further study is now underway to obtain the desired laser scribe widths and offset separation of the P2 scribe cell. This will ensure that the contact resistance and lateral conduction of the TCO do not significantly alter the solar cells performance at the module level.

ACKNOWLEDGEMENT

This work has been funded by the ERDF Convergence funding (SPARC Cymru project) along with HEFCW funding for the Low Carbon Research Institute (LCRI). The supply of pastes from DuPont is also gratefully acknowledged.

REFERENCES

[1]   A. Bosio, D. Menossi, S. Mazzamuto, N. Romeo, *Thin Solid Films* 519 (**2011**) 7522–7525.
[2]   A. J. Clayton, S. L. Rugen-Hankey, W. S. M. Brooks, G. Kartopu, V. Barrioz, D. A. Lamb, S. D. Hodgson & S. J. C. Irvine, Proceedings of the PVSAT-8 Conference, April 2012, Northumbria University
[3]   Z. Jingquan, F. Lianghuan , L. Zhi, C. Yaping, L. Wei, W. Lili, L. Bing, C. Wei, Z. Jiagui, *Sol. Energ. Mat. Sol. C.,* 93 (**2009**) 966–969
[4]   P. Westin, J. T, Wätjen, U. Zimmermann, M. Edoff, *Sol. Energ. Mat. Sol. C.,* 98 (**2012**) 172–178
[5]   J. Perrenoud, B. Schaffner, S. Buecheler, A.N. Tiwari, Sol. Energ. Mat. Sol. C., 95 (**2011**) S8– S12.
[6]   S. J. C. Irvine, V. Barrioz, D. Lamb, E. W. Jones and R. L. Rowlands-Jones, *J. Cryst. Growth*, 310 (**2008**) 5198.
[7]   R.L. Rowlands, S.J.C. Irvine, V. Barrioz, E.W. Jones and D.A. Lamb, *Semicond. Sci. Technol.*, 23 (**2008**) 15017
[8]   T. Aramoto, F. Adurodija, Y. Nishiyama, T. Arita, A. Hanafusa, K. Omura, A. Morita, *Sol. Energ. Mat. Sol. C.,* 75 (**2003**) 211-217
[9]   L.R. Cruz, W.A. Pinheiro, R.A. Medeiro, C.L. Ferreira, J.N. Duenow, *Vacuum*, 87 (**2013**) 45-49

Compound Semiconductors

Mater. Res. Soc. Symp. Proc. Vol. 1538 © 2013 Materials Research Society
DOI: 10.1557/opl.2013.842

Integration of GaAs on Ge/Si towers by MOVPE

A. G. Taboada,[1] T. Kreiliger,[1] C. V. Falub,[1] M. Richter,[2] F. Isa,[3] E. Müller,[5] E. Uccelli,[2] P. Niedermann,[4] A. Neels,[4] G. Isella,[3] J. Fompeyrine,[2] A. Dommann,[4] H. von Känel[1]

[1] Laboratory for Solid State Physics, ETH Zürich, Schafmattstr. 16 CH-8093 Zürich, Switzerland
[2] IBM Research GmbH, Zurich Research Laboratory, Säumerstrasse 4, CH-8803 Rüschlikon, Switzerland
[3] L-NESS and Dipartamento di Fisica-Politecnico di Milano,Via Anzani 42, I-22100 Como, Italy
[4] Microsystems Technology, CSEM, Jaquet Droz 1, CH-2002 Neuchâtel, Switzerland
[5] Electron Microscopy ETH Zürich (EMEZ), Wolfgang-Pauli-Str. 16, CH-8093 Zürich, Switzerland

ABSTRACT

We report on the maskless integration of micron-sized GaAs crystals on patterned Si substrates by metal organic vapor phase epitaxy. In order to adapt the mismatch between the lattice parameter and thermal expansion coefficient of GaAs and Si, 2 μm tall Ge crystals were first grown as virtual substrate by low energy plasma enhanced chemical vapor deposition. We investigate the morphological evolution of the GaAs structures grown on top of the Ge crystals at the transition towards full pyramids with energetically stable {111} facets. A substantial release of strain is shown in GaAs crystals with a height of 2 μm and lateral sizes up to 15×15 μm^2 by both X-ray diffraction and photoluminescence.

INTRODUCTION

New semiconductor functionalities imply mainstream Si technology to be extended to other semiconducting materials with optical and electrical properties beyond those of Si. In order to exploit the advantages of the III-V/Si system, significant challenges will need to be met. Defects due to the lattice or thermal expansion coefficients mismatch (4.1% and ~ 60 %, respectively at 300 K for GaAs/Si), and anti-phase domains (APDs) must be avoided.

Several approaches were attempted in order to optimize the epitaxial integration of GaAs on Si, such as: low temperature nucleation by both metal organic vapor phase epitaxy (MOVPE),[1] and molecular beam epitaxy (MBE),[2, 3, 4] use of Ge or GeSi virtual substrates,[5, 6] and the incorporation of intermediate graded GaAsP or InGaP layers.[7]

In order to solve the inherent issues related to the integration of GaAs on Si, the most promising III-V/Si combination, we chose to adapt the highly successful concept of 3-dimensional (3D) hetero-epitaxy recently reported.[8] Our approach involves Ge crystals grown by low energy plasma enhanced chemical vapor deposition (LEPECVD)[9] on top of μm-sized Si pillars as a substitute for Ge substrates, characterized by a very small lattice mismatch (~0.07%) and comparable thermal expansion coefficients (~15%) with respect to GaAs. Moreover, growth on patterned substrates has additional advantages compared to the planar growth. First, if the crystals exhibit sufficiently large aspect ratios, threading dislocations can escape laterally at the crystal sidewalls. This so-called aspect ratio trapping (ART) mechanism has been previously used for the growth of GaAs and GaAs/Ge epilayers on both micro- and nano-structured Si patterns. [10, 11, 12]

In this work we report on the integration of strain-free GaAs by MOVPE on Si substrates deeply patterned at the micron scale. The excellent crystalline quality of the GaAs crystals is confirmed by high-resolution X-ray diffraction (HRXRD) and low-temperature photoluminescence (PL) measurements.

EXPERIMENT

Si (001) substrates off-cut 6° towards [110] were patterned by conventional photolithography and deep reactive ion etching (DRIE) based on the Bosch process.[13] Patterns consisting of pillars with widths ranging between 2 μm and 40 μm were used for these experiments. Two micron tall Ge crystals were grown on the patterned Si substrates by LEPECVD at T=500°C. GaAs growth was performed by MOVPE. We have applied the most widely used growth method for the integration of III-V compounds on Si and bulk Ge substrates, usually referred to as the two-step growth method.[14] After a thin seed layer (7 nm) grown at T= 500°C, GaAs with different nominal thickness ranging between 2 μm and 6 μm was grown onto the Ge/Si crystals at T= 680° C, growth rate 28 nm/ min and reactor pressure 100 mbar. The morphology of GaAs/Ge/Si crystals was characterized by scanning electron microscopy (SEM) (Zeiss ULTRA 55 digital field emission). Crystal cross-sections were prepared by a dual beam FIB/SEM, Zeiss NVision 40 with the Ga liquid metal ion source operated at 30 kV, imaging currents 10 pA, and milling currents up to 13 nA. High resolution X-ray diffraction (HRXRD) was performed with Cu Kα1 radiation using a PANalytical X'Pert Pro-MRD diffractometer equipped with a 4-bounce Ge(220) crystal monochromator on the incident beam, and an analyzer crystal and a Xe point detector on the diffracted beam. A 532 nm continuous wave laser line attenuated to 5 mW was used as excitation source to investigate the PL at T=5 K. The PL spectrum was recorded with an InGaAs line detector attached to a 0.3 m focal length spectrograph.

DISCUSSION

GaAs truncated pyramids on Ge/Si patterned substrates

A top SEM view of 2-μm-tall GaAs structures grown by MOVPE onto an array of Ge/Si crystals is displayed in Fig. 1.a. The substrate pattern consists of 8-μm-tall and 2-μm-wide Si pillars separated by 3-μm-wide trenches arranged in 10×10 blocks. Before the GaAs growth, the Si pillars were coated with 2 μm of Ge by LEPECVD. The GaAs crystals exhibit regular pyramidal top morphologies with well-defined facets. The pyramidal structures are formed by the intersection of {111} facets. Figure 1 (b.-d.) shows the top view and cross section SEM micrographs of a 4-μm-tall GaAs crystal grown on a 15-μm-wide Ge/Si pillar. The asymmetrical distribution of {111}, {113} facets and the top (001) facet results from the substrate miscut of 6° (Fig.1.b). This asymmetry is evident also in the (1-10) and (110) FIB cross sections displayed in Fig. 1(c, d). Along the miscut direction, just a small part of the surface remained parallel to the offcut substrate plane, while the exact [001] surface orientation was almost fully restored. A rotation of the growth axis towards the GaAs natural directions can hence be identified for GaAs crystals grown on miscut patterned substrates. Basically, GaAs crystals are tilted in the (110) plane, corresponding to the miscut direction [110]. The tilt magnitude of 6° corresponds to the off-cut angle of the Si substrate.

Top view SEM images of 2-, 4- and 6-μm-tall GaAs crystals grown by MOVPE on square Si pillars 5-, 10- and 15-μm-wide, respectively, are shown in figure 2. (a-c). Prior to the GaAs growth, 2 μm of Ge were deposited by LEPECVD. GaAs crystal heights and Si pillar base dimensions were selected in order to keep the height-to-base aspect ratio constant. The ratio between the equivalent {111} and {113} facet surface areas belonging to the pillars of different sizes is constant within 4.5%.

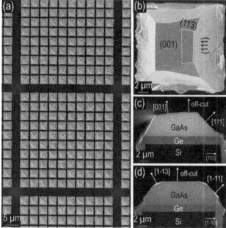

Fig.1.(a) Top SEM view of GaAs/Ge pyramids grown by MOVPE on top of 8-μm-tall and 2-μm-wide Si pillars arranged in 10×10 blocks. (b) Detail of the top surface of 4-μm-tall GaAs crystals grown on 15-μm-wide Si pillars. Prior to the GaAs growth, 2 μm of Ge were deposited by LEPECVD. The FIB cross sections parallel to the (1-10) (c) and (110) (d) planes display the strong morphological asymmetry of GaAs/Ge crystals grown on offcut Si patterned substrates.

Therefore, the different morphological stages of the GaAs crystals before the pyramid formation may be traced to a good approximation following the facet distribution of GaAs/Ge crystals with constant thickness grown on Si pillars of different width. Figure (2.d-g) shows 2-μm-tall GaAs crystals grown by MOVPE on 2-, 5-, 9- and 15-μm-wide Ge/Si pillars. As mentioned before, the GaAs full pyramids are terminated by {111} facets. Nevertheless, in the initial stages of the growth a distribution of facets belonging to the {113} and {001} family planes coexist with the main {111} facets in the GaAs truncated pyramids (Fig. 2. d-g). The GaAs facet distribution suggests an evolution towards the pyramidal shape due to the lower surface energy of {111} surfaces with respect to the (001) surface. This is in good agreement with the higher growth rate by a factor of 20 along [001] with respect to [111], estimated from the thicknesses along these two directions (Fig. 1.c.).

Elastic strain release of GaAs/Ge crystals on patterned Si substrates.

In order to quantify the strain relaxation for the different patterns, the tetragonal distortions of the cubic GaAs/Ge lattice were evaluated. With this aim, the HRXRD symmetrical (004) measurements in the different patterns were complemented with their corresponding scans in the asymmetrical (224) azimuth. Figure 3.a. compares the HRXRD 2θ/ω scans measured around the (004) reflection for a GaAs/Ge layer grown at T=680° C on a planar

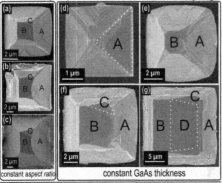

Fig.2. Figures 2.(a- c) correspond to GaAs crystals 2-, 4- and 6-μm-tall, respectively, grown on top of Ge coated Si pillars 5-, 10- and 15-μm-wide. The Si pillar sizes and GaAs crystal thicknesses were chosen in order to keep the height-to-base aspect ratio constant. Figures 2.(d-g) show the top SEM view of 2-μm-tall GaAs crystals grown on top of Ge/Si pillars 2-, 5-, 9- and 15-μm-wide, respectively. The different facets {111}, {001}, {113} are labeled with the capital letters A, B and C, respectively. The area labeled with a D corresponds to the surface misoriented 6° with respect to the [001] direction.

Si(001) substrate, and GaAs/Ge crystals grown under similar conditions on Si pillars with widths ranging from 2 to 40 μm. We relate the broadening of the GaAs and Ge HRXRD peaks compared with the one of the Si substrate both with the presence of defects and with the tilts of the individual GaAs/Ge/Si microstructures.
A clear shift towards lower Bragg angles can be identified for GaAs crystals grown on patterned substrates. That is because the biaxial thermal strain can relax elastically due to the high aspect ratio of the GaAs/Ge/Si heterostructures. The Ge and GaAs layers grown on the Si pillars up to 5×5 μm² show a complete release of the strain. For structures of 15×15 μm² the GaAs strain has increased to about 0.05 % (reference lattice constant (GaAs)= 5.6532 Å). A uniform increment of the strain was found for larger structures up to a relatively high tensile strain of 0.19 % observed for the GaAs growth on Ge on the non-patterned Si.
Figure 3.b shows the PL spectra of GaAs crystals grown by MOVPE on Ge-coated Si pillars with base widths ranging between 2 and 15 μm compared with the spectrum of GaAs layers grown on a planar Ge/Si substrate. The PL signal collected at T= 5 K from the GaAs grown on the different patterned substrates contains three different components. The corresponding optical transitions indicate a slight strain-driven shift depending on the microstructure size. The PL emission energy maxima collected from GaAs crystals grown on 2-μm-wide Ge/Si pillars is centered at 1.519 eV, in good agreement with the bulk GaAs bandgap at T=5 K, and with the PL measured in GaAs homoepitaxially grown on a GaAs substrate, used as a reference. Further increasing the Si pillar size leads to gradual redshifts of the PL energy, e.g., by about 6 meV for GaAs crystals grown on Ge coated 15-μm-wide Si pillars.

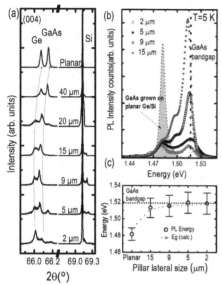

Fig.3 (a) 2θ-ω XRD scans measured on different GaAs/Ge/Si microstructures with widths from 2 μm to 40 μm, compared with control GaAs layers grown on a planar Ge/Si substrate. (b) Low temperature PL spectra of GaAs/Ge microstructures grown on 2-, 5-, 9- and 15-μm-wide Si pillars. They are compared with the PL spectrum of a 2-μm-thick GaAs film grown on a planar Ge/Si substrate. (c) PL energy obtained by Gaussian fits of the high energy peaks from the luminescence spectra (Fig.3.a.) and the calculated bandgap shifts using the $\varepsilon_{//}$ measured by HRXRD at room temperature as an input to Eq.(1). The strain values were corrected taking into account the thermal strain induced by cooling down from RT to T=5 K. Error bars correspond to the FWHM of the fitted PL spectra.

The intermediate peak, centered at ~1.498 eV (with small variations for the different sizes) may be related to a free electron-acceptor transition (e-C^0, 1.494 eV) due to carbon impurities incorporated during the MOVPE growth.[15] The low energy optical transition at 1.484 eV coincides with the main line collected from the GaAs layers grown on non-patterned substrates which can therefore be related with the strained material in the trenches. GaAs grown on planar Ge/Si substrates exhibits a similar two-peak structure, but strain-shifted by ~20 meV to the red. The change induced in the GaAs bandgap as a result of strain can be calculated using the expression:[16]

$$\Delta E_g = \left[-2a\left(\frac{C_{11} - C_{12}}{C_{11}}\right) + b\left(\frac{C_{11} + 2C_{12}}{C_{11}}\right)\right]\varepsilon_{\parallel} \qquad Eq.(1)$$

where DE_g is the change in the bandgap caused by the biaxial strain $\varepsilon_{//}$, C_{11} and C_{12} are the stiffness coefficients and a and b are the hydrostatic and shear deformation potentials respectively ($C_{11}=119$ GPa, $C_{12}=53.4$GPa, $a=-9.8$ eV and $b=-2.0$ eV for GaAs). A comparison of the PL maxima obtained by Gaussian fits to the different luminescence spectra, and the calculated bandgap shifts using the $e_{//}$ measured by HRXRD as an input to Eq. (1) can be found in Fig. 3.c. The thermal strain induced during the cooling down from the HRXRD measurement temperature (RT) and the PL characterization temperature (T=5 K) was considered. The good agreement between the calculations and the PL peaks settle the previously commented size dependent elastic strain relaxation of GaAs crystals grown on Ge/Si.

An increment on the low temperature PL integrated intensity normalized to the microstructure volume (up to a factor 8 for structures with height to base aspect ratio ~0.1) was found in the patterned substrates when comparing with GaAs grown on planar Ge/Si and GaAs homoepitaxialy grown on a commercial GaAs substrate (up to a factor 4 for structures with height to base aspect ratio ~0.1).

CONCLUSIONS

In summary, maskless integration of high quality epitaxial GaAs on Ge/Si patterned substrates by MOVPE was demonstrated. GaAs microstructures evolve to pyramidal morphologies formed by energetically stable {111} facets. GaAs microstructures grown on misoriented patterned Si substrates exhibit a rotation of their growth axis towards the GaAs natural directions. This allows obtaining well oriented GaAs (001) surfaces from misoriented Si substrates. Independent analysis of both the HRXRD diffraction and low temperature PL spectra reveals a gradual release of the elastic strain relaxation of GaAs/Ge/Si microstructures with decreasing size. Thus, GaAs crystals grown on 15x15 μm^2 wide Ge/Si pillars exhibit a 75% release of the strain compared with the GaAs layers grown on planar substrates.

ACKNOWLEDGMENTS

Financial support by the Swiss Federal Program Nano-Tera through projects NEXRAY and COSMICMOS is gratefully acknowledged. We acknowledge FIRST Center of ETH Zürich, Pilegrowth Tech Srl, helpful discussions with E. Gini and K. Mattenberger.

REFERENCES

1. M Akiyama, Y. Kawarada and K. Kaminishi, J. Cryst. Growth 68, 21-26 (1984).
2. K. Nozawa, Y. Horikoshi, Jap. J. Appl. Phys. 29, L540-L543 (1990).
3. R. Fischer, N. Chand, W. Kopp, H. Morkog, L. P. Erickson and R. Youngman, Appl. Phys. Lett. 47, 397 (1985).
4. Y. González, L. González and F. Briones, Jap. J. Appl. Phys. 31, L816-L819 (1992).
5. S. M. Ting and E. A. Fitzgerald J. Appl. Phys. 87, 2618 (2000).
6. J. A. Carlin, S. A. Ringel, E. A. Fitzgerald, M. Bulsara, B. M. Keyes Appl. Phys. Lett. 76, 1884 (2000).
7. Yuji Komatsu, Keiji Hosotani, Takashi Fuyuki and Hiroyuki Matsunami, Jap. J. of Appl. Phys., 36 5425 (1997).
8. C. V. Falub, H. von Känel, F. Isa, R. Bergamaschini, A. Marzegalli, D. Chrastina, G. Isella, E. Müller, P. Niedermann , L. Miglio, Science 335, 1330 (2012).
9. C. Rosenblad, H. von Känel, M. Kummer, A. Domman, and E. Muller, Appl. Phys. Lett , 76, 427, (2000).
10. E. A. Fitzgerald, N. Chand, Journal of Electronic Materials 20, 10 (1991).
11. J. Z. Li,J. Bai, J.-S. Park, B. Adekore, K. Fox, M. Carroll, A. Lochtefeld, and Z. Shellenbarger, Appl. Phys. Lett. 91, 021114 (2007).
12. M. Richter, E. Uccelli, A.G. Taboada, D. Caimi, N. Daix, M. Sousa, C. Marchiori, H. Siegwart, C.V. Falub, H. von Känel, F. Isa, G. Isella, A. Pezous, A. Dommann, P. Niedermann, J. Fompeyrine, J. Cryst. Growth, in press, (2013) http://dx.doi.org/10.1016/j.jcrysgro.2012.12.111
13. F. Laermer, A. Schilp, U.S. Patent 5501893_3, 26 (1996).
14. W. I. Wang, Appl. Phys. Lett. 44, 1149 (1984).

288

15. B. G. Yacobi, D. B. Holt, "Cathodoluminescence Microscopy of Inorganic Solids", Springer, (1990).
16. G. E. Pikus and G. L. Bir, Sov. Phys. Solid State 1, 136 (1959).

Mater. Res. Soc. Symp. Proc. Vol. 1538 © 2013 Materials Research Society
DOI: 10.1557/opl.2013.585

Development of high k/III-V (InGaAs, InAs, InSb) structures for future low power, high speed device applications

Edward Yi Chang,[1,2,*] Hai-Dang Trinh[1], Yueh-Chin Lin[1], Hiroshi Iwai[3], and Yen-Ku Lin[1]

[1]Department of Materials Science and Engineering, National Chiao Tung University, Taiwan.
[2]Department of Electronic Engineering, National Chiao Tung University, Taiwan.
[3]Frontier Research Center, Tokyo Institute of Technology, Tokyo , Japan
[*] Email: edc@mail.nctu.edu.

ABSTRACT

III-V compounds such as InGaAs, InAs, InSb have great potential for future low power high speed devices (such as MOSFETs, QWFETs, TFETs and NWFETs) application due to their high carrier mobility and drift velocity. The development of good quality high k gate oxide as well as high k/III-V interfaces is prerequisite to realize high performance working devices. Besides, the downscaling of the gate oxide into sub-nanometer while maintaining appropriate low gate leakage current is also needed. The lack of high quality III-V native oxides has obstructed the development of implementing III-V based devices on Si template. In this presentation, we will discuss our efforts to improve high k/III-V interfaces as well as high k oxide quality by using chemical cleaning methods including chemical solutions, precursors and high temperature gas treatments. The electrical properties of high k/InSb, InGaAs, InSb structures and their dependence on the thermal processes are also discussed. Finally, we will present the downscaling of the gate oxide into sub-nanometer scale while maintaining low leakage current and a good high k/III-V interface quality.

INTRODUCTION

High-k/III-V structure has been extensively studied recently in order to realize the 16 nm node and beyond complementary metal-oxide-semiconductor (MOS) technology due to their high carrier mobility and drift velocity. Since the devices have been getting smaller, the requirement for the devices is not only high performance but also low power consumption. In this context, the III-V based devices emerge as a potential candidate which could allow to have very high performance at low supply voltage. Regardless of long term study, many issues on high k/III-V structure still need to be addressed. The density of state at high k/III-V interface is away high due to the poor III-V native oxide quality as well as high defect (vacancies, dangling bonds, like-atoms bonds) at III-V surfaces. The defect in high k oxide is also a problem which affects the performance of devices. Besides, the downscaling the equivalent oxide thickness into sub-nanometer with appropriate leakage current is also an important issue. In this paper, we present our effort including surface treatment, annealing treatment processes to deal with those issues. InGaAs, InAs and InSb based MOSCAP structures with Al_2O_3 and HfO_2 high k oxides are focused in the study.

EXPERIMENTAL PROCEDURE

The MOSCAP structures used in this study is shown in Fig. 1. For InGaAs and InAs based devices, the wafers consist of 5×10^{17} cm^{-3} Si-doped n-type 100nm $In_{0.53}Ga_{0.47}As$ and 5nm $InAs/3nm\ In_{0.7}Ga_{0.3}As/100nm\ In_{0.53}Ga_{0.47}As$ multilayer stacks grown by molecular beam epitaxy

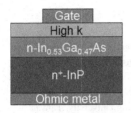 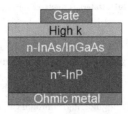 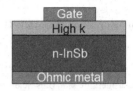

Figure 1. InGaAs, InAs and InSb based MOSCAP structures used in this study

(MBE) on n^+-type InP substrates were used. For InSb based devices, wafer is naturally n-type behavior with donor concentration of 2.2×10^{16} cm^{-3} at room temperature, determined by Hall measurement. Samples were firstly degreased in acetone and iso-propanol followed by wet chemical treatment steps. There are two kinds of wet chemical treatments were used for comparison including HCl and sulfide cleans. The HCl treatment was used by simply dipping samples in HCl 4% solution for 1-2min. For the sulfide treatment, samples were first underwent HCl treatment and were then dipped in $(NH_4)_2S$ 7% solution for 20 min. After that, samples were loaded in to the ALD chamber for the high k deposition. In the ALD chamber, ten pulses of trimethyl aluminum (TMA) were employed for *in situ* TMA self-cleaning before the deposition of Al_2O_3 (or HfO_2) using TMA (or tetrakis ethylmethylamino hafnium, TEMAH) and H_2O as precursors [1-4]. After oxide deposition, the samples were underwent a post deposition annealing (PDA) step. The PDA temperature and gas were modulated for comparison. The gate metal was then formed via photolithography/e-beam evaporation/lift-off processes. Finally, back side ohmic contact was deposited followed by post metal annealing step. The measurements including multi frequency capacitance-voltage (C-V), quasi-static C-V, current-voltage (I-V) and X-ray photoelectron spectroscopy (XPS) were performed to characterize properties of samples.

RESULTS AND DISSCUSSION

Effect of surface treatment and gas annealing condition on the electrical properties of Al_2O_3/InGaAs structure

The electrical properties of Al_2O_3/$In_{0.53}Ga_{0.47}As$ with TMA-only and sulfide plus TMA treatments in conjunction with PDA step in N_2 or H_2 gases are compared. The samples marks and corresponding process are shown in table 1. The C-V multi-frequency responses (1kH-1MHz) are shown in Fig. 2. At the gate voltage ranges from -4V to -1V, sample S_1 exhibits inversion bumps at high frequency and low frequency-like behavior at frequency smaller than 10 kHz as shown in Fig. 2(a). This C-V response is similar to previous reports [5-9] and it is believed this behavior dominated by the high interface trapping states [5, 6, 9]. The C-V curves of sample S_2 shows low frequency behavior with inversion carrier layer occurred at a frequency around 4 kHz. The observed inversion layer originates from the abrupt change of C-V curves

Table 1. Al_2O_3/InGaAs samples with different surface treatments and gas annealing condition

Sample	Surface treatment	ALD AL_2O_3	PDA condition	Gate metal	Ohmic metal	PMA
S_1	TMA-only	18nm, 300°C	500°C, 10min, N_2	Ti/Pt/Au	Au/Ge/Ni/Au	400°C, N_2, 30s
S_2	Sulfide plus TMA	18nm, 300°C	500°C, 10min, N_2	Ti/Pt/Au	Au/Ge/Ni/Au	400°C, N_2, 30s
S_3	Sulfide plus TMA	18nm, 300°C	500°C, 10min, H_2	Ti/Pt/Au	Au/Ge/Ni/Au	400°C, N_2, 30s

292

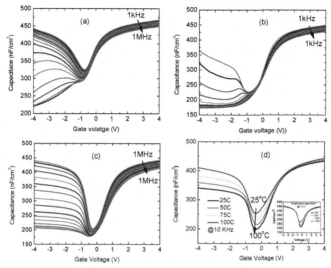

Figure 2. Multi-frequency C-V responses in (a)- TMA treated-; (b)- sulfide + TMA treated Al_2O_3/n-$In_{0.53}Ga_{0.47}As$ MOSCAPs, with post deposition annealing in N_2 gas; (c)- sulfide + TMA treated Al_2O_3/n-$In_{0.53}Ga_{0.47}As$ MOSCAPs, with post deposition annealing in H_2 gas; (d)- the temperature dependent C-V responses at 10 kHz of the same sample. The inset in figure 1(d) shows the temperature dependent C-V responses at 1 kHz

from depletion region to inversion. The appearance of the inversion layer implies the reduction of trapping states at lower half-part of the bandgap. Figure 2(c) shows the multi-frequency C-V responses of sample S_3. "Low-frequency" C-V behavior with "flat", no "bump" inversion response at frequency as high as 1 MHz indicates that the minority carriers (holes) could be generated and freely moving to form an inversion layer. The "free" inversion generation implies large decrease of interface trapping effect i.e. a significant reduction of interface traps density at lower half-part of InGaAs band gap. High inversion capacitance equal to oxide capacitance (C_{ox}) was obtained at frequency around 1 kHz (Fig. 2(c)), implying that at this value of frequency, minority carriers response freely to signal and forms a fully inversion layer. This result is consistent with previous report which estimated the response time, τ_R of minority carrier in n-$In_{0.53}Ga_{0.47}As$ is about 10^{-3} s [7].

Figure 2(d) shows the C-V characteristics measured at 10 kHz at different temperatures for sample S_3. The increase of minority carrier response time, τ_R versus temperature is clearly seen as evidenced by the increase of inversion capacitance. At frequency of 1 kHz, a fully inversion layer is formed, thus, the inversion capacitance becomes independent of temperature (the inset in Fig. 2(d)). As show in Fig. 2(d), the accumulation capacitance is almost unchanged and the C-V curves do not shift horizontally with the temperature. The inset in Fig. 2(d) shows clearly the identical C-V stretch-out at different temperatures. These characteristics further confirm the reduction of interface charge trap density.

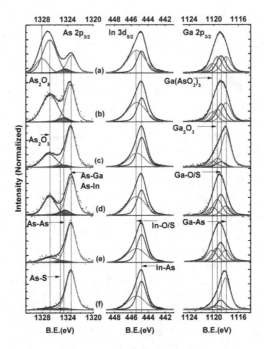

Figure 3. The As 2p$_{3/2}$, In 3d$_{5/2}$, Ga 2p$_{3/2}$ XPS spectra of (a)- Native oxide-covered InGaAs surface; (b)- TMA treated sample, with ALD Al$_2$O$_3$, as deposited; (c)- sulfide + TMA treated sample, with ALD Al$_2$O$_3$, as deposited; (d)- TMA treated sample, with ALD Al$_2$O$_3$, after PDA in N$_2$; (e)- Sulfide + TMA treated sample, with ALD Al$_2$O$_3$, after PDA in N$_2$; (f)- Sulfide + TMA treated sample , with ALD Al$_2$O$_3$, after PDA in H$_2$.

The As 2p$_{3/2}$, In 3d$_{5/2}$ and Ga 2p$_{3/2}$ XPS spectra of the native oxide-covered InGaAs surface, the Al$_2$O$_3$/InGaAs interface as deposited and after post deposition annealing in N$_2$ and H$_2$ gases are shown in Fig. 3(a) - 3(f), respectively. Although TMA treatment is effective in the reduction of As-O and Ga-O bonds as shown in Fig. 3(b), the amount of native oxides is still significant. A strong surface cleaning effect was made by using sulfide treatment followed by TMA pretreatment as indicated by the decrease of the native oxides signals in Fig. 3(c). By using TMA pretreatment, further removal of the native oxides is expected after sulfide treatment [3]. From Fig. 3(d) and Fig. 3(e), it is clearly seen that the decrease of As$_2$O$_3$ oxide results in the increase relative signal of the Ga-related oxides after annealing in N$_2$. For the sample with TMA treatment only, significant amount of As$_2$O$_3$ oxide is still remaining (Fig. 3(d)). In contrast, as shown in Fig. 3(f), by using H$_2$ annealing, the As$_2$O$_3$ was almost completed removed while the reduction of Ga-related oxides also occurred but with a slightly increase of Ga-O/S bonds. The In-related oxides seem to be stable after both N$_2$ and H$_2$ annealing as indicated by the very similar In 3d$_{5/2}$ spectrum of the samples before and after annealing.

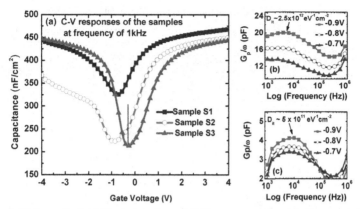

Figure 4. (Color online) (a)-The comparison of C-V responses of sample S_1, sample S_2 and sample S_3 at frequency of 1 KHz; (b) and (c)- The G_p/ω - f curves of sample S_1 and sample S_2, where G_p is the parallel conductance and ω is the measured angular frequency.

Figure 4(a) shows the C-V responses of samples at frequency of 1 kHz. The simulation results of oxide/n-$In_{0.53}Ga_{0.47}As$ MOSCAPs shows that the C-V curve of ideal device without any interface state density has an asymmetrical sharp with higher slope at negative voltage side and low minimum capacitance value, C_{min} [10]. According to this result, in our case, the C-V characteristic of sample S_3 indicates that this curve approach closest to the ideal curve as compared to either S_1 or S_2 (Fig. 4). By comparison, the evidence of high interface state density in sample S_1 exhibited by the observation of highest value of Cmin as well as accumulation capacitance, C_{acc} [10]. The conductance method with the application limited to the depletion region is used to estimate the interface trap density near midgap of sample S_1 and sample S_2 [11]. From the G_p/ω versus frequency curves shown in Fig. 4(b) - 4(c), the values obtained are about 2.5×10^{12} $eV^{-1}cm^{-2}$ and 5×10^{11} $eV^{-1}cm^{-2}$, respectively for sample S_1 and S_2. For sample S_3, the inversion layer occurred at frequency as high as 1 MHz and the use of conductance method would be inaccurate. The D_{it} value of this sample, however, is smaller than that of sample S_2, i,e, $< 5\times10^{11}$ $eV^{-1}cm^{-2}$.

Effect of surface treatments on the electrical properties of Al_2O_3/InAs structure

In this section, samples with TMA-only, HCl plus TMA and sulfide plus TMA treatments were performed and comparison. The list of samples and process are shown in table 2. Fig. 5 shows the In $3d_{5/2}$ and As $2p_{3/2}$ XPS spectra of the InAs native-oxide-covered surface, the Al_2O_3/HCl plus TMA treated InAs interface, and the Al_2O_3/sulfide plus TMA treated InAs interface. For both surface treatments, As-related oxides were removed to below the XPS detection level (Fig. 5, As $2p_{3/2}$ spectra). In $3d_{5/2}$ spectra indicate a similar effect of two kinds of treatments though a significant reduction of In_2O_3 oxide. In^{1+} chemical state signal seems to be slightly increased in both samples but more significantly in the sulfide plus TMA treated sample due to the contribution of In-S bonds.

Table 2. Al₂O₃/InAs samples with different kind of surface treatments

Sample	Surface treatment	ALD AL₂O₃	PDA condition	Gate metal	Ohmic metal	PMA
Control	TMA-only	18nm, 300°C	400°C, 30s, N₂	Ti/Pt/Au	Au/Ge/Ni/Au	400°C, N₂, 30s
HCl+TMA	HCl plus TMA	18nm, 300°C	400°C, 30s, N₂	Ti/Pt/Au	Au/Ge/Ni/Au	400°C, N₂, 30s
Sulfide + TMA	Sulfide plus TMA	18nm, 300°C	400°C, 30s, N₂	Ti/Pt/Au	Au/Ge/Ni/Au	400°C, N₂, 30s

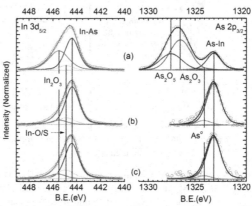

Figure 5. The In $3d_{5/2}$ and As $2p_{3/2}$ XPS spectra of (a) native-oxide-covered InAs surface; (b) 1.5 nm ALD Al₂O₃/HCl +TMA treated InAs interface; (c) 1.5 nm ALD Al₂O₃/sulfide+TMA treated InAs interface. Native oxides including In₂O₃, As₂O₃, and As₂O₅ were significantly reduced after the use of surface treatments

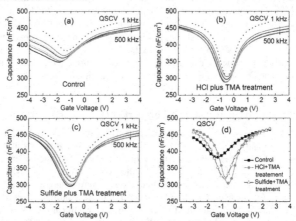

Figure 6. Multi-frequency C-V responses (solid lines) and QSCV curves (dashed lines) in (a) control sample, (b) HCl plus TMA sample, and (c) sulfide plus TMA treated sample of 18 nm ALD Al₂O₃/InAs MOSCAPs; (d) QSCV curves of all three samples,

Figure 6 shows the multi-frequency C-V responses and quasi-static C-V (QSCV) curves of 18 nm Al_2O_3/n-InAs MOSCAP samples. In the accumulation regime, the multi-frequency responses do not show the obvious difference in frequency dispersion between samples. As shown in Figs. 6(a)-6(c), the values of frequency dispersion of samples are small, in the range of 0.65-0.75% per decade. These low frequency dispersions including the control sample indicate that surface treatments do not seem to affect significantly the C-V responses in the accumulation regime.

Low frequency-like C-V behavior is observed for all samples in the whole range of measured frequencies. This behavior originates from the short minority carrier response time (τ_R) in very low band gap, high intrinsic density materials as InAs. As shown in Figs. 6(a) and 6(d), the control sample exhibits high value of depletion capacitance (C_{dep}) in depletion regime which reveals large values of semiconductor capacitance (C_S) and/or interface trap capacitance (C_{it}). Large frequency dispersion in inversion regime in this sample as shown in Fig. 6(a) also implies high contribution of interface traps. In chemicals plus TMA treated samples, nice C-V curves with small frequency dispersion in inversion regime are observed [Figs. 6(b)-6(c)]. In Fig. 6(d), smaller C_{dep} of these two samples compared to the control sample indicates that the contribution of C_{it} and/or C_S is reduced. Out of the two chemical plus TMA treated samples, the HCl plus TMA sample exhibits better electrical characteristics as compared to the sulfide plus TMA treated sample, with smaller frequency dispersion in inversion regime and smaller stretch out [Figs. 6(b)-6(d)]. This result seems contradictory to most of reports on high k/ GaAs (InGaAs) but it is consistent with the report on HfO_2/InAs structure [12].

Low frequency CV-simulations were performed by full numerical solution of the Poisson equation taking into account the complete multilayer structure as well as the interface states at

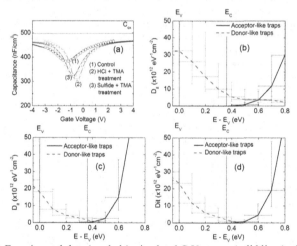

Figure 7. (a) Experimental data (symbols), simulated C-V curves (solid lines) of ALD 18 nm Al_2O_3/n-InAs MOSCAP structures with various surface treatments. Interface state density profiles of all three samples, extracted from simulation, are shown as well: (b) control sample, (c) HCl plus TMA treated sample, (d) sulfide plus TMA treated sample.

the InAs/high-k oxide interface, similar to approach in [10]. The interface state density (D_{it}) at the InAs/high-k interface was varied, until good fit to experimental data was obtained. All experimental QSCV curves (symbols) were well fitted by the simulations (solid lines). D_{it} profiles of samples extracted from simulation are shown in Figs. 7(b)-7(d), where the estimated error bars of the extracted D_{it} are shown as well. Errors were estimated and taken into account due to the following reasons: (i) error on metal work function, (ii) charge quantization effects and non-parabolicity in the conduction band which were not included in the simulation and (iii) uncertainty on absolute value of oxide capacitance, C_{ox}. The derived D_{it} profiles present a U-shape with minimum in D_{it} profile located around the conduction band minimum (E_C) for all samples. The interface state density shows strong similarities with the $In_{0.53}Ga_{0.47}As/Al_2O_3$ D_{it} profile [10]. It can be clearly seen that the two different surface treatments significantly reduce the donor-like traps over the full energy region as compared to the control sample [Figs. 7(b)-7(d)]. The D_{it} of the sulfide treated plus TMA sample shows slightly higher values of donor-like traps as compared to the HCl plus TMA treated sample at, in agreement with the comparison of C-V characteristics between these two samples.

Influence of PDA temperatures on electrical properties of Al_2O_3/InSb structure

Electrical properties of the Al_2O_3/InSb MOSCAP structure are very sensitive to the temperatures because InSb has low thermal budget. Table 3 shows the process of samples with different post deposition (PDA) temperatures for studying. Fig. 8 shows the $In3d_{3/2}$ and $Sb3d_{3/2}$ X-ray photo electron spectroscopy (XPS) spectra of InSb native oxides surface and 2nm Al_2O_3/HCl plus TMA treated InSb interface. The use of HCl treatment, and TMA pretreatment before the deposition of Al_2O_3 resulted in significant reduction of both In-O and Sb-O bonds.

Figure 9 shows typical multifrequency C-V responses of the MOSCAPs. Strong inversion responses are observed in the whole range of measured frequencies (100 Hz - 1 MHz) due to very short minority carrier response time in InSb. The frequency dispersion in conduction band side is always lager that in valence band side for all samples [see also Fig. 10(b)]. This indicates lager amount of boder traps located in conduction band side. Figure 10(a) shows the reduction of maximum capacitance at the PDA temperature of 300°C and above. The hysteresis in conductance band side increased from 70 mV to 100 mV while PDA temperature increased to 350°C and above. The frequency dispersion in both side of InSb bandgap also increase with the increase of PDA temperatures as shown in Fig 10(b).

Figure 11 shows the comparison of $(C_{max}-C_{min})/C_{max}$ and the C-V stretch-out values of the samples. The $(C_{max}-C_{min})/C_{max}$ value decreases from 47.7% for the sample without PDA to 37.7% for the sample PDA at 400°C. This indicates that samples without or with low PDA temperatures are easily to get inversion state as compared to samples with higher PDA temperatures. The C-V stretch-out value increases from 755 mV (w/o PDA sample) to 1030 mV (400°C PDA sample) indicate the increase of border traps inside the oxide when PDA temperature increased. The performing degradation of the MOSCAP samples with increasing

Table 3. Al_2O_3/InSb samples with different PDA temperatures

Sample	Surface treatment	ALD AL_2O_3	PDA condition	Gate metal	Ohmic metal	PMA
w/o PDA	HCl plus TMA	7.5nm, 250°C	without	Ni/Au	Au/Ge/Ni/Au	250°C, N_2, 30s
300°C PDA	HCl plus TMA	7.5nm, 250°C	300°C, 30s, N_2	Ni/Au	Au/Ge/Ni/Au	250°C, N_2, 30s
350°C PDA	HCl plus TMA	7.5nm, 250°C	350°C, 30s, N_2	Ni/Au	Au/Ge/Ni/Au	250°C, N_2, 30s
400°C PDA	HCl plus TMA	7.5nm, 250°C	400°C, 30s, N_2	Ni/Au	Au/Ge/Ni/Au	250°C, N_2, 30s

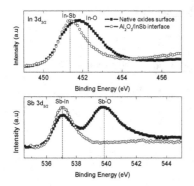

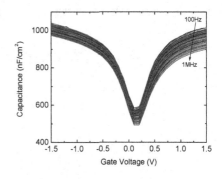

Figure 8. In 3d$_{3/2}$ and Sb 3d$_{3/2}$ XPS spectra of bare InSb native oxide surface and as deposited 2nm Al$_2$O$_3$/HCl-treated InSb interface show the significant reduction of InSb native oxides after HCl treatment and Al$_2$O$_3$ deposition

Figure 9. A typical multi-frequency C-V responses of Al$_2$O$_3$/InSb MOSCAPs

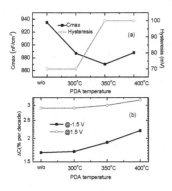

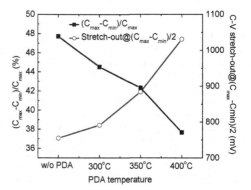

Figure 10. (a) Maximum capacitance, hysteresis, and (b) frequency dispersions vary PDA temperature of Al2O3/InSb structures

Figure 11. Maximum capacitance, hysteresis, and frequency dispersions vary PDA temperature of Al$_2$O$_3$/InSb structures

PDA temperatures could attribute to the interdiffusion between Al$_2$O$_3$ and InSb during thermal process. In fact, the transmission electron microscopy (TEM) graphs and energy-dispersive x-ray spectroscopy (EDX) exhibit the extension of interdiffusion regions (data not shown). The interdiffusion would result in the degradation of both gate oxide as well as InSb layer near the Al$_2$O$_3$/InSb interface

299

Scaling down the equivalent oxide thickness into sub-nanometer

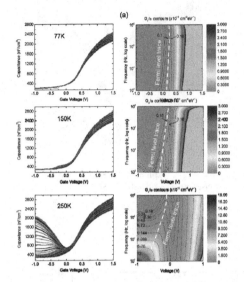

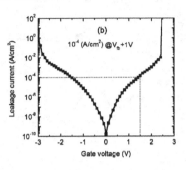

Figure 12. (a) Multifrequency C-V behavior and conductance maps at different temperatures and (b) leakage current of 3.5 nm HfO₂/-InGaAs MOSCAP strutures

For scaling down the EOT into sub-nm, we used the deposition of HfO₂. 3.5 nm ALD HfO₂ was deposited on n-InGaAs after the use of HCl plus TMA for surface treatment. Figure 12(a) shows the multifrequency C-V behaviors and corresponding conductance maps of the sample at different temperatures (77K, 150K and 250K). The C-V curves exhibit a nice behavior with distinct accumulation, depletion, and inversion regions. The value of maximum capacitance is 2800 nF/cm² at 1.5 V and 250K which corresponds to about 1.1 nm capacitance equivalent thickness (EOT~ 0.95 nm) [13]. The conductance map shows the variation of conductance peaks with frequency and gate voltage which implies the free movement of Fermi level (dash-white lines). A low leakage current of 10^{-4} A/cm² at V_{th}+1 V is also obtained for this sample. The result made a good progress for future study of the MOSFET devices.

CONCLUSIONS

The effects of various surface treatments and different gas annealing conditions on the electrical characteristics of ALD Al₂O₃/n-In₀.₅₃Ga₀.₄₇As MOS capacitors were studied. We report the true inversion channel in Al₂O₃/n-In₀.₅₃Ga₀.₄₇As MOS capacitor structure by using the combination of ex-situ sulfide treatment and in-situ TMA pretreatment and post deposition annealing in pure H₂ gas. Both C-V and XPS data show a strong effect of H₂ annealing on the reduction of interface trapping states. A true inversion behavior supports an evidence of the free movement of Fermi level at lower half-part band gap. For Al₂O₃/InAs structure, the effect of interface states on accumulation capacitance behavior is small and does not depend on the surface treatments. In contrast, the C-V characteristics of Al₂O₃/n-InAs in depletion and inversion region were significantly improved by surface treatments. The D_{it} profiles extracted from simulation shows a significant reduction of donor-like traps after surface treatments in complete InAs bandgap, as well as in the lower part of conduction band. Results also revealed that HCl plus TMA treatment has stronger effect on the reduction of donor-like traps than sulfide plus TMA treatment. For Al_2O_3/InSb structure, the electrical properties of the MOSCAPs with different PDA temperatures have been studied. XPS showed the significant reduction of InSb native oxides by using HCl plus TMA treatment before the deposition of Al_2O_3. Nice C-V responses with strong inversion behavior in whole range of measured frequency were observed. However, the PDA temperature of above 300°C would result in the significant degradation of electrical properties of the MOSCAPs. The degradation could attribute to the interdiffusion between Al_2O_3 and InSb during thermal treatment process. For down scaling, a low EOT of 0.95 nm, leakage current of 10^{-4} A/cm^2 at $V_{th}+1$ V and free movement of Fermi level at interface has been obtained for 3.5nm HfO_2/HCl plus TMA treated InGaAs structure.

ACKNOWLEDGMENTS

This work was supported by Taiwan National Science Council under Contract Nos. 101-2923-E-009-002-MY3 and 99-2221-E-164-MY3.

REFERENCES

1. H.-S. Kim, I. Ok, M. Zhang, F. Zhu, S. Park, J. Yum, H. Zhao, J. C. Lee, P. Majhi, N. Goel, W. Tsai, C. K. Gaspe, and M. B. Santos, *Appl. Phys.Lett.* **93**, 062111 (2008).
2. H. Ko, K. Takei, R. Kapadia, S. Chuang, H. Fang, P. W. Leu, K. Ganapathi, E. Plis, H. S. Kim, S. Y. Chen, M. Madsen, A. C. Ford, Y. L. Chueh, S. Krishna, S. Salahuddin, and A. Javey, *Nature* **468**, 286 (2010).
3. M. Milojevic, C. L. Hinkle, F. S. Aguirre-Tostado, H. C. Kim, E. M. Vogel, J. Kim, and R. M. Wallace, *Appl. Phys.Lett.* **93**, 252905 (2008).
4. H. D. Trinh, E. Y. Chang, P. W. Wu, Y. Y. Wong, C. T. Chang, Y. F. Hsieh, C. C. Yu, H. Q. Nguyen, Y. C. Lin, K. L. Lin, and M. K. Hudait, *Appl.Phys.Lett.* **97**, 042903 (2010).
5. E. O'Connor, S. Monaghan, R. D. Long, A. O'Mahony, I. M. Povey, K. Cherkaoui, M. E. Pemble, G. Brammertz, M. Heyns, S. B. Newcomb, V. V. Afanas'ev, and P. K. Hurley, *Appl. Phys. Lett.* **94**, 102902 (2009).
6. E. O'Connor, R. D. Long, K. Cherkaoui, K. K. Thomas, F. Chalvet, I. M. Povey, M. E. Pemble, P. K. Hurley, B. Brennan, G. Hughes and S. B. Newcomb, *Appl. Phys. Lett.* **92**, 022902, (2008).
7. Y. C. Chang, M. L. Huang, K. Y. Lee, Y. J. Lee, T. D. Lin, M. Hong, J. Kwo, T. S. Lay, C. C. Liao, and K. Y. Cheng, *Appl.Phys. Lett.* **92**, 072901 (2008).
8. N. Goel, P. Majhi, W. Tsai, M. Warusawithana, D. G. Schlom, M. B. Santos, J. S. Harris and Y. Nishi, *Appl. Phys. Lett.* **91**, 093509 (2007).
9. Y. Hwang, M. A. Wistey, J. Cagnon, R. Engel-Herbert, and S. Stemmer, *Appl. Phys.Lett.* **94**, 122907 (2009).
10. G. Brammertz, H.-C. Lin, M. Caymax, M. Meuris, M. Heyns, and M. Passlack, *Appl.Phys. Lett.* **95**, 202109 (2009).
11. D. K. Schroder, *Semiconductor Material and Device Characterizatic*, (John Wiley and Sons, Inc.,2006) pp. 321-323.

12. D. Wheeler, L.-E. Wernersson, L. Fröberg , C. Thelander, A. Mikkelsen, K.-J. Weststrate, A. Sonnet, E.M. Vogel, A. Seabaugh, *Microelectron. Eng.* **86**, 1561-1563 (2009).
13. R. Suzuki, N. Taoka, M. Yokoyama, S. Lee, S. H. Kim, T. Hoshii, T. Yasuda, W. Jevasuwan, T. Maeda, O. Ichikawa, N. Fukuhara, M. Hata, M. Takenaka, and S. Takagi, *Appl. Phys. Lett.* **100**, 132906 (2012).

Mater. Res. Soc. Symp. Proc. Vol. 1538 © 2013 Materials Research Society
DOI: 10.1557/opl.2013.550

Growth and Characteristics of a-Plane GaN/ZnO/GaN Heterostructure

Chiao-Yun Chang[1], Huei-Min Huang[1], Yu-Pin Lan[1], Tien-Chang Lu[1] *, Hao-Chung Kuo[1],

Shing-Chung Wang[1] and Li-Wei Tu[2], Wen-Feng Hsieh[1]

[1]Department of Photonics & Institute of Electro-Optical Engineering, National Chiao Tung University, 1001 University Road, Hsinchu 30050, Taiwan
[2] Department of Physics, National Sun Yat-Sen University, Kaohsiung 80424, Taiwan

Abstract

The crystal structure of a-plane GaN/ZnO heterostructures on r-plane sapphire was investigated by using the XRD and TEM measurment. It was found the formation of (220) $ZnGa_2O_4$ and crystal orientation of semipolar $(10\bar{1}3)$ GaN at GaN/ZnO interface. The epitaxial relation of normal surface direction are the sapphire $(1\bar{1}02)$ // a-GaN $(11\bar{2}0)$ and $ZnGa_2O_4$ (220) // semi-polar GaN $(10\bar{1}3)$. Beside, the emission peak energy of ZnO appears shift about 60 meV in the GaN/ZnO/GaN heterostructures due to the re-crystallization of ZnO layer with Ga or N atom and the formation of the localized state.

Introduction

The ZnO has attracted extended attention because of the superior material characteristics in optoelectronics and piezoelectronics[1]. Furthermore, the physical properties of ZnO greatly resemble GaN, which include the lattice constant and thermal expansion coefficient, crystal structure and energy bandgap. The lattice mismatches between the wurtzite structure of GaN and ZnO are 0.4% for the a-axis direction and 1.9% along the c-axis direction. Based on their similar lattice constant, the related GaN/ZnO -based heterojunctions are suitable to realize high performance optoelectronic devices[2]. Therefore, it has the potential to grow along the non-polar orientation direction such as a-plane $(11\bar{2}0)$ and m-plane $(1\bar{1}00)$ for the growth of GaN epilayer on ZnO layer. However, the growth mechanism and the optical properties of the GaN/ZnO interface are still not yet clear. This work investigates the growth and characteristics of *a*-plane GaN/ZnO epitaxial structures, which can be applied to various optoelectronic devices.

Experimental procedure

The *a*-plane GaN/ZnO heterostucture was grown on *r*-plane sapphire. First of sample, a 1.5-μm-thick *a*-plane GaN layer was grown on r-plane sapphire at the temperature of 1100 °C by using the metal organic chemical vapor deposition (MOCVD). Next, *a*-plane ZnO epitaxial layer was deposited at the temperature of 550 °C by using the plasma-assisted molecular beam epitaxy (PAMBE) system. Afterwards, the GaN was grown on the ZnO template at the temperature of 620 °C by molecular beam epitaxy (MBE). The optical properties of GaN/ZnO/GaN heterostructure were investigated by the photoluminescence (PL) measurment and time-resolved photoluminescence (TRPL) measurment. The crystal orientation and the interface elements of composition for GaN/ZnO heterostucture were determined using high solution x-ray diffraction (XRD) measurement and TEM measurements.

Results and discussion

The PL spectrum of GaN/ZnO heterostucture on r-plane sapphire at room temperature were shown in Fig. 1(a). It was appeared the near-band-edge (NBE) transition of ZnO at 3.28 eVon the ZnO template. However, the PL spectrum of GaN/ZnO heterostucture on *r*-plane sapphire was obviously observed two emission peaks at 3.42 and 3.22 eV. The emission peak of 3.22 eV was related to the shallow donor-acceptor-pair (DAP) transitions or free-to-bound transitions (A°X) in ZnO epitaxial layer [3, 4], which shift almost 60 meV compared to ZnO templet. As we known that the growth temperature of GaN epitaxial layer was higher than that of ZnO epitaxial layer and the single-step growth was used without any buffer layer, leading to the decomposition of ZnO epitaxial layer. Simultaneously, the Ga and N atoms could be interacted with the Zn and O atoms. It is believed that the regrowth process lead to form the ZnO co-doping with Ga and N (ZnO:Ga, N). Besides, the bandstructure of GaN/ZnO interface was type-II band alignment, which also formed the localization state to affect the emission energy. Therefore, the lifetime of GaN NBE was about 270 ps, and GaN/ZnO interface transited has longer lifetime about 440 ps. It implied that the interface states of the type-II GaN/ZnO heterostructure could trap the parts of carriers transitioning from higher levels to lower levels and delay the carrier lifetime. Fig. 1(b) exhibited the θ-2θ XRD spectrums of GaN/ZnO heterostucture and ZnO template on r-plane sapphire. It can be found the additional diffraction peaks of $(10\bar{1}3)$ GaN and the (220) $ZnGa_2O_4$. It could result in the formation of ZnO:GaN alloys at GaN/ZnO interface due to the diffusion effects of ZnO and the recombinated process during regrowth GaN. And the GaN/ZnO heterostructure not only formed the ZnO:GaN alloys of $ZnGa_2O_4$ but also created the new crystal orientation of semipolar $(10\bar{1}3)$ GaN[5].

The crstal structure of the GaN/ZnO heterostructure can be studied by using TEM, EDS mapping and the SAD pattern, showed in Fig 2. Compared TEM-image with the EDS mapping image, it was found that an intermediate spinel $ZnGa_2O_4$ layer distributed among GaN/ZnO interface. Beside,the original wurtzite a-plane ZnO epitaxial layer indeed has been transformed into the spinel $ZnGa_2O_4$ to induce the appearance of semi-polar orientation GaN, and the crystallographic relationship between GaN and $ZnGa_2O_4$ was established as $(110)_{ZnGa_2O_2} \parallel (10\bar{1}3)_{GaN}$. Therefore, the TEM results have a good agreement with the XRD results. The formation of the intermediate spinel $ZnGa_2O_4$ changes the crystal orientation of GaN from non-polar $(11\bar{2}0)$ GaN to semi-polar $(10\bar{1}3)$ GaN during the regrowth.

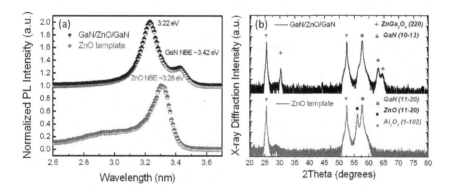

Fig.1 (a) Room temperature photoluminescence spectra of the epitaxial GaN/ZnO heterostructure and the related optical transitions were marked clearly. (b) 2θ-ω scans of GaN/ZnO heterostructure identified the surface orientation and indicated the crystalline

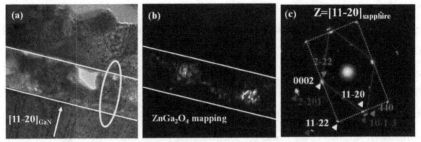

Fig. 2. (a) TEM image shows the interface of GaN/ZnO/GaN structure on r-plane sapphire. (b) EDS mapping image for ZnGa$_2$O$_4$ element. (d) The SAD pattern of the GaN/ZnO/GaN structure on *r*-plane sapphire.

Conclusion

In summary, the optical and structural characteristics of GaN/ZnO heterostructure were influenced by the interaction between GaN and ZnO due to the existence of the interface states and additonal ZnO:GaN alloys. The unexpected spinel ZnGa$_2$O$_4$ and semi-polar orientation GaN formed at the interface of GaN/ZnO were verified by the XRD and TEM. The formation of the interface states induced the lower transition energy and longer carriers life time in the GaN/ZnO heterostructure. Consequently, these results describe the growth mechanism and the optical properties of non-polar ZnO/GaN heterostructures, which provides important information for the epitaxial technique.

Acknowledgments

This work was supported by the MOE ATU program and in part by the National Science Council of Republic of China (ROC) in Taiwan under contract NSC-99-2120-M-009-007 and NSC 99-2221-E-009-035-MY3.

Reference

[1] F.D. Auret, S. Goodman, M. Legodi, W.E. Meyer, D. Look, Applied physics letters, 80, 1340 (2002).

[2] H. Xu, Y. Liu, Y. Liu, C. Xu, C. Shao, R. Mu, Applied Physics B: Lasers and Optics, 80, 871 (2005).

[3] M. Kumar, T.H. Kim, S.S. Kim, B.T. Lee, Applied physics letters, 89,   112103-1 (2006).

[4] L. Gorbatenko, O. Novodvorsky, V.Y. Panchenko, O. Khramova, Y.A. Cherebilo, A. Lotin, C. Wenzel, N. Trumpaicka, J. Bartha, Laser physics, 19, 1152 (2009).

[5] T.J. Baker, B.A. Haskell, F. Wu, P.T. Fini, J.S. Speck, S. Nakamura, Japanese journal of applied physics, 44, L920 (2005).

Devices and LEDs

Mater. Res. Soc. Symp. Proc. Vol. 1538 © 2013 Materials Research Society
DOI: 10.1557/opl.2013.504

Nickel Foam as a Substrate for III-nitride Nanowire Growth

Michael A. Mastro[1], Neeraj Nepal[1], Fritz Kub[1], Jennifer K. Hite[1], Jihyun Kim[2], and Charles R. Eddy, Jr.[1]

[1]U.S. Naval Research Laboratory, 4555 Overlook Ave., SW, Washington, D.C. 20375

[2]Department of Chemical and Biological Engineering, Korea University, Seoul, South Korea

ABSTRACT
This article presents the use of flexible metal foam substrates for the growth of III-nitride nanowire light emitters to tackle the inherent limitations of thin-film light emitting diodes as well as fabrication and application issues of traditional substrates. A dense packing of gallium nitride nanowires were grown on a nickel foam substrate. The nanowires grew predominantly along the a-plane direction, normal to the local surface of the nickel foam. Strong luminescence was observed from undoped GaN and InGaN quantum well light emitting diode nanowires.

INTRODUCTION
The GaN-based light emitting diode (LED) market has grown into a multi-billion dollar market in just the last two decades. Despite this rapid progress, certain restrictions are inherent to the thin-film on a planar substrate design. These constraints can be generalized into light extraction, defectivity, substrate cost, and processing cost limitations as well as a lack of mechanical flexibility.[1]
Sapphire is the predominant substrate for epitaxy of III-nitride light emitting diode thin films. Sapphire is non-conductive and presents a large lattice mismatch with the III-nitride material system. Silicon carbide substrates are expensive but have a closer lattice mismatch to GaN and can be supplied in a conductive state.[2-4] Silicon is available in larger diameters but GaN epitaxy on silicon suffers from thermal stress constraints.[5] The cost of the wafer is significant but often overstated when compared to the processing and balance of system costs.[6] A move to larger wafers allows a significant increase in back-end processing throughput and commensurate decrease in the cost per die.[7]
An ongoing idea is to produce III-nitride LEDs on low-cost, large-area substrates such as glass akin to the thin film photovoltaic technologies.[8,9] Despite some progress for producing III-nitride LEDs on glass,[10] a poly-crystal type growth is inherently produced when the underlying substrate, such as glass, does not present crystalline order to which reactant atoms can align to form an ordered thin film. Poly-crystal or fine-grain III-nitride material presents an exceedingly large number of dislocations and other defects that effectively destroy operation of the pn junction.[11]
Even for high-quality thin film growth on sapphire, lattice mismatch leads to the formation of dislocations with densities greater than 10^8 cm^{-2}, which limit the internal quantum efficiency of LEDs.[12] Furthermore, green LEDs require high indium content in

the $In_xGa_{1-x}N$ quantum wells that under planar lattice stress can encourage the formation of V-pits. Moreover, $In_xGa_{1-x}N$ stability is reduced at the elevated growth temperature needed for thin film metal organic chemical vapor deposition (MOCVD).

Another intrinsic issue with a standard LED structure is the low external extraction efficiency of light owing to the index of contrast difference with air.[13] A majority of the light generated in a semiconductor thin film on a planar substrate suffers from total internal reflection. Complex procedures such as die shaping, photonic crystals, micro-cavities, and surface roughening are used to extract light that would normally be trapped in the semiconductor and substrate slab.[14-19]

Nanowire III-nitride light emitters have been suggested and demonstrated to avoid many of the deleterious issues associated with thin film structures.[20,21] The dimensions of the nanowire are on the order of the optical wavelength; therefore, light is easily scattered out of the semiconductor into the surrounding air. The growth of nanowires via vapor-liquid-solid mechanism proceeds at temperature lower than that used for thin-film MOCVD easing the incorporation of indium into the active region.[22] The removal of the in-plane lattice constraint allows nanowires to grow with a greatly reduced defect level relative to its thin-film equivalent. It is known that the vapor-liquid-solid growth of GaN nanowires is known to have a lesser dependence on the underlying substrate.[23]

Recently, a two-step reactive vapor / MOCVD approach was used to grow GaN nanowires with an InGaN shell on a stainless steel substrate.[24] In general, a metal substrate can be scaled to any reasonable process tool size or shape.

This article presents a nanowire LED design based on a metal foam substrate. In contrast to a solid metal substrate, the foam form further lowers the amount and cost of material used for a given surface area. Furthermore, the suppleness of the foam substrate extends the application space to other areas including coiled piezoelectric energy harvesting devices and flexible light sources. It is expected from simple ray tracing that the vast majority of light generated from the nanowires will not be reabsorbed somewhere else in the architecture. This includes light generated deep in the structure that can be expected to directly escape through the micron-scale pores of the foam substrate.

EXPERIMENTAL

A 0.05 M nickel nitrate solution was repeatedly dripped onto a nickel foam substrate and blown dry in N_2 then loaded into a vertical impinging-flow MOCVD reactor. A 50-Torr, N_2/H_2 mixed atmosphere was used during the ramp to growth temperature. Trimethylgallium was flowed for 2 sec prior to the onset of NH_3 flow to prevent nitridation of the nickel seeds. The GaN nanowire core was grown at a temperature of 850°C, a pressure of 50 Torr and a V/III ratio of 50.[25] Under proper growth conditions, the metal catalyst particle captures reactants and enhances the growth rate perpendicular to the substrate, thus creating a pseudo-one-dimensional semiconductor wire. The LED nanowires continued with growth of an InGaN quantum well shell at 600°C and a V/III ratio of 150. Immediately thereafter, the GaN:Mg shell was grown to avoid decomposition of the InGaN. The introduction of Mg dopant atoms is known to encourage a higher lateral growth rate relative to growth of undoped GaN nanowires under equivalent conditions.[26,27] The samples were cooled from growth temperature in pure nitrogen ambient to avoid rapid decomposition in hydrogen and, in the case of the samples with a p-type shell, to activate the acceptor dopant (Mg). Structural characterization was performed with a LEO FE Scanning Electron Microscope (SEM). Photoluminescence (PL) measurements were carried out using a HeCd laser at 325 nm and an Ocean Optic QE6500 spectrometer.

RESULTS AND DISCUSSION

Images of the GaN nanowires on the nickel foam framework are observable in a series of scanning electron micrographs in Fig 1. The nanowires grow uniformly over the entire foam surface. This particular foam has pores with an average diameter of 200 μm although a similar uniform coating was achieved for pore diameters down to 5 μm. At smaller pore volumes, it is possible to form a coalesced film.

The direction of GaN growth is generally perpendicular to the local surface. Growth of III-nitride nanowires via a VLS mechanism under this growth condition tends to proceed in the <11-20> a-direction with an isosceles triangular cross-section exhibiting a distinct set of facets of (0001), (1-10-1), and (-110-1).

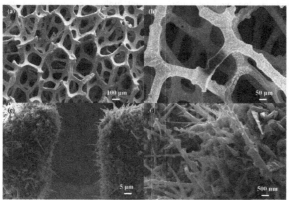

Figure 1. Electron micrograph of GaN nanowires on a nickel foam substrate at increasing level of magnification.

The GaN wire length is proportional to growth time with 1 hr of growth yielding wires of approximately 10 μm in length. The GaN nanowires have an approximate 200 nm diameter over a length of several microns. A slight tapering is evident owing to growth at an elevated nanowire growth temperature although these conditions are known to produce higher quality GaN nanowires.[28]

While the nanowires grow along the a-direction, the curvature of the underlying foam presents a range of diffracting planes at the nominal surface. A 2θ-θ X-ray diffraction pattern in Fig. 2 presents sharp diffraction from the (10-10), (0002), and (10-10) planes as well as weaker diffraction from higher index planes.

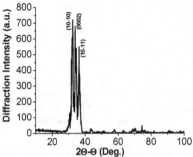

Figure 2. X-ray diffraction pattern of GaN nanowires on a nickel foam substrate. The nickel foam form intrinsically possesses a continuum of surface normals. Thus, nanowires grown on this curved surface present a spectrum of GaN crystal planes. Low index GaN diffraction planes dominate the pattern although higher index planes are also evident.

The room temperature PL of the GaN nanowires in Fig. 3 displays two major bands in the spectrum corresponding to the band-edge and the donor-acceptor luminescence transitions. The dominant band at 3.4 eV is associated with recombination processes involving the annihilation of free-excitons and a band-to-band transition. The enhancement of the higher energy side of the band-edge PL emission is typically attributed to the high-excess of free electron carriers in high-quality GaN.[21]

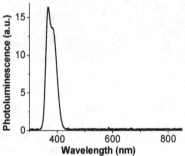

Figure 3. Photoluminescence intensity of GaN nanowires on a nickel foam substrate. The near band-edge emission dominates the spectrum.

Similar samples as described above were fabricated with the addition of n-type doping to the GaN:Si core followed by the formation of an InGaN-well/GaN:Mg shell around the core. The thickness of the InGaN shell layer is approximately 5 nm and, the outer thickness of the GaN:Mg layer is 200 nm after 5 min of growth. The thickness of the InGaN/GaN:Mg sheath was directly proportional to growth time. The structure of the nano-wires was designed to create a thin InGaN well for quantum confinement of the injected carriers, and a thicker GaN:Si core and GaN:Mg sheath for optical confinement of the optical mode.

Fig. 4 shows the PL spectrum from GaN:Mg / InGaN well (sheath) / GaN:Si (core) nano-wires. The nano-wires display the expected GaN band-edge (near 3.4 eV) and near band-edge dopant-based transitions as well as the 542 nm InGaN luminescence from the quantum well. Likely, the emission of the InGaN is slightly blue-shifted by the onset of quantum confinement in the 5-nm well.

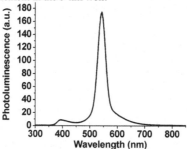

Figure 4. Photoluminescence spectrum of GaN:Mg / InGaN well / GaN:Si core-shell nanowires on a nickel foam substrate.

CONCLUSION

The focus of this investigation was to demonstrate high-quality, reproducible group-III nitride nano-wire emitters on a conductive, low-cost, and flexible metal foam substrates.

ACKNOWLEDGEMENTS

Research at the US Naval Research Lab is partially supported by the Office of Naval Research.

REFERENCES

1 M.A. Mastro, J.D. Caldwell, R.T. Holm, R.L. Henry, C.R. Eddy Jr., Adv. Mater., **20**-1, 115 (2008)
2 M.A. Mastro, R. T. Holm, C.R. Eddy, Jr., J-H. Kim, J. Cer. Proc. Res. **9**, 1-5 (2008)
3 M.A. Mastro, C. R. Eddy Jr.; N. D. Bassim; M. E. Twigg; R. L. Henry; R. T. Holm; A. Edwards, Solid State Electronics, 49-2, 251
4 M. A. Mastro, M. Fatemi, D.K. Gaskill, K-K Lew, B.L. Van Mil, C.R. Eddy Jr., C.E.C. Wood, J. Appl. Phys. 100, 093510 (2006)
5 M.A. Mastro, C.R. Eddy Jr., D.K. Gaskill, N.D. Bassim, J. Casey, A. Rosenberg, R.T. Holm, R.L. Henry, M.E. Twigg, J. Crystal Growth, 287, 610 (2006)
6 M.A. Mastro, R.T. Holm, N.D. Bassim, C.R. Eddy, Jr., R.L. Henry, M.E. Twigg, A. Rosenberg, Jpn. J. Appl. Phys. **45**-31, L814 (2006)
7 M.A. Mastro, R.T. Holm, N.D. Bassim, C.R. Eddy Jr., D.K. Gaskill, R.L. Henry, M.E. Twigg, Appl. Phys. Lett., 87, 241103 (2005)

8 M. Jiang, Y. Li, R. Dhakal, P. Thapaliya, M.A. Mastro, J.D. Caldwell, F. Kub, X. Yan, J. Photon. Energy 1, 019501 (2011)

9 M.A. Mastro, J.A. Freitas, Jr., C.R. Eddy Jr., F. Kub, J.-H. Ahn, H.-R. Kim, J. Kim, Physica E : Low-dimensional Systems and Nanostructures, 41, 487 (2008)

10 J. H. Choi, A. Zoulkarneev, S. I. Kim, C. W. Baik, M. H. Yang, S. S. Park, H. Suh, U. J. Kim, H. B. Son, J. S. Lee, M. Kim, J. M. Kim, K. Kim, Nature Photonics 5, 763–769 (2011)

11 Y.N. Picard, M.E. Twigg, J.D. Caldwell, C.R. Eddy Jr., M.A. Mastro, R.T. Holm, Scripta Materialia 61, 773 (2009)

12 J.D. Caldwell, M.A. Mastro, N.D. Bassim, M.E. Twigg, K.D. Hobart, O.J. Glembocki, C.R. Eddy, Jr., M.J. Tadjer, R.T. Holm, R.L. Henry, F. Kub, P.G. Neudeck, A.J. Trunek, J.A. Powell, ECS Trans. 3, 189 (2006)

13 M.A Mastro, B.-J. Kim, Y. Jung, J. Hite, C.R. Eddy, Jr., J. Kim, Current Applied Physics, 11, 682 (2011)

14 H.Y. Kim, M.A. Mastro, J. Hite, C.R. Eddy, Jr., J. Kim, J. Crystal Growth 326, 58(2011)

15 M.A. Mastro, C.S. Kim, M. Kim, J. Caldwell, R.T. Holm, I. Vurgaftman, J. Kim, C.R. Eddy Jr., J.R. Meyer, Jap. J. Appl. Phys. 47, 7827 (2008)

16 B.-J. Kim, Y. Jung, M.A. Mastro, N. Nepal, J. Hite, C.R. Eddy, Jr., J. Kim, J. Vacuum Science and Technology B 29, 021004 (2011)

17 M. A. Mastro, E.A. Imhoff, J.A. Freitas, J.K. Hite, C.R. Eddy, Jr, Solid Sate Comm. 149, 2039 (2009)

18 Hong-Yeol Kim, Younghun Jung, Sung Hyun Kim, Jaehui Ahn, Michael A. Mastro, Jennifer Hite, Charles R. Eddy, Jr., Jihyun Kim, J. Crystal Growth 326, 65 (2011)

19 M.A. Mastro, L. Mazeina, B.-J. Kim, S.M. Prokes, J. Hite, C.R. Eddy, Jr., J. Kim, Photonics and Nanostructures, 9, 91 (2011)

20 Y. Li, F. Qian, J. Xiang and C.M. Lieber, Materials Today 9 (10), 18 (2006).

21 M.A. Mastro, S. Maximenko, M. Murthy, B.S. Simpkins, P.E. Pehrsson, J.P. Long, A.J. Makinen, J.A. Freitas, Jr., J. Hite and C.R. Eddy, Jr., Journal of Crystal Growth, 310, 2982 (2009)

22 F. Qian, Y. Li, S. Gradecak and C.M. Lieber, Nano Lett. 5, 2287 (2005).

23 F. Qian, Y. Li, S. Gradecak, D. Wang, C.J. Barrelet and C. M. Lieber, Nano Lett. 4, 1975 (2004)

24 C. Pendyala, J.B. Jasinski, J.H. Kim, V.K. Vendra, S. Lisenkov, M. Menon, M.K. Sunkara, Nanoscale, 4, 6269 (2012)

25 M.A. Mastro, H.-Y. Kim, J. Ahn, B. Simpkins, P. Pehrsson, J. Kim, J.K. Hite, C.R. Eddy, Jr., IEEE Trans. Elec. Dev. 58, 3401(2011)

26 Z. Zhong, F. Qian, D. Wang and C.M. Lieber, Nano Lett. 3, 343 (2003).

27 Michael A. Mastro, Blake Simpkins, Mark Twigg, Marko Tadjer, R. T. Holm, Charles R. Eddy, Jr., ECS Trans. 13-3, 21 (2008)

28 A. A. Talin, G. T. Wang, E. Lai et al., Appl. Phys. Lett. 92(9), 093105-1

Mater. Res. Soc. Symp. Proc. Vol. 1538 © 2013 Materials Research Society
DOI: 10.1557/opl.2013.548

Non radiative recombination centers in ZnO nanorods

D. Montenegro[1], V. Hortelano[2], O. Martínez[2], M. C. Martínez-Tomas[1], V. Sallet[3], V. Muñoz[1] and J. Jiménez[2]

[1]Departamento de Física Aplicada y Electromagnetismo, Universitat de Valencia, Dr. Moliner 50, 46100 Burjassot, Spain
[2]GdS-Optronlab, Departamento Física Materia Condensada, Edificio I+D, Universidad de Valladolid, Paseo de Belén 1, 47011, Valladolid, Spain
[3]Groupe d'Etude de la Matière Condensée (GEMAC), CNRS-Université de Versailles St-Quentin, 45 avenue des Etats-Unis, 78035 Versailles Cedex, France

ABSTRACT

Nowadays, the nature of the non radiative recombination centres in ZnO is a matter of controversy; they have been related to extended defects, zinc vacancy complexes, and surface defects, among other possible candidates. We present herein the optical characterization of catalyst free ZnO nanorods grown by atmospheric MOCVD by microRaman and cathodoluminescence spectroscopies. The correlation between the defect related Raman modes and the cathodoluminescence emission along the nanorods permits to establish a relation between the non radiative recombination centers and the defects responsible for the local Raman modes, which have been related to Zn interstitial complexes.

INTRODUCTION

One dimensional ZnO nanostructures have attracted a great deal of attention because of its potential application in UV light emitting devices [1, 2]. The properties of ZnO, as the direct band-gap energy (3.37 eV at room temperature), and the large free exciton binding energy (60 meV), make from it a candidate for highly efficient UV lasers. However, the attractive properties of ZnO are hindered by the difficulty of stable p-doping, and the insufficient knowledge about the electro-optic role played by native defects and their complexes [3,4]. The quantum efficiency of the UV emission is reduced by the presence of deep levels, which are responsible for a broad luminescence band in the yellow-orange spectral window, and the non radiative recombination centers (NRRCs). While the origins of the deep level emission (DLE) have received a great deal of attention, captured in a huge literature, little is known about the NRRCs. It is usually assumed that a high UV/DLE emission ratio means a good crystalline quality; however, this criterion does not take account of NRRCs, which reduce the internal quantum efficiency of both UV and visible emissions. There are non negligible differences in the luminescence emission between samples prepared under different growth methods, giving similar UV/DLE emission ratio, but with very different overall emission levels.

It is usually assumed that the non radiative recombination is concerned with extended defects and/ or surface defects. However, good quality hydrothermal crystals are almost dislocation free; also, nanorods present a good crystalline quality, in spite of a few stacking faults. Chichibu el al. [5] argued that the recombination lifetime was governed by the occurrence

317

of defect complexes associated with Zn vacancies, V_{Zn}, while isolated point defects were discarded as NRRCs [6]. The work by Chichibu et al. [5] was based on time resolved PL experiments combined with positron annihilation spectroscopy (PAS). Both measurements were averaged over extended volumes; therefore, one cannot establish a correlation between the local light emission efficiency and the vacancy defects detected by PAS, and identified as the NRRCs in ZnO crystals.

Advances in the understanding of the DLEs have been achieved by modifying the growth conditions and by thermal treatments in selected atmospheres. We present herein a combined analysis using microscopic characterization tools, cathodoluminescence (CL), and microRaman spectroscopy, of catalyst free ZnO nanorods grown under different conditions. The measurements were carried out in individual nanorods, in order to establish a reliable spatial correlation between the defects revealed by Raman spectroscopy and the luminescence emission revealed by CL, along the nanorods. The NRRCs in these structures are related to Zinc interstitial (Zn_i) complexes.

EXPERIMENTAL AND SAMPLES

Vertically well aligned ZnO nanorods were grown on c-sapphire substrates with and without a ZnO buffer layer in an atmospheric MOCVD reactor. ZnO nanorods arrays were synthesized using helium as a carrier gas and N_2O as an oxygen precursor, and DMZn-TEN for Zn. The VI/II flow ratio was varied between 80 and 300. Two sample series were prepared in which one of the precursors flow was fixed and the other was varied and vice versa. The morphology of the samples and the growth details were described in a previous work [7]. The growth time for nanorods grown on the buffer layer was varied from 10 to 40 minutes, while for nanorods grown on bare c-sapphire substrates it was varied from 5 to 15 min. All the ZnO nanorods were grown at 800 °C.

The Raman spectra were acquired at room temperature using the 532 nm line of a frequency doubled Nd-YAG laser. The scattered light was analyzed by means of a Raman spectrometer (Labram HR800 UV from Horiba-Jobin-Yvon) equipped with a LN2-cooled charge-coupled device (CCD) detector. CL measurements were carried out at 80 K with a Gatan MonoCL2 system attached to a field emission scanning electron microscope (FESEM) (LEO 1530). The acceleration voltage of the e-beam was varied between 3 and 20 kV. The CL spectra were acquired using a Peltier cooled CCD detector.

RESULTS AND DISCUSSION

The Raman spectra recorded on ZnO nanorods grown with and without the buffer layer are shown in Fig.1. Besides to the normal modes E_2(low) (E_{2l}) (99 cm^{-1}), $2E_2$ (333 cm^{-1}), and E_2(high) (E_{2h}) (437 cm^{-1}) [8], additional peaks at 275, 510, 580 and 643 cm^{-1} appear only in the nanorods grown on the c-sapphire substrate without the buffer layer, which do not correspond to wurtzite ZnO normal modes. These peaks are labelled as additional modes associated with defects [9,10].

The CL spectra are shown in Fig. 2. The luminescence spectrum of ZnO is generally separated in three spectral windows: E > 3.33 eV, which corresponds to the near band edge

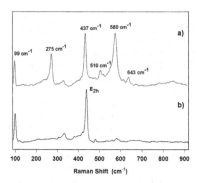

Figure 1. Raman spectrum of ZnO nanorods grown on c-sapphire substrate (a), and grown on the ZnO buffer layer (b).

(NBE) emission governed by the excitonic transitions, 3.33 eV> E > 3.00 eV, associated with free to bound, DAP transitions, and phonon replicas, and E < 3.0 eV, which basically refers to the DLEs. The general aspect of the spectrum depends on the impurities and defects present and their respective concentrations. The incorporation of the impurities and defects is very sensitive to the growth process. The nature of the DLE is still under debate; however, one can summarize the main hypothesis about it as follows: the green luminescence in absence of Cu doping, is related to oxygen deficiency (V_O) [3]; the yellow orange emission is most probably due to Zn deficiency, namely, V_{Zn}, or interstitial oxygen, O_i [11]. The red emission is more unusual; it has been related to Fe and N-impurities [12]. Zn_i has also been claimed to be involved with the red luminescence [13]; therefore, the presence of a dominant red band in the DLE spectrum, could support evidence of the presence of Zn_i.

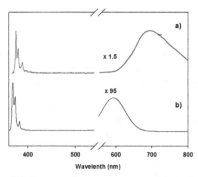

Figure 2. CL spectrum of ZnO nanorods grown on c-sapphire substrate (a), and grown on the ZnO buffer layer (b).

319

The DLE in the nanorods grown on the c-sapphire substrate consists of an unstructured band peaking in the red around 1.75 eV, without green or yellow emission. The amplitude of the red emission lies in the order of magnitude of the NBE emission. However, the spectrum of the nanorods grown on the buffer layer presents in addition to the NBE emission, a weak orange band, without the presence of the red band, together with an overall higher emission with respect to the other nanorods. These differences account for the presence of different defects and concentrations in both types of nanorods.

If one looks at the panchromatic CL images, the differences between the two types of nanorods concern the distribution of the CL emission along the nanorods, Fig.3. The CL of nanorods grown on the c-sapphire substrate presents an emission decrease from tail to tip, looking dark contrasted in about one third of the nanorod length close to the tip; while the tail presents bright contrast, accounting for an emission efficiency decreasing from tail to tip, Fig.3a. Instead of this CL emission distribution, the nanorods grown with the buffer layer do not show emission variation along the nanorod length, if any, it seems to increase along the nanorod, Fig.3b.

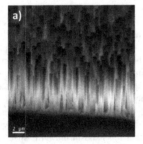 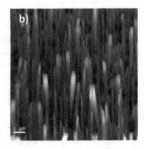

Figure 3. Panchromatic CL images of ZnO nanorods grown on c-sapphire substrate (a), and grown on the ZnO buffer layer (b).

If one compares this CL distribution along the nanorods with the Raman spectra acquired along individual nanorods, one observes a relevant difference between the two processes as well; in the nanorods grown on the buffer layer only the normal modes are observed all along it; however, in the nanorods grown on the c-sapphire substrate the additional Raman peaks are observed. The Raman spectra acquired at regularly spaced points along the nanorod show a continuous evolution from tail to top, evidencing a progressive increase of the intensity of the defect related peaks from tail to tip, Fig.4. These peaks are absent in the nanorod tail, where the CL emission is higher, and become more intense, even more than the normal modes, close to the tip, where the CL is quenched.

The defect related Raman peaks were long time associated with N impurities [9,10]. However, they were also reported to occur with other dopants, namely, Fe, Sb, Al, and Ga [14]. Recent work by Friedrich et al. [15] in natural and isotopically pure (^{68}Zn) ZnO layers grown by pulsed laser deposition, provides of the role of Zn_i complexes in the occurrence of the defect related Raman peaks. CL follows a reverse spatial correlation with these peaks. This spatial anticorrelation between both, CL and defect related Raman peaks, might permit to establish a correlation between the NRRCs and the defects responsible for the Raman peaks, and according to the Friedrich work [15], it confers a relevant role to the Zn_i complexes as NRRCs.

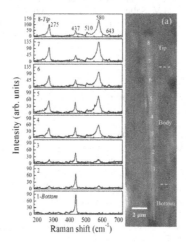

Figure 4. Raman spectra acquired along an individual nanorod grown on the c-sapphire substrate (see the optical image).

The two series of samples, grown under either rich Zn (increasing the DMZn-TEN flow rate), or rich O (increasing the NO_2 flow rate) in c-sapphire substrate, share the idea exposed above about the role of the excess Zn with respect to the NRRCs. In fact, the CL emission was observed to increase with increasing NO_2 flow rate, while it decreased for increasing the DMZn-TEN flow rate; simultaneously, the defect related Raman peaks decreased with increasing the NO_2 flow, and increased with increasing the DMZn-TEN flow. All this is highly consistent with our previous discussion on individual nanorods.

CONCLUSION

CL and microRaman measurements in ZnO nanorods allowed establishing a relation between the luminescence efficiency and defects identified by its Raman signature. According to recent interpretation of the defect related Raman peaks, one can tentatively associate the NRRCs in ZnO nanorods with Zn_i complexes.

ACKNOWLEDGMENTS

The authors gratefully acknowledge Spanish Government for financial support under the Projects MAT2007-66129, MAT-2010-20441-C02, MAT-2010-16116 and TEC2011-28076-C02-02, and Generalitat Valenciana under the projects Prometeo/2011-035 and ISIC/2012/008, Institute of Nanotechnologies for Clean Energies of the Generalitat Valenciana. The authors, D.N. Montenegro, V. Hortelano and V. Sallet thank specially to Spanish MICINN, the European Social Fund (ESF) and Universitat de Valencia, respectively, for the financial support.

REFERENCES

1. M. Willander, et al., *Nanotechnology* **20**, 332001 (2009).
2. A.B. Djurišić, A.M.C. Ng, and X.Y. Chen, *Prog. Quant. Electron.* **34**, 191 (2010).
3. K. Vanheusden, W.L. Warren, C.H. Seager, D.R. Tallant, J.A. Voigt, and B.E. Gnade, *J. Appl. Phys.* **79**, 7983 (1996).
4. A.B. Djurišić and Y.H. Leung, *Small* **2**, 944 (2006).
5. S.F. Chichibu, T. Onuma, M. Kubota, A. Uedono, T. Sota, A. Tsukazaki, A. Ohtomo, and M. Kawasaki, *J. Appl. Phys.* **99**, 093505 (2006).
6. X. Wen, J.A. Davis, L. Van Dao, P. Hannaford, V.A. Coleman, H.H. Tan, C. Jagadish, K. Koike, S. Sasa, M. Inoue, and M. Yano, *Appl. Phys. Lett.* **90**, 221914 (2007).
7. D.N. Montenegro, A. Souissi, M.C. Martínez-Tomás, V. Muñoz-Sanjosé, and V. Sallet, *J. Cryst. Growth* **359**, 122 (2012).
8. R. Cuscó, E. Alarcón-Lladó, J. Ibáñez, L. Artús, J. Jiménez, B. Wang, and M.J. Callahan, *Phys. Rev. B* **75**, 165202 (2007).
9. F. Friedrich and N.H. Nickel, *Appl. Phys. Lett.* **91**, 111903 (2003).
10. A. Kaschner, U. Haboeck, M. Strassburg, G. Kaczmarczyk, A. Hoffmann, and C. Thomsen, *Appl. Phys. Lett.* **80**, 1909 (2002).
11. M. Gomi, N. Oohira, K. Ozaki, and M. Koyano, *Jpn. J. Appl. Phys.* **42**, 481 (2003).
12. T. Monteiro, C. Boemare, and M.J. Soares, *J. Appl. Phys.* **93**, 8995 (2003).
13. Y.F. Mei, G.G. Siu, R.K.Y. Fu, K.W. Wong, P.K. Chu, C.W. Lai, and H.C. Ong, *Nuclear Instruments and Methods in Phys. Research B* **237**, 307 (2005).
14. C. Bundesmann, N. Ashkenov, M. Schubert, D. Spemann, T. Butz, E.M. Kaidashev, M. Lorenz, and M. Grundmann, *Appl. Phys. Lett.* **83**, 1974 (2003).
15. F. Friedrich, M.A. Gluba, and N.H. Nickel, *Appl. Phys Lett.* **95**, 141903 (2009).

Mater. Res. Soc. Symp. Proc. Vol. 1538 © 2013 Materials Research Society
DOI: 10.1557/opl.2013.657

Assessment of Homogeneity of Extruded Alumina-SiC Composite Rods Used in Microwave Heating Applications by Impedance Spectroscopy

Justin R. Brandt[1] and Rosario A. Gerhardt[1]
[1]School of Materials Science and Engineering, Georgia Institute of Technology
Atlanta, GA 30332-0245

ABSTRACT

Composite rods consisting of Alumina (Al_2O_3) and Silicon Carbide whiskers (SiC_w) are used to fabricate microwave cooking racks because they effectively act as a microwave intensification system that allows cooking at much faster rates than conventional microwave ovens. The percolation behavior, electrical conductivity and dielectric properties of these materials have been reported previously. However, it has been observed that the electrical response of the extruded bars is a function of the rod length and that long rods show substantially different behavior than thinner disks cut from them. A percolation model has been proposed that describes the effect of the alignment of the semiconducting SiC whiskers and the quality of the interfaces present in the composite rods: SiC-SiC and SiC-Al_2O_3-SiC for example. This study was undertaken with the goal of testing out whether the response of the individual sections could be used to generate the response of the full length rods and to assess the importance of the homogeneous distribution of the SiC fillers on the resultant impedance response.

INTRODUCTION

The electrical properties of aluminum oxide (Al_2O_3) containing silicon carbide whiskers (SiC_w) made by different processing methods: hot pressing of circular disks, and pressureless sintering of circular disks and extruded rods have previously been investigated.[1-5] Silicon carbide whiskers are 0.5 μm in diameter and can have aspect ratios as high as 20. The different processing methods therefore result in different whisker alignment, and consequently they have different electrical responses. These composite materials, especially the extruded rods, are of interest because of the fast heating rates they experience in the microwave frequency range, making them useful for making microwave oven inserts for faster and more homogeneous cooking.[6]

One of the previous papers demonstrated that the long rods behave differently than thin sections cut from them.[3,4] Detailed impedance measurements of the full length rods and thinly sliced samples were studied in detail. In the large rods, it was found that the Schottky contact electrode effect[2] was buried under the large magnitude of the bulk impedance, whereas in the thin slices, the bulk and electrode contributions were more comparable. The presence of the electrode effect was confirmed by applying a DC bias.[2] The electrode effect was successfully minimized in the large rods by the 6.7V mark. Any subsequent increase in DC bias had a minimal effect up to 40V. Similarly in the thin slices, 5V DC bias was sufficient to eliminate the low frequency electrode semicircle while having minimal impact on the bulk response.[5] Increasing the voltage also had a secondary effect of shrinking the high frequency bulk semicircle in the thin slices. This was suggested to be caused by the non-linear response at the SiC_w-SiC_w interfaces in the percolated SiC_w clusters.[5] A model that takes into account the quality of the contacts between individual whiskers and the network they form has been compared to river geography.[5] The purpose of this study was to conduct a more detailed study of long rods, and different thickness sections cut from them to try to understand what led to the differences observed.

EXPERIMENT

In order to bridge the gap between the full rod and thin slice extremes, the samples were cut into smaller equally sized pieces down to 1/8th sized slices. Since the full rods were roughly 25.60 cm in length, the 1/8th sized slices were around 3.2 cm in length. The samples were also cut into 1/2th slices and 1/4th slices. Each of these samples was measured using the Solartron 1260/1296 impedance analyzer mainly in the extruded direction. Impedance measurements were taken using a frequency range from 10MHz to 10mHz. A constant AC voltage of 0.5 V was applied for each measurement. The rods themselves are anisotropic because the whiskers align along the extrusion direction. Figure 1 shows schematics of the expected alignment of SiC_w for both along and perpendicular to the extrusion direction. Most of the measurements were made along the extrusion direction but some were also obtained perpendicular to the extrusion direction.

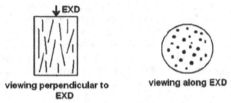

viewing perpendicular to EXD　　　viewing along EXD

Figure 1: SiC_w shown in the alumina matrix in parallel and perpendicular directions with respect to the extrusion direction[4]

The halves, fourths and eighths were measured in sequence so that the impedance values throughout the bar could be added and then compared to the response of the intermediate steps at the different lengths. It was found that this procedure worked well when the extruded rods were equally homogeneous at different positions along the length of the rods. However, it was quite easy to quickly determine if a section of a rod contained a more or less inhomogeneous distribution of the semiconducting fillers, thus proving that the impedance measurements are superb at distinguishing the response of the insulating alumina and the semiconducting SiC. Because of this, multiple rods were analyzed in order to assess the inhomogeneity from sample to sample after extrusion. Eight 20% SiC_w:Alumina whole rods were measured. The reason the 20% concentration of whiskers was chosen was because this would set the samples in a region well beyond the percolation threshold for extruded alumina- SiC_w ceramic composite rods.[4] This buffer would provide a cushion of security making sure the differences in conductivity before and after the percolation threshold do not become an additional variable when investigating the homogeneity of the rods themselves. The magnitude of the impedance and phase angle plotted versus log frequency for the eight rods is presented in Figure 2. It is clear that the response from rod to rod is very similar, although clear differences can be discerned by looking at the phase angle data. This illustrates that during the extrusion process, in addition to the variations possible within the rods, different rods can vary from one another to a certain degree. Similar behavior has been observed for extruded polymer composite wires.[7]

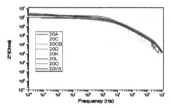 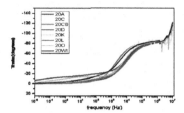

Figure 2: (a) Log Magnitude of the Impedance vs. Log Frequency and (b) Comparison of Phase Angle vs. Log Frequency for eight 20% $SiC_w Al_2O_3$ Rods Measured.

Data obtained from the measurements of the different slices from one of the rods that was cut into various sized pieces was organized by size and plotted in Nyquist plot form, where the imaginary part lies on the y-axis, while the real component occupies the x-axis. Figure 3 showcases some of the data for one rod that had been cut in half and subsequently into eighths. Figure 3(a) clearly shows that one half of the rod is much more insulating than the other. This is indicated by the fact that one has two relatively small semicircles while the other shows two large semicircles. In contrast, the differences among the different $1/8^{th}$ pieces measured, shown in Fig. 3(b), are not as drastic, but one can definitely tell that each $1/8^{th}$ piece has a slightly different behavior.

(a)   (b)

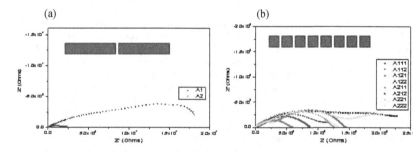

Figure 3: Complex Impedance Graphs for one 20% SiC_w:Alumina Extruded Rod Cut Into: (a) Two Sections and (b) Cut Into Eight Sections

DISCUSSION

The impedance data for the various rods and slices measured was then analyzed using a modeling and analysis program called Zview.[8] The data points supplied by the Solartron were used to fit an equivalent circuit matching the response displayed by the measured samples. Figure 4 depicts the proposed equivalent circuit and an example fit for one of the full sample bars. The figure also presents a schematic of how the alignment and variation of the whisker alignment may result in different behavior along the length of the rods.

(a)

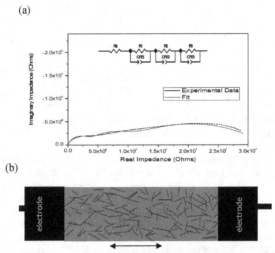

(b)

Figure 4: (a) Proposed equivalent circuit with experimental data and fit for the 20L full bar (b) schematic for the whiskers dispersed in the sample between the two electrodes.

The proposed equivalent circuit includes multiple parallel RC elements in series combined with a standalone resistor. The RC elements show non-ideal response and therefore the capacitor C is represented by a CPE,[9] which allows for a distribution in the values of C. The RC elements represent varying levels of whisker concentration and alignment within the rods, as schematically illustrated in Fig. 4(b). The electrode effect is also responsible for one of the RC element semicircles. Similarly to the long rods and thin slices, by applying a DC bias, this can be confirmed for different sized samples as well (not shown). Detailed analysis of our results show that fitting the complex impedance data alone often does not accurately represent the sample behavior completely. Admittance, permittivity, and modulus need to be used simultaneously in order to be able to obtain the equivalent circuit elements that will most accurately represent the sample microstructure and electrical properties being measured.[10] More details about the fitting procedure for these samples will appear in future publications.

The fitted values for one of the full length rods and all its individual halves, fourths, and eighths are shown in Table 1 on the next page (the distribution parameter for each C is not included here for brevity sake). Looking over the values in Table 1, it can be seen that the standard deviation value in relation to the average value for each sample size gives a good indicator of the spread in values. In this study, the fact that the standard deviations are around the mean value going both above and below in some cases shows that a certain level of variation does in fact exist, but the values are not all over the place. In other words, the level of variation is still contained within a reasonable level indicating overall repeatability.

On the other hand, we must not forget that high variability within one single rod can give rise to very different responses along the length of the rod, as was demonstrated by Figure 3(a), where the two halves gave very different responses. There exists roughly an order of magnitude difference between the two halves, where the majority of resistance was contained within a single half. Similar behavior occurred in the 1/4th sized samples as well. The effect is minimized when the samples are cut down into 1/8th sized samples as shown in Fig. 3(b). The increasing amount of sample that is lost through the cutting process may be a contributing factor to this homogenization behavior. Considering that some material is lost in each cutting step as the samples become smaller, it is possible for vital points within the whisker matrix degree of connectivity to become lost and thus contribute to removal of some of the variability observed. Nevertheless, the main message to carry from these results is that since the electrical response is highly sensitive to the amount and distribution and orientation of the semiconducting whiskers, this proves that using impedance spectroscopy is an excellent method that can be used as a way to establish quality control in multicomponent materials such as ceramic and polymer composites, especially for compositions well beyond the percolation threshold where the conducting network provides a good way to pin the response and use it to our advantage.[11] Much more work is needed to expand on this work in order to establish what links if any exist with other properties and specific microstructures of these heterogeneous materials.

Table 1: Average values and standard deviations for equivalent circuit elements obtained by fitting the impedance response of one long extruded rod and all of its parts

20L	R4	R1	C1	R2	C2	R3	C3
full	-1.31×10^{4}	3.36×10^{6}	1.90×10^{-11}	2.03×10^{7}	2.89×10^{-8}	6.64×10^{6}	1.41×10^{-9}
Average Halves	-1.35×10^{4}	4.00×10^{6}	1.10×10^{-8}	7.31×10^{6}	7.02×10^{-7}	3.83×10^{5}	2.20×10^{-9}
St Dev	8.10×10^{3}	3.68×10^{6}	1.60×10^{-9}	8.79×10^{6}	9.00×10^{-7}	2.57×10^{5}	1.25×10^{-9}
Average Fourths	-7.55×10^{3}	2.41×10^{6}	1.31×10^{-8}	3.71×10^{6}	1.58×10^{-8}	4.76×10^{5}	1.12×10^{-5}
St Dev	4.06×10^{3}	2.52×10^{6}	5.68×10^{-9}	6.62×10^{6}	3.05×10^{-8}	5.51×10^{5}	1.70×10^{-5}
Average Eighths	-2.37×10^{3}	1.16×10^{5}	8.83×10^{-9}	7.56×10^{5}	1.83×10^{-8}	6.75×10^{5}	5.60×10^{-7}
St Dev	1.64×10^{3}	8.52×10^{4}	1.02×10^{-8}	4.78×10^{5}	7.91×10^{-9}	5.13×10^{5}	1.17×10^{-6}

It should be pointed out that it is not unusual at all for extruded objects to show variability in their properties along the extruded direction. In fact, it was shown in a prior study of extruded CB/ABS wires that they were also fairly inhomogeneous along the length of the extruded wires[7] Conducting a second extrusion was found to be beneficial to making the wires more homogeneous. Unfortunately, a similar process would be too difficult to carry out on a ceramic composite such as alumina rods filled with SiC_w.

CONCLUSIONS

As a result of the extrusion process, even though the SiC whiskers are fairly well aligned, the properties vary throughout the length of each rod because the exact alignment can vary. This behavior can be deduced from data obtained from multiple rods and their cut sections. By comparing both the raw impedance data in a Nyquist plot as well as seeing the fitted circuit element values, it is easy to arrive at the same conclusion, which is that inhomogeneity across sample rods is a consistent outcome and needs to be considered when extruded materials are used in vital applications.

ACKNOWLEDGMENTS

Funding for this work was provided by a National Science Foundation DMR-1207323 grant. JRB would also like to thank fellow lab workers Rachel Muhlbauer, Tim Pruyn, and Salil Joshi for teaching me how to operate new lab equipment and stimulating intellectual discussion.

REFERENCES

1. D.S. Mebane and R.A. Gerhardt, "Interpreting Impedance Response of Silicon Carbide Whisker/Alumina Composites Through Microstructural Simulation," J.Am.Ceram.Soc. **89**[2], 538-543, 2006.
2. Brian D. Bertram And Rosario A. Gerhardt, "Room Temperature Properties Of Electrical Contacts to Alumina Composites Containing Silicon Carbide Whiskers,"J. Applied Phys. **105**, 074902,2009.
3. B.D. Bertram and R.A. Gerhardt, "Frequency-dependent dielectric properties and percolation behavior of alumina filled with SiC whiskers," J. Am. Ceram. Soc. **94**[4],1125-1132,2011.
4. B. D. Bertram, R. A. Gerhardt, and J. W. Schultz, "Extruded and Pressureless-Sintered Al2O3–SiCw Composite Rods: Fabrication, Structure, Electrical Behavior, and Elastic Modulus" J. Am. Ceram. Soc. **94**, 4391,2011.
5. Brian D. Bertram, Rosario A. Gerhardt, John W. Schultz, "Impedance response and modeling of composites containing aligned semiconductor whiskers: Effects of dc-bias partitioning and percolated cluster length, topology, and filler interfaces" J. Appl. Phys. 111, 124913, 2012.
6. T. E. Quantrille, "Novel Composite Structures for Microwave Heating and Cooking"; in 41st Annual Microwave Symposium. International Microwave Power Institute, Vancouver, BC, 2007.
7. R. Ou, R.A. Gerhardt, C. Marrett, A. Moulart, and J.S. Colton, "Assessment of percolation and homogeneity in ABS/carbon black composites by electrical measurements" Composites. Part B **34**, 607, 2003.
8. Zview Software, Scribner Associates, http://www.scribner.com/
9. Evgenij Barsoukov, J. Ross Macdonald, eds., Impedance Spectroscopy: Theory, Experiment, and Applications, John Wiley & Sons, 2005.
10. R. Gerhardt, "Dielectric and Impedance Spectroscopy Revisited: Distinguishing Localized Relaxation from Long Range Conductivity," J. Phys. Chem. Solids. **55**(12), 1491-1506, 1994.
11. R. Gerhardt, "Microstructural Characterization of Composites via Electrical Measurements," Ceram. Eng. Sci. Proc. **15**(5), 1174-1181,1994.

Mater. Res. Soc. Symp. Proc. Vol. 1538 © 2013 Materials Research Society
DOI: 10.1557/opl.2013.574

Improvement of Minority Carrier Lifetime in Thick 4H-SiC Epi-layers by Multiple Thermal Oxidations and Anneals

Lin Cheng[1], Michael J. O'Loughlin[1], Alexander V. Suvorov[1], Edward R. Van Brunt[1], Albert A. Burk[1], Anant K. Agarwal[1], and John W. Palmour[1]
[1]Cree, Inc. 4600 Silicon Drive, Durham, NC 27703

ABSTRACT

This paper details the development of a technique to improve the minority carrier lifetime of 4H-SiC thick ($\geq$ 100 μm) n-type epitaxial layers through multiple thermal oxidations. A steady improvement in lifetime is seen with each oxidation step, improving from a starting ambipolar carrier lifetime of 1.09 μs to 11.2 μs after 4 oxidation steps and a high-temperature anneal. This multiple-oxidation lifetime enhancement technique is compared to a single high-temperature oxidation step, and a carbon implantation followed by a high-temperature anneal, which are traditional ways to achieve high ambipolar lifetime in 4H-SiC n-type epilayers. The multiple oxidation treatment resulted in a high minimum carrier lifetime of 6 μs, compared to < 2 μs for other treatments. The implications of lifetime enhancement to high-voltage/high-current 4H-SiC power devices are also discussed.

INTRODUCTION

The quality of thick 4H-SiC n-type epitaxial layers has increased dramatically in recent years. 4 inch wafers with blocking layers thicker than 100 μm and doping concentrations of less than 2×10^{14} cm^{-3} can now be readily fabricated, making bipolar 4H-SiC power devices such as GTOs and PiN diodes a viable prospect for pulse power applications with voltage ratings of 10 kV and beyond [1]. In addition to supporting high voltage, these devices must also operate at high current densities. A key factor to achieve good on-state performance is a high ambipolar carrier lifetime in the device's drift region. This effect becomes more important for thicker epitaxial regions when aiming for much higher blocking voltages. For devices with $\geq$ 100 μm thick as-grown drift regions, typically observed lifetimes are in the range of 1 μs and insufficient to achieve full conductivity modulation in the drift region at high current density, leading to high on-state voltages and high conduction losses [1].

Several techniques to enhance the ambipolar lifetime beyond values present after crystal growth have been developed following the identification of an electron trap ($Z_{1/2}$) as the main lifetime killer in 4H-SiC epitaxial layers [2]. Growing epitaxy in a carbon-rich environment had been shown to provide higher ambipolar lifetime, but at the cost of a greater density of epitaxial defects [2]. Consequently, post-epitaxial growth lifetime enhancement techniques that involved the annihilation of the carbon vacancies with carbon interstitials were developed, with the interstitials either released through thermal oxidation [3] or introduced by carbon implantation [2].

This work examines the effects of repeated thermal oxidation cycles on the ambipolar lifetime of thick ($\geq$ 100 μm) 4H-SiC epitaxial layers. A series of 5 hour 1300°C oxidation cycles were performed, followed by a six minute post oxidation anneal at 1550°C in an inert ambient. A steady increase in lifetime was observed after each oxidation step; following the post-oxidation anneal, the lifetime increased to > 10x than its as-grown value.

EXPERIMENT

Three 4" silicon carbide wafers with identical 100 µm thick n-type epitaxial layers were used in this experiment. The as-grown lifetime was measured using the microwave photoconductivity decay technique (µPCD) to serve as a control value. Following the initial lifetime measurement, one wafer was oxidized at 1300 °C for 5 hours in a dry O_2 ambient, and then measured using µPCD again. This process was repeated several times, followed by 1550 °C anneals in an inert ambient after several oxidation steps were completed. The µPCD measured lifetime is shown in Fig. 1 as a linescan near the horizontal center of the wafer, along with the sequence of oxidations and anneals performed.

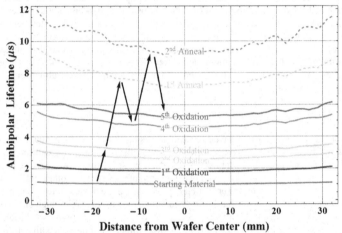

Distance from Wafer Center (mm)

Fig. 1: µPCD linescan across wafer horizontal centerline for multiple oxidation/anneal cycles

A steady increase in lifetime is visible after each oxidation step, implying that more carbon vacancies or $Z_{1/2}$ defect centers could be reduced by the carbon interstitials that were released from the subsequent thermal oxidations and driven in during the high-temperature anneals. The anneal steps increase the measured lifetime significantly, but also increase the variation in lifetime observed across a single wafer. Oxidations performed after a 1550 °C annealing step returned the variance to close to the previous values. A box plot of the treatment sequence is shown in Fig. 2. The maximum lifetime achieved was measured after 4 oxidations and two anneal steps, reaching a mean value of 11.2 µs. Of note is the fact that in addition to increasing the mean lifetime, the minimum lifetimes ($\tau_{A,min}$) observed on the wafer also increased after most of the oxidation steps.

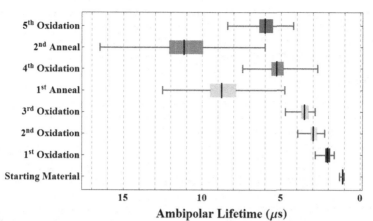

Ambipolar Lifetime (μs)

Fig. 2: Box chart of μPCD measured ambipolar lifetime for multiple-oxidation based lifetime enhancement treatment process. Means are indicated by a dark line in each distribution.

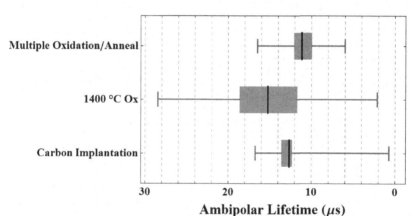

Ambipolar Lifetime (μs)

Figure 3: Box plot of alternate treatments for generating high ambipolar lifetime in 4H-SiC, compared to multiple oxidation based treatment.

This multiple oxidation/anneal approach was compared to a single, high temperature dry O_2 oxidation at 1400°C for 5 hours. The mean value of lifetime generated by this treatment was 15.25 μs, however, this treatment resulted in a large spread of lifetime values (IQR, interquartile range, = 6.96 μs, compared to IQR = 2.16 μs for the best multiple oxidation treatment). A carbon implantation/anneal based lifetime enhancement procedure was also performed for comparison. Although this treatment resulted in a high overall lifetime ($\tau_{A,mean}$ = 12.67 μs) as well as a tight distribution of lifetimes across the wafer (IQR = 1.26 μs), the resulting lifetime distribution is strongly positively skewed, indicating that significant portions of the wafer remain at low lifetime ($\tau_{A,min} = 0.74$ µs). A box plot of the resulting lifetime distributions for the treatments resulting in high lifetime is shown in Fig. 3. It can be seen that the multiple oxidation approach resulted in material having the highest minimum lifetime ($\tau_{A,min} = 6.01$ µs).

DISCUSSION

A large increase in the observed lifetime is seen after each oxidation or annealing step in the multiple oxidation based lifetime treatment. This has been previously attributed to the presence of a defect level possibly associated with carbon interstitials (HK0) [2]. A similar increase in lifetime has been observed for single-step oxidation or carbon implantation with anneal treatments [2]. The reduction in measured lifetime following the anneal steps is thus likely related to reintroduction of defects through the oxidation process, as the surface recombination velocity of oxide passivated SiC surfaces is known to be lower than that of bare SiC [4].

The high minimum lifetime of the multiple oxidation treatment is a desired result from a device fabrication standpoint. A high, but uniform, lifetime is preferred especially for large-area power device fabrication to ensure uniform distributions of current density across the device during bipolar device operation. Non-uniform current distributions caused by localized regions of high lifetime in a device are a potential failure mode for high-power bipolar switches [5]. Lifetimes in 4H-SiC can be subsequently controlled after enhancement using electron irradiation [4]. Thus, lifetime enhancement to raise $\tau_{A,min}$, followed by lifetime control (electron irradiation) to tighten the distribution of lifetime can result in a wafer that is suitable for fabrication of large area bipolar power devices. The high-temperature (1400°C, 5 hours) oxidation and carbon implantation based treatments had high observed mean lifetimes, but the large spread in distribution will result in either widely varying device specifications, or low yield if a requirement of uniform lifetime is imposed.

16 kV PiN diodes previously studied have shown decreasing differential on-resistance with increasing temperature, which is an effect attributed to an increase in lifetime [1]. The on-state J-V characteristic and measured $R_{ON,diff}$ are shown in Fig. 4. The measured room-temperature ambipolar lifetime of this wafer had an average of 1.22 µs. Taking the relationship [6] that

$$\tau(T) = \tau_{300K}\left(\frac{T}{300K}\right)^{3.2}$$

$\tau_{200°C}$ for the measured PiN diode is 5.2 µs. This relationship, combined with the decreasing relationship in the reduction of $R_{ON,diff}$ vs. temperature near 200 °C indicate that lifetimes above 5 µs are required for drift regions with thickness of 120 µm and above for optimal performance in a pulsed-power application.

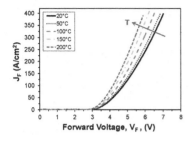

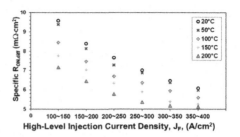

Figure 4 (a): Forward J-V characteristic of 16 kV PiN diode at 20 °C to 200 °C

Figure 4 (b): $R_{ON,diff}$ at various current densities and temperatures for PiN diode shown in Fig. 4 (a).

CONCLUSIONS

Enhancement of the ambipolar lifetime in 4H-SiC wafers through multiple oxidations shows promise for bipolar device fabrication due to the high minimum lifetimes obtained through this technique. The lifetimes obtained through multiple oxidations are high enough such that investigation into lifetime-killing techniques for 4H-SiC (such as electron irradiation) should be an area of further research for 4H-SiC material preparation.

REFERENCES

1. L. Cheng et al, Mat. Sci. Forum 740-742 (2013) pp. 895-898
2. P. B. Klein, Mat. Sci. Forum 717-720 (2012) pp. 279-284
3. T. Hiyoshi and T. Kimoto, Appl. Phys. Expr 2 (2009) 041101
4. T. Hayashi et al, J. Appl. Phys. 112 (2012) 064503
5. Y. Shimizu et al, IEEE Trans. Elec. Dev. 46 (1999) pp. 413-419
6. W. Sung et al, Proc. ISPSD (2009) pp. 271-274

Mater. Res. Soc. Symp. Proc. Vol. 1538 © 2013 Materials Research Society
DOI: 10.1557/opl.2013.573

Impact of Gate Metal on Surface States Distribution and Effective Surface Barrier Height in AlGaN/GaN Heterostructures

Nitin Goyal and Tor A. Fjeldly
Department of Electronics and Communication, Norwegian University of Science and Technology, Trondheim, Norway

ABSTRACT

A physics based model is presented to describe the surface donor density distribution for metal/AlGaN/GaN structures. This model partly relies on experimental observations to describe the reduction that takes place in surface donor density when the metal gate is deposited. This new model is based on our previous work on the bare surface barrier height for both unrelaxed and partially relaxed barrier layers. The model predictions are consistent with reported experimental data.

INTRODUCTION

GaN based High Electron Mobility Transistors (HEMTs) have shown great promise for future high-power and high frequency applications due to their wide band gap and high electron mobility. Wide bandgap III-V based HEMTs also possess excellent thermal properties, which make them suitable candidates for high-temperature applications. The most interesting feature of these devices is the presence of a high mobility, two-dimensional electron gas (2DEG) with a sheet density of the order of 10^{13} cm^{-2} at the AlGaN/GaN interface, even in the absence of both AlGaN barrier layer doping and a gate metal (bare surface). This phenomenon is attributed to the strong piezoelectric as well as spontaneous polarization effects in the structure[1].

In the last few years, considerable efforts have been made to explain the mechanism and source of electrons for the formation of this 2DEG. It has been shown that the presence of distributed donor states in the forbidden gap of the barrier AlGaN surface is the main source of the 2DEG [2-6]. We have previously shown how the bare surface (Schottky) barrier height (SBH) relates to polarization, thickness, and Al content of the barrier, as well as to the surface donor state distribution. This model applies well to bare surface AlGaN/GaN [7,8], but the understanding of what happens when a gate metal is deposited on top AlGaN layer is still lacking. Understanding of metal contacts on AlGaN/GaN heterostructures is important, since these contacts are required for an efficient gate control in practical HEMTs. A large SBH implies low leakage current as well as high breakdown voltage.

THEORY

The SBH of a metal-semiconductor contact is defined as the energy difference between the conduction band minimum (CBM) and the Fermi level E_F at the interface. The SBH has been a constant topic of research for several decades. In their basic model, Schottky and Mott assumed ideal interfaces with no surface states, where SBH is given by $q\Phi_b = q\Phi_m - \chi$, where Φ_m is the metal work function and χ is the electron affinity of semiconductor [9] . Bardeen later introduced a model based on the assumption of a very high density of surface states in order to explain an observed pinning of the Fermi level. A part of this problem is that the metal can give rise to so-called metal induced gap states (MIGS) in the forbidden gap of semiconductor, and that defects present at the interface can introduce defect induced gap states (DIGS) [9]. These surface states can be of acceptor or donor type depending on the position in bandgap. Cowley and Sze explained the variation observed in the barrier height for non-polar materials with different metal contacts in terms of a corresponding change in the surface state properties post metal deposition [10]. But this theory is not applicable to AlGaN/GaN heterostructures. The underlying reason is that the inherent polarization present in a thin AlGaN layer fully depletes this layer and that the thickness of the AlGaN layer also strongly affects the barrier height. In reality, for the AlGaN/GaN heterostructures, the surface state density will be somewhere in between, where the Mott model does not quite apply and where the Fermi level is not fully pinned.

Experimentally, it has been shown that deposition of metals to form Schottky contacts reduces both the 2DEG and the SBH by partially neutralizing the surface states, but it also causes a redistribution of the surface states [11]. In turn, this modifies both the SBH and the charge carrier density n_s in the 2DEG. The density and distribution of the surface states are strongly dependent on the surface treatment technique used and the Al content in the AlGaN layer [12]. However, to date no satisfactory theoretical model has been proposed to explain the change in the 2DEG and the SBH in the presence of the metal gate. This requires a careful consideration of surface donor density and its distribution post metal deposition. Other experiments show a large unexplained difference between SBH measured on bare $Al_{0.29}Ga_{0.71}N$ surfaces (about 2.4 eV) and on surfaces with deposited metal (1.4-1.6 eV found using internal photoemission spectroscopy) [13-15]. An interfacial oxide is usually etched away prior to the metal deposition to improve reliability and reproducibility. The effect of the metal on this post-deposition distribution is found to depend on the metal work-function, which, in effect, rules out the existence of high-density surface states that would otherwise have pinned the Fermi level. On the other hand, the Mott theory applies to semiconductor surfaces with no surface states and therefore gives inaccurate results for practical semiconductors [9].

To simplify the development of an analytical model for the modifications resulting from the metal gate deposition on the AlGaN/GaN heterostructures, we assume that basic nature of surface states remain unchanged, i.e., that they consist of donor states with an intermediate density and a fairly flat distribution in the relevant energy range about the Fermi level. We base the present modeling on our recent work on the SBH and 2DEG sheet density in the bare surface AlGaN/GaN structure for strained thin and partially relaxed thick AlGaN layers [7,8]. As indicated above, the deposition of a metal gate has been observed to diminish both the SBH and the 2DEG density which, in turn, corresponds to a reduction in both the surface donor density n_o and in the surface donor level E_d. These changes are schematically illustrated in Figure 1, where E_{db} and E_{dm} are the donor level for the bare and the metalized surface, respectively, ΔE_c is the conduction band offset at the hetero-interface, and d is the AlGaN layer thickness.

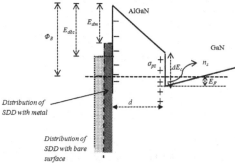

Figure 1. Schematic illustration of the band diagram in the AlGaN/GaN hetero-structure, with indication of surface density and energy distribution with a bare and a metallized surface.

The bare surface model for the unrelaxed AlGaN/GaN hetero-structure gives the following expression for the SBH and the 2DEG charge density

$$q\Phi_b = \frac{\left(n_o E_d + \sigma_{pz}/q\right)d + \varepsilon_{AlGaN}\,\Delta E_c/q^2}{n_o\left(d + \varepsilon_{AlGaN}/q^2 n_o\right)} \tag{1}$$

$$qn_s = n_o d\,\frac{\sigma_{pz} - \varepsilon_{AlGaN}(E_d - \Delta E_c)/q}{n_o d + \varepsilon_{AlGaN}/q^2} \tag{2}$$

where σ_{pz} is the polarization charge density, ε_{AlGaN} is the permittivity of the barrier layer, and ΔE_c is the conduction band offset at the heterointerface.
Based on the assumptions above, (1) and (2) also apply after deposition of the metal gate, but with modified values of the parameters n_o and E_d. Since n_o and E_d depend on the metal work function, so do $q\Phi_b$ and qn_s.

DISCUSSION

As indicated, the reduction in the surface barrier height and the 2DEG density, observed after deposition of the metal gate, occurs because of a reduction in both the surface donor level and surface donor density. We show in the Figure 2 the experimentally observed values of the SBH for a 25nm thick AlGaN barrier layer versus the Al mole fraction. These values are considerably lower than those reported for the similar bare surfaces [4-7]. Similar trends for the 2DEG density for bare and metallized surfaces are also shown in Figures 3 and 4,

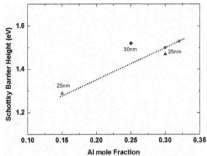

Figure 2. Experimental values of SBH versus the Al mole fraction in the barrier of an AlGaN/GaN heterostructure with deposited metal gate (Ni). AlGaN thickness: 25 nm.

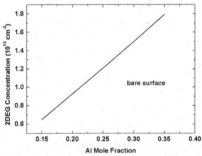

Figure 3. 2DEG density at the AlGaN/GaN heterointerface versus Al mole fraction in the barrier with a bare surface (no metal), calculated from the bare surface model formulation [7,8]. AlGaN thickness: 25 nm.

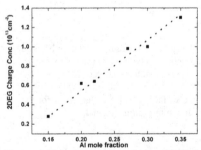

Figure 4. Experimental data for the 2DEG density at the AlGaN/GaN heterointerface post metal gate (Ni) deposition. AlGaN thickness: 25 nm.

To determine the surface donor density n_o, we extract from Figure 4, $n_s = (5.14x - 0.49) \times 10^{13}$ cm^{-2} and $q\Phi_b = (1.41x + 1.07)$ eV for the 25nm thick AlGaN barrier layer. By substituting these

338

values in equation (1) and (2) and solving, we obtain n_o for post metal gate deposition for any Al mole fraction. The result is shown in Figure 5 and compared with n_o of the bare surface case.

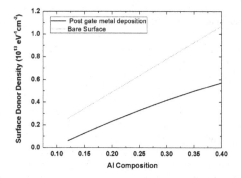

Figure 5. Modeled surface donor density versus Al mole fraction for bare surface and after deposition of metal (Ni) on top AlGaN barrier layer.

As shown in the Figure 5, the surface donor density reduces considerably when the metal gate is deposited. These calculations are consistent with the experimental observations and the proposal by Shin *et al.* that the gate metal causes a reduction in 2DEG by partial neutralization of the surface donor states [11]. This also shows the necessity for understanding the surface physics of metal AlGaN interface, which determines the 2DEG at the AlGaN/GaN interface. A theoretical understanding of this surface is also important for design of efficient gate control and design of Schottky contacts with different metals. To the best of our knowledge, this is first such theoretical attempt in this direction and more theoretical and experimental efforts are needed. This result can be used to derive a threshold voltage model and to develop advanced concepts related to these surface properties of the GaN HEMTs.

CONCLUSIONS

In this paper, we have developed a model framework for the effect of the metal gate deposition on the top surface of an AlGaN/GaN heterostructure. The model predicts how this leads to a reduction in the donor density and distribution at the top surface. This model can be useful in understanding how various physical phenomena in AlGaN/GaN HEMT devices are related to the surface donor states and can also be used for the calculation of threshold voltage. The model agrees well with experimental observations.

ACKNOWLEDGMENTS

The work was carried out with support by the European Commission under Grant Agreement 218255 (COMON) and the Norwegian Research Council under contract 970141669 (MUSIC).

REFERENCES
1. 1. Begin typing text here. O. Ambacher, J. Smart, J. R. Shealy, N. G. Weimann, K. Chu, M. Murphy, W. J. Schaff, L. F. Eastman, R. Dimitrov, L. Wittmer, M. Stutzmann, W. Rieger, and J. Hilsenbeck, " Two-dimensional electron gases induced by spontaneous and piezoelectric polarization charges in N- and Ga-face AlGaN/GaN heterostructures", J. Appl. Phys. 85, 3222 (1999)
2. J. P. Ibbetson, P. T. Fini, K. D. Ness, S. P. DenBaars, J. S. Speck, and U. K. Mishra, "Polarization effects, surface states, and the source of electrons in AlGaN/GaN heterostructure field effect transistors", Appl. Phys. Lett. 77, 250 (2000)
3. B. Jogai, "Influence of surface states on the two-dimensional electron gas in AlGaN/GaN heterojunction field-effect transistors", J. Appl. Phys. 93, 1631 (2003)
4. Masataka Higashiwaki, Srabanti Chowdhury, Mao-Sheng Miao, Brian L. Swenson, Chris G. Van de Walle, and Umesh K. Mishra, "Distribution of donor states on etched surface of AlGaN/GaN heterostructures", J. Appl. Phys. 108, 063719 (2010)
5. G. Koley and M. G. Spencer, "On the origin of the two-dimensional electron gas at the AlGaN/GaN heterostructure interface", Appl. Phys. Lett. 86, 042107 (2005)
6. Luke Gordon, Mao-Sheng Miao, Srabanti Chowdhury, Masataka Higashiwaki, Umesh K Mishra and Chris G Van de Walle, "Distributed surfaces donor states and the two-dimensional electron gas at AlGaN/GaN heterojunctions". J. Phys. D: Appl. Phys. 43 505501
7. N. Goyal, B. Iniguez, and T. A. Fjeldly, "Analytical modeling of bare surface barrier height and charge density in AlGaN/GaN heterostructures" Appl. Phys. Lett. 101, 103505 (2012).
8. N. Goyal, T. A. Fjeldly, "Effects of Strain Relaxation on Bare Surface Barrier Height and Two-Dimensional Electron Gas in AlxGa1−xN/GaN Heterostructures" Journal of Applied Physics,113, 014505 (2013).
9. S. Noor Mohammad, "Contact mechanisms and design principles for alloyed ohmic contacts to n-GaN," J. Appl. Phys. 95, 7940 (2004)
10. A. M. Cowley and S. M. Sze ,"Surface States and Barrier Height of Metal-Semiconductor Systems", J. Appl. Phys. 36, 3212 (1965)
11. Jong Hoon Shin, Young Je Jo, Kwang-Choong Kim, T. Jang, and Kyu Sang Kim, "Gate metal induced reduction of surface donor states of AlGaN/GaN heterostructure on Si-substrate investigated by electro reflectance spectroscopy", Appl. Phys. Lett. 100, 111908 (2012)
12. S. T. Bradley, S. H. Goss, J. Hwang, W. J. Schaff, and L. J. Brillson, "Surface cleaning and annealing effects on Ni/AlGaN interface atomic composition and Schottky barrier height," Appl. Phys. Lett. 85, 1368 (2004)
13. L. S. Yu, Q. J. Xing, D. Qiao, S. S. Lau, K. S. Boutros, and J. M. Redwing, "Internal photoemission measurement of Schottky barrier height for Ni on AlGaN/GaN heterostructure," Appl. Phys. Lett. 73, 3917 (1998)
14. S. T. Bradley, S. H. Goss, J. Hwang, W. J. Schaff, and L. J. Brillson, "Pre-metallization processing effects on Schottky contacts to AlGaN/GaN heterostructures," J. Appl. Phys. 97, 084502 (2005)
15. E. T. Yu, X. Z. Dang, L. S. Yu, D. Qiao, P. M. Asbeck, S. S. Lau, G. J. Sullivan, K. S. Boutros, and J. M. Redwing, "Schottky barrier engineering in III–V nitrides via the piezoelectric effect," Appl. Phys. Lett. 73, 1880 (1998)

Mater. Res. Soc. Symp. Proc. Vol. 1538 © 2013 Materials Research Society
DOI: 10.1557/opl.2013.505

Effect of Growth Pressure and Gas-Phase Chemistry on the Optical Quality of InGaN/GaN Multi-Quantum Wells

E.A. Armour[1], D. Byrnes[1], R.A. Arif[1], S.M. Lee[1], E.A. Berkman[1], G.D. Papasouliotis[1], C. Li[2], E.B. Stokes[2,5], R. Hefti[3], and P. Moyer[4,5]

[1] Veeco Instruments, Turbodisc Operations, 394 Elizabeth Avenue, Somerset, NJ 08873, U.S.A.

[2] Department of Electrical and Computer Engineering, University of North Carolina at Charlotte, 9201 University City Blvd, Charlotte, NC 28223, U.S.A.

[3] Nanoscale Science Program, University of North Carolina at Charlotte, 9201 University City Blvd, Charlotte, NC 28223, U.S.A.

[4] Department of Physics & Optical Science, University of North Carolina at Charlotte, 9201 University City Blvd, Charlotte, NC 28223, U.S.A.

[5] Center for Optoelectronics and Optical Communications, University of North Carolina at Charlotte, 9201 University City Blvd, Charlotte, NC 28223, U.S.A.

ABSTRACT

Blue light-emitting diodes (LED's), utilizing InGaN-based multi-quantum well (MQW) active regions deposited by organometallic chemical vapor epitaxy (OMVPE), are one of the fundamental building-blocks for current solid-state lighting applications. Studies [1,2] have previously been conducted to explore the optical and physical properties of the active MQW's over a variety of different OMVPE growth conditions. However, the conclusions of these papers have often been contradictory, possibly due to a limited data set or lack of understanding of the fundamental fluid dynamics and gas-phase chemistry that occurs during the deposition process.

Multi-quantum well structures grown over a range of pressures from typical low-pressure production processes at 200 Torr, up to near-atmospheric growth conditions at 700 Torr, have been investigated in this study. At all growth pressures, clear trends of gas-phase chemical reactions are observed for increased gas residence times (lower gas speeds from the injector flange and lower rotation rates) and increased V/III ratios (higher NH_3 flows).

Confocal microscopy, excitation-dependent PL (PLE), and time-resolved photo-luminescence (TRPL) have been employed on these MQW structures to investigate the carrier lifetime characteristics. Confocal emission images show spatially-separated bright and dark regions. The bright regions are red-shifted in wavelength relative to the dark regions, suggesting microscopic spatial localization of high indium content regions. As the growth pressure and gas residence times are reduced, a larger difference in band-gap between bright and dark regions, longer lifetimes, and higher average PL intensities can be obtained, indicating that higher optical quality material can be realized. Optimized MQW's grown at high pressure exhibit higher PLE slope intensities and IQE characteristics than lower pressure samples. Results on simple LED structures indicate that the improvement in MQW optical quality at high pressures translates to higher output power at a 110 A/cm^2 injection current density.

INTRODUCTION

As InGaN-based LED epi-structure brightness and electrical performance continues to be improved, primarily through structural and process recipe optimization, there exists a desire to more completely understand the fundamental material and gas-phase chemistry limitations influencing these parameters. Current production-level LED structures are typically grown at low-pressure conditions for the MQW active region ($\leq$ 200 Torr), where the reactor designs can maintain extremely reproducible and uniform performance. However, there have been limited reports on the chemical mechanisms that occur during the InGaN OMVPE process [1,3-6] and the process optimization needed to achieve high quality material. Of particular interest is the effect that growth pressure has on the optical and morphological properties of the InGaN material.

To understand these properties over a wide range of process conditions, we have employed a simple 4-period thin InGaN MQW structure. The growths were performed in a Veeco K465i Turbodisc reactor, utilizing vertical geometry and high-speed rotation. All growths were performed with equivalent plug-flow stability parameters, whereby the total flow into the reactor matched the rotationally-driven pumping speed of the disk to achieve laminar and stable fluid flow streamlines. More details of a high speed rotating disk reactor operation can be found in [7].

EXPERIMENTAL

The MQW samples in this study were grown on planar c-plane oriented (0001) sapphire substrates, miscut 0.2° towards the m-plane. The underlying GaN buffer layers consisted of a 2.2 μm nominally undoped GaN buffer, followed by a 2 μm n-type GaN layer with approximately 7×10^{18} cm^{-3} silicon doping level, as measured by SIMS. The XRD FWHM of the GaN templates were controlled to a range of 250-270" for (002)ω and 260-280" for (102)ω rocking curves, which translates into a threading dislocation density of approximately 4.4×10^8 cm^{-2} [8]. The surface morphology of the n-GaN templates was measured using AFM with a 5 x 5 μm scan area, exhibiting a typical roughness of 0.4 nm RMS. Following the n-GaN buffer layers, the MQW structure was deposited, which consisted of 4 pairs of a 2.7 nm InGaN quantum well (QW) sandwiched between 10 nm GaN quantum barriers (QB). Triethylgallium (TEGa) and trimethylindium (TMIn) with nitrogen (N$_2$) as a carrier gas were used for the QW's. The period growth rate was controlled at 0.9 nm/min. by changing the overall TEGa and TMIn fluxes, keeping a fixed molar ratio of 1.7 to obtain identical structures between the different growth conditions. The same TEGa flow rate was used for both QW and QB layers. Ammonia (NH$_3$) was used as the group V source. The deposition temperature was controlled using a fixed emissivity pyrometer measuring the wafer carrier, and was adjusted for each growth sample to achieve approximately 450 nm peak PL wavelength.

Our experimental dataset explored the morphological and optical quality of the MQW's as a function of growth pressure, V/III ratio, and total flow of the gases. The total flow is determined by the amount of N$_2$ carrier gas and NH$_3$ injected into the reactor. In these experiments, relatively small changes in NH$_3$ flows were used (in a range of 30 to 70 slm), such that the N$_2$ carrier gas flow usually dominated the total flow ratio. Low-power photoluminescence (PL) maps were acquired and averaged over two 4" diameter full wafer samples, using a Nanometrics RPM Blue with a pump wavelength of 377 nm and an incident power of ~127 W/cm^2.

To more accurately assess the optical quality of the films closer to the carrier densities employed in typical LED structures, power-dependent PL excitation (PLE) spectroscopy was utilized. For our PLE experiments, the center point of the 4" wafer was pumped with a frequency doubled Ti:sapphire femtosecond laser operating at 76 MHz with an illumination wavelength of 375 nm. The output of the second harmonic generator was coupled to a fiber and focused on the sample with a spot size of approximately 1 mm diameter. The PLE measurements were conducted over an average incident power in the range of 6 mW to 160 mW, which corresponds to an estimated injection current density of 6 to 150 A/cm^2. The integrated intensity as a function of illumination power was plotted, and the slope of this curve was used as a metric of "brightness" of each MQW sample.

Confocal laser scanning microscopy (CLSM) was also employed to characterize these samples. A 405 nm pulsed laser beam, with 10 MHz frequency and 300 ps pulse width, was used as an excitation source for the CLSM measurements. The laser was coupled into a single mode optical fiber after which a 405/20 nm band pass filter (BPF) was placed to block any fluorescence emission from the fiber. The laser beam was then reflected by a 430 nm dichroic beam splitter (DBS) to a 100× objective lens whose numerical aperture is 1.25. The objective lens focused the laser beam onto the samples, and also collected PL emitted from the samples. PL collected by the objective lens passed through the DBS and a pinhole, and then was detected using a time-correlated single photon counting (TCSPC) avalanche photodiode (APD), which was preceded by a 445/40 nm BPF so that only near-band-edge emission light contributed to the imaging. The optical probe was stationary relative to the instrument frame, while the sample was mounted to a high speed scanning lead zirconium titanate (PZT) stage with nanometer precision control along three axes and a translation range of 70 μm by 70 μm by 50 μm in x-, y-, and z-directions, respectively. The microscope had a lateral spatial resolution of about 200 nm, and vertical spatial resolution of about 100 nm. Localizing features to better than 5 nm is possible using deconvolution with a point spread function. With suitable synchronization and time-counting software, the system could also be used to measure time-resolved PL (TRPL) of micrometer or nanometer scale structures. For spectral measurements, the 445/40 nm BPF and TCSPC APD were replaced by a spectrometer equipped with a thermoelectrically-cooled CCD camera.

RESULTS AND DISCUSSION

At 200 and 450 Torr growth pressures, higher total flows and V/III ratios for a given chamber pressure generally resulted in stronger PL intensities. As the growth pressure was raised, the PL intensity ratio between low flow and high flow conditions increased, with low total flow conditions at 700 Torr having very weak PL signals. For the 700 Torr samples, there was also a substantial improvement in surface morphology as the total flow was increased, as illustrated in Figure 1. With low total flows (i.e. long gas residence time in the chamber), samples under these conditions exhibited 3-dimensional indium inclusions and a Stranski-Krastanov type of growth mode. As the total flow increases, we observed a transition through a spiral growth mode pattern towards a two-dimensional step-structure morphology.

At 700 Torr pressure, increasing the V/III ratio did not result in improved PL intensity, but instead reduced the intensity, unlike the cases for 200 Torr and 450 Torr. To achieve equivalent

PL intensities between 50 slm NH₃ and 70 slm NH₃ flows, the total flow for these conditions needed to be increased by approximately 20%

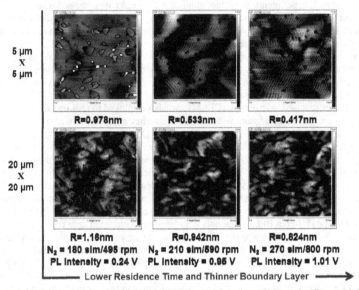

Figure 1. Surface morphology for 700 Torr MQW's as a function of N_2 carrier flow with NH_3 flow at 50 slm.

These characteristics reinforce the observations that we previously published [9], suggesting that parasitic gas-phase reactions occur between TEGa, TMIn, and NH_3, and are particularly amplified at higher process pressure. The gas-phase reactions are likely changing the ratio of chemical adducts adsorbing onto the growth surface, which affect the adatom surface diffusion characteristics. This explanation endorses the proposal of [1], which suggested that as pressure is increased, the adatom surface diffusion may be impeded by either adsorbed reactants or carrier gas species. However, from our total flow experiments where only the N_2 carrier gas is being modulated, it does not appear that adsorbed carrier gas species (N_2) are necessarily suppressing adatom surface diffusion.

Two effects occur when total flow is increased: (a) residence time of the chemical species is reduced between injection at the top flange down to the thermal and momentum boundary layers, which is proportional to the gas speed, and (b) the boundary layer is compressed due to increased rotation required to balance the higher total flow. It should be pointed out, however, that continually increasing the N_2 flow rate to achieve even larger total flows results in a reduced partial pressure of the NH_3 species, which is not desirable do to its negative effect on MQW (InGaN) quality.

To more easily understand the optimized process parameter space, we have developed two metrics to describe the gas residence time to the growth surface and the effective amount of NH_3 that is available for growth:

$$S^* = \text{(gas velocity at injector plate)} \times \text{(wafer carrier rotation speed)}$$

$$V/III^* = \frac{\text{# of moles of NH}_3 * X_{NH3}}{\text{# of moles of Alkyls}}$$

with

$$X_{NH3} = \frac{NH_3 \, flow}{NH_3 \, flow + N_2 \, flow}$$

where S^* is defined as the effective gas speed, and V/III^* is the diluted V/III ratio, which is influenced by the NH_3 partial pressure ratio. Using these two parameters, for a fixed QW growth rate and TEGa/TMIn molar ratio, we can easily compare optical qualities from one condition to another in a normalized process space.

As can be seen from the 200 Torr and 450 Torr contour plots of Figure 2, as the effective gas speed S^* is increased simultaneously with V/III^* (diagonal upwards indicated by the arrows), the PL intensity increases for the samples. The highest intensities achieved at both pressures occur at a gas speed $S^* > 90$. For 700 Torr conditions, very high brightness samples can be obtained, although the increased S^* vs. V/III^* trend observed at the lower pressures is obfuscated. At this high growth pressures, the process space becomes more limited due to the smaller achievable gas velocities and higher total flows (i.e. N_2 carrier gas flow) necessary to achieve stable plug flow. At 700 Torr, laminar streamlines without recirculation cells are hard to obtain due to geometrical limitations in the injector design, which limits the run-to-run repeatability needed to fully characterize this process space.

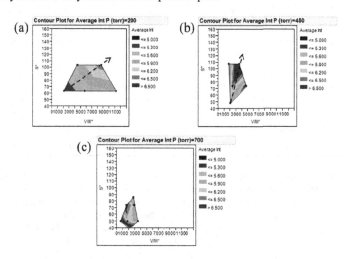

Figure 2. PL Intensity as a function of effective Gas Speed (S*) and effective V/III ratio (V/III*) for (a) 200 Torr, (b) 450 Torr, and (c) 700 Torr growth pressures.

Another observation is that when pressure is increased, a higher wafer carrier temperature is needed to achieve similar indium concentrations in the MQW and hence maintain the target wavelength of 450 nm. Typically, about 5-8° C increase is necessary between similar 200 Torr and 450 Torr conditions, and another 8-12° C increase is needed to achieve the same wavelength at 700 Torr conditions. Measurements using an in-situ 405 nm pyrometer (Veeco BluTemp^{TM}) indicate that for these controlled structures, the surface temperature remains the same, and that the differential temperature between the wafer carrier and wafer enlarges with increasing temperature [10]. It is proposed that convective surface cooling at the wafer relative to the wafer carrier from the higher density of gas and higher total flows at the higher pressures may be the driving forces for this difference. The higher gas temperature in the vicinity of the wafer carrier could also result in increased pyrolysis of the NH_3, resulting in a higher effective V/III ratio at the growth surface for the high pressure conditions.

Table I. Summary of PLE, CLSM, and TRPL measurements for a subset of the explored growth conditions.

Sample ID	Growth pressure (torr)	V/III*	S*	PLE slope Efficiency	CLSM Images average intensity	Bandgap difference (meV)	PL peak intensity ratio	PL lifetime τ2 in bright region (ns)	PL lifetime τ2 in dark region (ns)
1	200	4457	104	44.84	70	44.2	3.63	22.94	18.97
2	200	2123	64	35.41	49	39.1	3.25	17.99	15.17
3	200	5778	64	34.32	54	25.4	3.01	20.11	16.77
4	200	3820	151	38.80	55	30.5	3.19	17.25	16.75
5	200	4504	104	36.60	60	37.5	3.34	21.41	18.88
6	200	3016	104	48.59	66	41.6	3.55	25.31	20
7	450	2463	74	35.44	60	35.3	3.27	17.86	14.82
8	450	1851	108	68.93	80	41.5	3.58	27.75	22.31
9	450	2092	48	30.78	44	22.8	2.93	15.72	11.31
10	450	8948	74	56.50	64	47.9	3.79	23.87	19.34
11	700	3572	50	50.84	71	36.9	3.3	26.85	24.17
12	700	684	50	42.70	52	27.1	3.09	23.85	22.57
13	700	2979	75	65.11	88	45.3	3.61	27.44	24.34
14	700	3668	81	39.84	57	27.6	3.11	23.67	22.63

Combined results of PLE, CLSM, and TRPL measurements for MQW's grown over a selection of different flow conditions are listed in Table I. PLE slope is determined by integrating the intensity over the pump laser power, and is suggested to be the best metric for translating to LED optical power efficiency, due to the high carrier injection intensities that are generated. The CLSM image average intensity was calculated by averaging the intensity of each pixel of the image obtained during the measurements. There is a rough correlation of PLE intensity to CLSM intensity, as indicated in Figure 3 (a). Figure 3 (b) shows the PLE slope efficiency as a function of effective gas speed S* plotted with identification of different pressures. In both measurement cases, the strongest PL intensity samples are obtained at higher pressure (450 Torr and 700 Torr) growth conditions. Similar levels of PL intensity could not be achieved at 200 Torr growth conditions, even after process optimization.

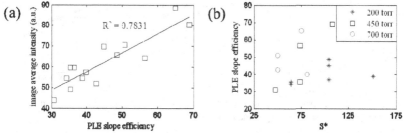

Figure 3. a) Correlation of PLE slope efficiency to CLSM image average intensity, (b) PLE slope efficiency as a function of effective gas speed (S*) for different growth pressures. Note the highest intensity samples are at 450 Torr and 700 Torr.

Figure 4 shows the CLSM image of near band-edge emission for a MQW sample grown at 700 Torr, in which an inhomogeneous distribution of PL intensity is observed. Bright regions, which are several micrometers in size, are surrounded by considerably dimmer dark regions. PL spectra measured at different places in the image show that the intensity at bright regions is > 3 times stronger than that in the dark regions, as indicated by the peak ratio in Table I. Furthermore, the PL peaks in the bright regions are red-shifted compared with the PL peak in the dark regions, as is shown in Figure 5 (a). This is consistent with smaller band-gap energy in the bright regions more effectually localizing carriers to recombine radiatively, and thus preventing them from migrating to regions with non-radiative recombination defect centers (NRRCs). The darker regions, on the other hand, exhibit higher band-gap energies and reduced potential well height, thus more easily allowing thermal diffusion into the NRRC sections. The average image PL intensity increases when the band-gap energy difference between these regions becomes larger, as shown in Figure 6 (a). As the CLSM image average intensity increases, the image becomes less uniform, as the image standard deviation, which is defined as how much variation or exists from the average, is larger for samples with larger image average intensity. This is shown in Figure 6 (b) and is to be expected, as samples with higher average intensity have a larger band-gap energy difference and stronger carrier localization effect in its narrow-band-gap-region, resulting in a larger difference between bright and dark region PL intensities.

Figure 4. Confocal microscope image of 445 nm near band-edge emission for a 700 Torr MQW sample. The image uses a 40 nm band-pass filter, illustrating the bright and dark regions. The image covers a 20 μm × 20 μm region in the center of the sample wafer.

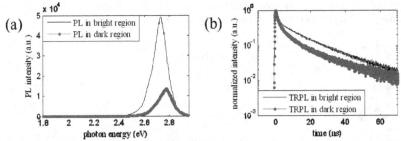

Figure 5. (a) Confocal PL spectra at bright and dark regions on the image in Figure 4, and (b) TRPL lifetime decay plots of the bright and dark regions.

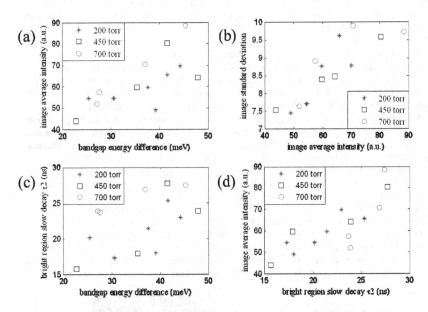

Figure 6. (a) CLSM image average intensity as a function of band-gap energy difference between bright and dark regions of the MQW samples from this study, (b) CLSM image average intensity compared to the contrast between bright and dark regions, (c) CLSM image average intensity as a function of τ2 lifetime of the bright regions in the sample, and (d) τ2 lifetime plotted against band-gap energy difference for the sample MQW's.

Confocal TRPL measurements were performed in both bright and dark regions on each sample. Figure 5 (b) shows the TRPL spectra for the sample that was pictured in Figure 4. A bi-exponential decay equation was used to fit the TRPL spectra, and two lifetimes $\tau1$ and $\tau2$ were obtained, where $\tau1$ is a fast observed decay component (no more than 3 ns) and $\tau2$ is a slower observed decay component (around 20 ns). The biexponential function is consistent with two spatially distinct recombination systems in each probed region. In our analysis, $\tau2$ is attributed to recombination in the strongly localized states, while $\tau1$ is attributed to weakly localized states where radiative recombination is more quenched by NRRC and also competes with carrier transfer into strongly localized states. As a result, the slow decay lifetime $\tau2$ tends to dictate the overall optical quality of the MQW samples, as plotted in Figure 6 (b). The lifetime of $\tau2$ within the bright regions is longer than the $\tau2$ within the dark regions, indicating higher quenching of recombination in strongly localized states in dark regions. The shorter lifetime and lower intensity of dark regions is consistent with a higher level of recombination through NRRC in the dark regions, and/or carrier transport from dark-to-bright regions. The longer $\tau2$ lifetime in bright regions also coincides with a larger band-gap energy difference between the bright and dark regions, which is consistent with our explanations above, as evidenced in Figure 6 (c).

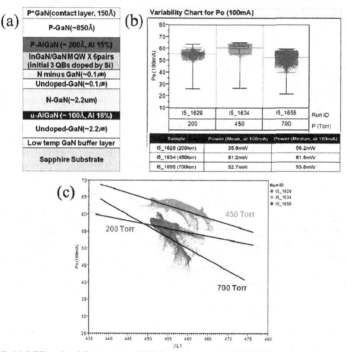

Figure 7. (a) LED epitaxial structure, (b) light output power (LOP) histogram measured for the 3 different pressures using optimized MQW flow conditions, and (c) output power plotted as a function of wavelength (nm) for the 3 LED wafers.

The flow conditions yielding the highest PLE slope efficiency for each pressure were selected to insert into the 6-pair MQW active region of a simple LED structure, which is shown in Figure 7 (a). For all 3 LED's tested, the same structure and doping levels were used, and only the active layer growth conditions were changed. The fabricated LED die were 300 μm x 300 μm in size, with a lateral electrode configuration, mesa-etched and passivated, but not separated. A production prober was used to map die throughout the entire wafer; no special temperature control was used during the measurements.

The LED devices with MQW's grown at 450 Torr were brighter than 200 Torr MQW LED's by ~ 9.7%, whereas the 700 Torr LED's were slightly dimmer, as shown in the histogram of Figure 7 (b), and the output power as a function of wavelength plots of Figure 7 (c). The brightness increase between 200 Torr and 450 Torr MQW optimized growth conditions has been confirmed through repeated growth and fabrication LED campaigns, averaging a 5-10% improvement in optical output power. Reference wafers during each fabrication campaign yielded batch-to-batch variations of < 5%. The drop in power for the 700 Torr LED was not entirely unexpected, as the 450 Torr conditions are within the plug-flow stability regime, but the 700 Torr MQW flow conditions have non-laminar streamlines, which can lead to large run-to-run variations. It should be noted that optimized higher pressure growths have also shown improvements in optical power output in customer structures and other LED studies [11].

CONCLUSIONS

Multi-quantum well structures have been studied over a variety of growth conditions to examine the morphology and optical quality of the materials grown. At increased growth pressures, parasitic chemical reactions can dominate, reducing material quality. To lessen the effect of these chemical pathways, decreasing gas residence time by increasing total flow can improve overall PL intensity. To further understand the relationship between growth conditions and PL intensity, we have developed two metrics: effective gas speed S* and diluted V/III* which allow us to compare growth conditions over a wide process space. For long gas residence time (low S*) and high V/III* conditions, gas-phase chemical reactions are amplified, resulting in a reduced parameter space to achieve high-quality MQW layers. This trend is particularly clear at high MQW growth pressures, where suppressing gas phase reactions with high total flows for high V/III ratio material results in higher PL intensity structures. Under optimized conditions, our brightest samples were achieved at elevated growth pressures.

Optical characterization of the MQW samples using power-dependent PL excitation, confocal laser scanning microscopy, and time resolved PL have shown that the highest emission samples have stronger localization with a greater difference in band-gap between bright and dark regions, and exhibit longer TRPL lifetimes. Initial LED studies incorporating equivalent MQW structures and optimized flow conditions have exhibited higher optical output power for 450 Torr samples when compared to 200 Torr samples.

ACKNOWLEDGEMENTS

The authors wish to acknowledge Frank Lu for assistance on TRPL and PLE optical measurements and Steve Ting for helpful suggestions.

REFERENCES

1. R.A. Oliver, M. J. Kappers, C.J. Humphreys, and G.A.D. Briggs, J. Appl. Phys. **97**, 013707 (2005).
2. W.V. Lundin, E.E. Zavarin, M.A. Sinitsyn, A.V. Sakharov, S.O. Usov, A.E. Nikolaev, D.V. Davydov, N.A. Cherkashin, and A.F. Tsatsulnikov, Fiz. Tekh. Poluprovodnikov **44**, 126, (2010) [Semiconductors **44**, 123 (2010)].
3. J.R. Creighton, G.T. Wang, W.G. Breiland, and M.E. Coltrin, J. Crystal Growth **261**, 204 (2004).
4. J.R. Creighton, M.E. Coltrin, and J.J. Figiel, Appl. Phys. Lett. **93**, 171906 (2008).
5. G.B. Stringfellow, J. Crystal Growth **312**, 735 (2010).
6. A, Demchuk, J. Porter, and B. Koplitz, J. Phys. Chem. A **102**, 8841 (1998).
7. B. Mitrovic, A. Gurary, and L. Kadinski, J. Crystal Growth **287**, 656 (2006).
8. S.R. Lee, A.M. West, A.A. Allerman, K.E. Waldrip, D.M. Follstaedt, P.P. Provencio, D.D. Koleske, and C.R. Abernathy, Appl. Phys. Lett. **86**, 241904 (2005).
9. E.A. Armour, B. Mitrovic, A. Zhang, C. Ebert, M. Pophristic, and A. Paranjpe in *Compound Semiconductors for Generating, Emitting and Manipulating Energy*, edited by T. Li, M. Mastro, A. Dadgar, H. Jiang, and J. Kim, (Mater. Res. Soc. Symp. Proc. **1396**, Boston, MA, 2011) pp. 3-14.
10. Alexander Manasson, private internal Veeco communication (2013).
11. J.H. Na, S.K. Lee, H.S. Lim, H.K. Kwon, S. Son, and M.S. Oh, presented at the 16th International Conference on Metal Organic Vapor Phase Epitaxy, Busan, Korea, 2012 (unpublished).

Mater. Res. Soc. Symp. Proc. Vol. 1538 © 2013 Materials Research Society
DOI: 10.1557/opl.2013.619

Hexagonal Pyramids Shaped GaN Light Emitting Diodes Array by N-polar Wet Etching

Jun Ma*, Liancheng Wang, Zhiqiang Liu, Guodong Yuan, Xiaoli Ji, Ping Ma , Junxi Wang, Xiaoyan Yi, Guohong Wang and Jinmin Li

Institute of Semiconductors, Chinese Academy of Sciences, Beijing 100083, P R China

ABSTRACT

In this work, we investigated the influence of N-polar wet etching on the properties of nitride-based hexagonal pyramids array (HPA) vertical-injection light emitting diodes (V-LEDs). The cathodeluminescence images showed the randomly distribution of hexagonal pyramids with isolated active regions. The transmission electron microscopy images demonstrated the reduced density of threading dislocations. The IQE was estimated by temperature dependence of photoluminescence, which showed 30% increase for HPA V-LEDs compared with broad area (BA) V-LEDs. The improved extraction efficiency was verified by finite difference time domain simulation, which was 20% higher than that of roughened BA V-LEDs. The electrical properties of HPA V-LEDs were measured by conductive atomic force microscopy (CAFM) measurements. HPA V-LEDs exhibited much lower leakage current due to the improved crystal quality.

INTRODUCTION

Nitride-based vertical-injection light emitting diodes (V-LEDs), of which the insulated sapphire substrates were substituted by the Cu substrates, have been believed to be a promising candidate for high power applications.[1,2,3] Due to the thermal, electrical conductivity of Cu submount and the vertical current injection, V-LEDs have the advantages of better current injection, excellent heat dissipation, superior electrostatic discharge protection, and simple packaging process.[4,5] However, the development of V-LEDs for high power applications still suffers from severe obstacle, such as large reversed leakage current, serious efficiency droop and limited light extraction. Recently, remarkable advances in the low-dimensional nanostructure LEDs seem to provide an inspiring solution. In particular, the nanowire/nanorod GaN LEDs have brilliant quantum efficiency and improved efficiency droop. [6,7] On the other hand, N-polar (000$\bar{1}$) surface wet etching process has been commonly adopted to acquire the textured surface in order to improve the extraction efficiency in the fabrication of conventional GaN broad area (BA) V-LEDs.[8] The hexagonal pyramid-shaped morphology produced on the N-polar face during the wet etching process has been reported by several groups.[9,10,11] It is believed that threading dislocations (TDs) serve as nucleation points for the etch process, and other defects play an important role in the wet etching of N-polar (000$\bar{1}$) surface.[12,13]

In this paper, we presented the fabrication of hexagonal pyramids array (HPA) V-LEDs. The hexagonal pyramids with isolated active regions were produced by the N-polar (000$\bar{1}$) face wet etching. The eliminating of TDs in hexagonal pyramids was realized by the wet etching process. The electrical and optical properties of HPA V-LEDs were investigated. Compared with conventional BA V-LEDs, the efficiency droop and output power of the exhibited a significant improvement, which could be due to the better crystal quality and the amelioration of extraction efficiency as the inherent advantage of hexagonal pyramid shape.

EXPERIMENT

The scanning electron microscopy (SEM) images of key procedures for the fabrications of HPA V-LEDs are shown in Figure 1(a)-(c) using Hitachi 4800. After the GaN epitaxial growth,

Ni/Ag/Pt/Au metallization p-contact with high reflectivity was deposited on p-GaN using electron-beam evaporator. The electroplating of Cu was performed on the metallization p-contact as the new substrate, followed by the removal of sapphire substrate through laser lift-off (LLO). Then the remaining epitaxial film was subjected to a 2.5 h wet etching to produce the hexagonal pyramids. Dilute potassium hydroxide (KOH) solution with a concentration of 6 mol/L served as the etch electrolyte at a constant temperature of 70 °C. BA V-LEDs sample with 10 min wet etching in the same KOH solution for a roughed N-polar face was fabricated as reference. The epilayer for HPA V-LEDs was separated into hexagonal pyramids by wet etching, as shown in Figure 1(a). The hexagonal pyramids are distributed randomly on the Ni/Ag/Pt/Au reflective metalized contact. Right following the wet etching process, the cover of silicone gel on isolated pyramids was implemented by spin-coating in order to protect the active regions of hexagonal pyramids. After that, the epilayer was subjected to O₂ plasma etching to expose the apexes of the pyramids. As seen in Figure 1(b), the dark areas in SEM image which submerge the bottom part of the pyramids indicate the cover of silicone gel because of its insulation to electron beam, while the top parts of pyramids that exhibit the bright hexagonal shape were exposed as a result of the O₂ plasma etching. Then, a 200 nm layer of indium tin oxide (ITO) was deposited on the N-polar surface of epilayer, which serve as transparent electrode and current spreading layer. As displayed in Figure 1(c), the interconnection of hexagonal pyramids is realized by ITO film. Cr/Pt/Au contacts were deposited on the ITO film as n-type electrode. Finally, the HPA V-LEDs and BA V-LEDs devices were fabricated into a 1200 μm×1200 μm chip.

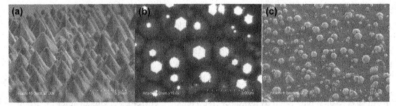

Figure 1 SEM images of key procedures for the fabrications of HPA V-LEDs. (a) Hexagonal pyramids array produced by wet etching. (b) coverage of spin-coated silicone gel after O2 plasma. (c) deposition of ITO film.

DISCUSSION

A FEI F20 microscope with acceleration voltage of 200 kV was used to obtain bright-field transmission electron microscopy (TEM) images. As seen in Figure 2(a) during the wet etching process, most TDs were connected with the grooves between the pyramids and terminated at the etched region during the wet etching process. Furthermore, it is noteworthy that no obvious screw or edge dislocations are observed inside the isolated pyramid. Moreover, the active regions positioned in the bottom part of the hexagonal pyramids are separated by the wet etching process. The InGaN/GaN MQWs shows abrupt interfaces over the whole pyramid, which indicates MQWs structure closed to the edge of the pyramid are not affected by the wet etching.

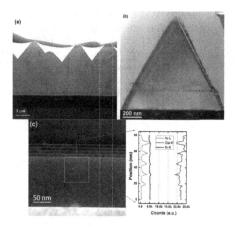

Figure 2 cross-sectional TEM images of (a) epilayer during the wet etching, (b) isolated hexagonal pyramid, (c) enlarged area of active region and the EDX line scanning profile.

Monochromatic cathodeluminescence (CL) images were collected by Gatan MONO CL3+ system. The top-view and bird-view monochromatic CL images of single and arrayed pyramids samples were shown in Figure 3. As seen in Figure 3(a), the top-view CL image of single pyramid displays the profile of active region, which is hexagonal shaped by the wet etching process. The bright area in bird-view CL image of single and arrayed pyramid clearly indicates the position of active region, the dark area above and below it corresponds to the n-type and p-type GaN, respectively, as seen in Figure 3(b)(c). No dark points or other wavelength are detected. Previously studies have revealed that the surface pits correlated with TDs, or regions containing high dislocation densities exhibit dark spots with reduced intensity in monochromatic CL images.[14,15,16] In this regard, these observations illustrate the formation of dislocation-eliminating hexagonal pyramids after wet etching process.

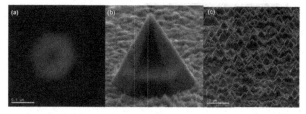

Fig. 3 (a) plan- and (b) tilt-view monochromatic CL images of single hexagonal pyramid and tilt-view hexagonal pyramids array

The crystal quality of HPA V-LEDs and BA V-LEDs are investigated by temperature dependence of photoluminescence (PL) spectrum. Figure 4 shows the Arrhenius plot of the normalized integral PL intensity of HPA V-LEDs and conventional BA V-LEDs. The measured internal quantum efficiency (IQE) of HPA V-LEDs increases by 30% compared with that of BA

V-LEDs, which suggests that the HPA V-LEDs has an improved crystallinity by the reduction of TDs density. Furthermore, the temperature at which the PL intensity starts to quench for HPA V-LEDs is much higher than that for BA V-LEDs. The quenching of PL intensity for HPA V-LEDs is well described using single activation energy. However, the PL intensity for BA V-LEDs first quenches with weak activation energy, then a second activation energy is dominated at high temperature. This observation indicates the larger activation energy for HPA V-LEDs, leading to the restricted activation of nonradiative recombination with increased temperature.[17,18]

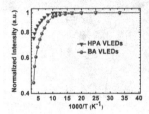

Figure 4 Arrhenius plot of the normalized integral PL intensity of HPA V-LEDs and conventional BA V-LEDs.

As shown in Figure 5(a), the 3D finite difference time domain (FDTD) computational domain for HPA V-LEDs is a 10-period hexagonal pyramids array with a diameter and pitch of 1 and 2 μm, respectively. The theoretically calculated extraction efficiency for HPA V-LEDs is about 71%, which is much higher than 52% for conventional roughened BA V-LEDs. As seen in Figure 5(b), the light intensity distribution of HPA V-LEDs spreads over the entire surface, suggesting that the light escape angle is significantly enlarged. Previous research revealed that the largest extraction efficiency enhancement was realized when the sidewall angle of nanostructure is approaching to 23.4°, which is the critical angle for total internal reflection (TIR) at the GaN/air interface. [19] The natural angle of sidewall (10$\bar{1}$1) for hexagonal pyramid formed by wet etching is fixed at 31.6°,[9] leading to the inherent advantage of high extraction efficiency for hexagonal pyramids shape.

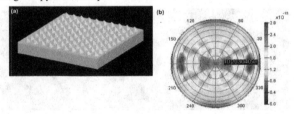

Figure 5 (a) Schematic of 3D FDTD computational domain for HPA V-LEDs and (b) simulated far-field emission pattern of HPA V-LEDs.

Localized I-V properties of single pyramid and BA V-LEDs were investigated using C-AFM at room temperature. Duty-cycle operation with cycle period of 20 ms and pulse width of 1ms was adopted at room temperature for electrical measurement. As shown in Figure 6(a)(b), the typical I-V property of single hexagonal pyramid was studied using conductive atomic force microscopy (C-AFM) in contact mode with an Au-coated tip. *I-V* properties of various contact points on the single pyramid and N-polar surface of BA V-LEDs were measured. As seen in Figure 6(c), the average leakage current for single pyramid has the constant value of 0.3 nA under reversed bias up to 10 V. It indicates that such a saturated current under reversed bias originate from the absence of defect induced trap levels within the band gap.[20] However, the reversed leakage current for BA V-LEDs significantly increases to 457 nA at 10 V. Such a large leakage current may attribute to the ICP etching damage and defects formed during the epitaxy. Moreover, the leakage current for BA V-LEDs exhibits a strong field dependent, indicating that defect-assisted tunneling is the dominant transport mechanism.

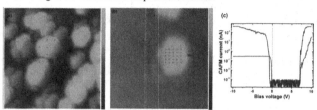

Figure 6 10 μm×10 μm AFM images of (a) typical hexagonal shaped morphology of roughened n-GaN surface by wet etching, data scale is 2 μm (b) isolated hexagonal pyramid, data scale is 4 μm and (c) localized I-V property measured by CAFM, red line for HPA V-LEDs, and blue line for BA V-LEDs.

CONCLUSIONS

In conclusion, we fabricated the high performance HPA V-LEDs through N-polar face wet etching. The TEM and CL analysis indicate that TDs in HPA V-LEDs are eliminated significantly by wet etching process. The estimated IQE for HPA V-LEDs by temperature dependence of PL showed an increase of 30% than that for BA V-LEDs due to the better crystal quality. The reduction of TDs in hexagonal pyramids is evidenced by the constant leakage current of 0.3 nA under reversed bias up to 10 V by C-AFM measurement. It is demonstrated that N-polar wet etching provided an effective approach to fabricate the micro- to nano-size V-LEDs array without detrimental ICP process and artificial lithography mask.

ACKNOWLEDGMENTS

This work was financially supported by the National High Technology Research and Development Program of China (No. 2011AA03A105).

REFERENCES

1. O. B. Shchekin, J. E. Epler, T. A. Trottier, T. Margalith, D. A. Steigerwald, M. O. Holcomb, P. S. Martin, and M. R. Krames. "High performance thin-film flip-chip InGaN–GaN light-emitting diodes," Appl. Phys. Lett. **89**, 071109 (2006).
2. Ch.-F. Chu, F.-I. Lai, J.-T. Chu, C.-C. Yu, C.-F. Lin, H.-C. Kuo, and S. C. Wang, "Study of GaN light-emitting diodes fabricated by laser lift-off technique," J . Appl. Phys. **95**, 3916 (2004).

3. W. Y. Lin, D. S. Wuu, K. F. Pan, S. H. Huang, C. E. Lee, W. K. Wang, S. C. Hsu, Y. Y. Su, S. Y. Huang, and R. H. Horng, "High-power GaN-mirror-Cu light-emitting diodes for vertical current injection using laser liftoff and electroplating techniques," IEEE Photonic. Tech. Lett. **17**, 1809-1811 (2005).

4. C.-F. Chu, C.-C. Yu, H.-C. Cheng, C.-F. Lin, and S.-C. Wang, "Comparison of p-Side Down and p-Side Up GaN Light-Emitting Diodes Fabricated by Laser Lift-Off," Jpn. J. Appl. Phys. **42**, L147–L150 (2003).

5. J.-T. Chu, H.-W. Huang, C.-C. Kao, W.-D. Liang, F.-I Lai, C.-F. Chu, H.-C. Kuo, and S.-C. Wang, "Fabrication of large-area GaN-based light-emitting diodes on Cu substrate," Jpn. J. Appl. Phys. **44**, 2509–2511 (2005).

6. H. P. T. Nguyen, K. Cui, S.f. Zhang, M. Djavid, A. Korinek, G. A. Botton, and Z. Mi, "Controlling electron overflow in phosphor-free InGaN/GaN nanowire white light-emitting diodes," *Nano Lett. 12*, 1317–1323(**2012**).

7. H.-M. Kim, Y.-H. Cho, H. Lee, S. I. Kim, S. R. Ryu, D. Y. Kim, T. W. Kang, and K. S. Chung, "High-brightness light emitting diodes using dislocation-free indium gallium nitride/gallium nitride multiquantum-well nanorod arrays," *Nano Lett. 4*, 1059–1062 (**2004**).

8. T. Fujii, Y. Gao, R. Sharma, E. L. Hu, S. P. DenBaars, and S. Nakamura, "Increase in the extraction efficiency of GaN-based light-emitting diodes via surface roughening," Appl. Phys. Lett. **84**, 855 (2004).

9. H. M. Ng, N. G. Weimann, and A. Chowdhury, "GaN nanotip pyramids formed by anisotropic etching," J. Appl. Phys. **94**, 650 (2003).

10. S. L. Qi, Z. Z. Chen, H. Fang, Y. J. Sun, L. W. Sang, X. L. Yang, L. B. Zhao, P. F. Tian, J. J. Deng, Y. B. Tao, T. J. Yu, Z. X. Qin, and G. Y. Zhang, "Study on the formation of dodecagonal pyramid on nitrogen polar GaN surface etched by hot H_3PO_4," Appl. Phys. Lett. **95**, 071114 (2009).

11. Y. Gao, T. Fujii, R. Sharma, K. Fujito, S. P. DenBaars, S. Nakamura, and E. L. Hu, "Roughening hexagonal surface morphology on laser lift-off (LLO) N-face GaN with simple photo-enhanced chemical wet etching," Jpn. J. Appl. Phys. **43**, L637–L639 (2004).

12. T. Fujii, Y. Gao, R. Sharma, E. L. Hu, S. P. DenBaars, and S. Nakamura, "Increase in the extraction efficiency of GaN-based light-emitting diodes via surface roughening," Appl. Phys. Lett. **84**, 855 (2004).

13. J.-H. Kim, C.-S. Oh, Y.-H. Ko, S.-M. Ko, K.-Y. Park, M. Jeong, J. Y. Lee, and Y.-H. Cho, "Dislocation-eliminating chemical control method for high-efficiency GaN-based light emitting nanostructures," *Cryst. Growth Des. 12*, 1292–1298 (**2012**).

14. J. S. Speck, and S. J. Rosner, "The role of threading dislocations in the physical properties of GaN and its alloys," Physica B **273-274**, 24-32 (1999).

15. T. J. Badcock, R. Hao, M. A. Moram, M. J. Kappers, P. Dawson, and C. J. Humphreys, "The effect of dislocation density and surface morphology on the optical properties of InGaN/GaN quantum wells grown on r-plane sapphire substrates," Jpn. J. Appl. Phys. **50**, 080201 (2011).

16. Y. B. Tao, T. J. Yu, Z. Y. Yang, D. Ling, Y. Wang, Z. Z. Chen, Z. J. Yang, G. Y. Zhang, "Evolution and control of dislocations in GaN grown on cone-patterned sapphire substrate by metal organic vapor phase epitaxy,"J. Cryst. Growth **315,** 183–187 (2011).

17. F. E. Williams, and H. Eyring, "The mechanism of the luminescence of solids," J. Chem. Phys. **15**, 289-304 (1947);

18. P. J. Dean, "Absorption and luminescence of excitons at neutral donors in gallium phosphide," Phys. Rev. **157**, 655–667 (1967)

19. J. H. Son , J. U. Kim , Y. H. Song , B. J. Kim , C. J. Ryu, and J.-L. Lee, "Design rule of nanostructures in light-emitting diodes for complete elimination of total internal reflection" Adv. Mater. **24**, 2259–2262 (2012).
20. P. Kozodoy, J. P. Ibbetson, H. Marchand, P. T. Fini, S. Keller, J. S. Speck, S. P. DenBaars, and U. K. Mishra, "Electrical characterization of GaN p-n junctions with and without threading dislocations," Appl. Phys. Lett. **73**, 975 (1998).

Optoelectronics

Mater. Res. Soc. Symp. Proc. Vol. 1538 © 2013 Materials Research Society
DOI: 10.1557/opl.2013.586

Hybrid III-V-on-Silicon Microring Lasers

Di Liang[1], Géza Kurczveil[1], Marco Fiorentino[1], Sudharsanan Srinivasan[2], David A. Fattal,
Zhihong Huang[1], John E. Bowers[2] and Raymond G. Beausoleil[1]
[1]Intelligent Infrastructure Lab, Hewlett-Packard Laboratories
Palo Alto, CA 94304, U.S.A.
[2]Department of Electrical and Computer Engineering, University of California
Santa Barbara, CA 93106, U.S.A.

ABSTRACT

Hybrid silicon laser is a promising solution to enable high-performance light source on large-scale, silicon-based photonic integrated circuits (PICs). As a compact laser cavity design, hybrid microring lasers are attractive for their intrinsic advantages of small footprint, low power consumption and flexibility in wavelength division multiplexing (WDM), etc. Here we review recent progress in unidirectional microring lasers and device thermal management. Unidirectional emission is achieved by integrating a passive reflector that feeds laser emission back into laser cavity to introduce extra unidirectional gain. Up to 4X of device heating reduction is simulated by adding a metal thermal shunt to the laser to "short" heat to the silicon substrate through buried oxide layer (BOX) in the silicon-on-insulator (SOI) substrate. Obvious device heating reduction is also observed in experiment.

INTRODUCTION

When faced with bandwidth, power, and signal integrity issues on conventional metal interconnects in silicon microelectronics, industry has committed to adopt high-capacity fiber-optic technology to build high-performance optical interconnect systems. Recently developed hybrid silicon platform has emerged as one of the most promising device platform for practical applications [1]. Among a variety of demonstrated hybrid silicon lasers, microring lasers were designed for their intrinsic advantages of small footprint, low power consumption and flexibility in wavelength division multiplexing (WDM), etc. Inset in Fig. 1 shows the schematic of a hybrid Si microring laser where III-V epitaxial layers are transferred to the (SOI) substrate to provide optical gain. A Si bus waveguide is used to extract a portion of optical power from the hybrid cavity [2]. The fabrication starts from transferring the thin InP-based III-V laser epitaxial layers (λ=1.515 µm) onto the SOI by a low-temperature O_2-assisted wafer bonding process. Upon blank depositing p-type metal contact and dielectric mask layer on the p-InGaAs layer, a self-aligned process is used to transfer microring pattern through p-type metal, III-V and all the way to silicon layer. Upon passivation the etched sidewall, another photolithography and similar but timed dry etching process is used to form the microring resonator mesa and expose the n-InP contact layer inside. Then n-type contact metal is placed inside the microring resonator, followed by probe-pad metal pad deposition. Detailed fabrication process and epitaxial layer structure can be referred to [2].

Directional bistability, the ability of a laser to operate either in the clock-wise (CW) or counter-clock-wise (CCW) mode due to structural symmetry, is a unique characteristic of ring lasers. While useful for some applications, bistability is undesirable for ring lasers used in optical interconnects. We demonstrate an unidirectional microring laser by without additional power consumption and chip complexity for the first time. Unidirectional emission is achieved by integrating a passive reflector that feeds laser emission back into laser cavity to introduce extra unidirectional gain. We show that the length of the passive reflector is a critical parameter in determining the lasing behavior [3].

To overcome the device heating bottleneck in hybrid silicon lasers, we also demonstrated the design and experimental evidence of device heating reduction by employing a metal thermal shunt. By etching through the buried oxide layer in silicon-on-insulator substrate and filling with high thermal conductive metal, device joule heating can be "shorted" to the silicon substrate. Simulation and direct experimental comparison in devices with and without thermal shunt clearly showed device heating reduction and device performance improvement [4].

UNIDIRECTIONAL HYBRID MICRORING LASER

Fig. 1 shows a typical light-current (LI) characteristic of a hybrid microring laser with a 50 μm diameter. Two integrated hybrid silicon photodetectors (inset of Fig. 1) capture CW and CCW laser emission separately. At low injection level (between Ith and 33 mA), the laser emits in both CW and CCW modes. At higher injection levels (between 33 and 40 mA), CW and CCW mode degeneracy breaks and the laser operates in the unidirectional bistability regime where one direction dominates in certain injection ranges [5]. The lasing direction can switch in a random fashion (e.g. ~38 mA in Fig 1).

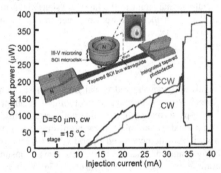

Fig. 1. LI characteristic of a diameter D=50 μm hybrid silicon microring laser. Inset: device schematic showing two integrated hybrid photodetectors.

A number of approaches have been demonstrated to achieve stable unidirectional lasing. "S-shape" ring resonator cavities are designed to introduce asymmetric round trip loss/gain [6]. Optical pulse injection from an external laser or light emitting diode (LED) has also been used to increase the net modal gain in one direction [5, 7, 8]. These approaches introduce additional optical loss or require an external light source, which either degrades laser performance or increases total system complexity and power consumption. Other designs that work for some cavity structures are not applicable to our devices [9].

We fabricated the unidirectional ring laser shown schematically in Fig. 2(a), in which an optical reflector at one end of the bus waveguide induces the laser to emit light toward the other end. In this structure, light emitted in the CW mode is partially coupled back to the CCW mode. The power circled back into the cavity leads to a photon density increase and unidirectional lasing in the desired direction (CCW here). Compared with external injection from another laser this approach does not require additional power and is free of wavelength/mode mismatch. Injection from an on-chip LED is easier to implement. However, because of the LED broad emission spectrum and low output power, it is not very effective in guaranteeing unidirectional operation.

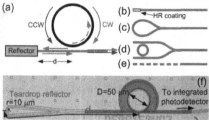

Fig. 2. (a) Schematic of a ring laser with an external reflector integrated on the bus waveguide. (b-e) Schematic of various passive reflector designs, (f) Image of a fabricated hybrid silicon microring laser with tear-drop reflector.

Different types of waveguide reflectors can be used to provide feedback. Figs. 2(b)-(e) show various possible implementations: a cleaved waveguide facet with high reflection coating, a teardrop reflector, a passive ring reflector, and a distributed Bragg reflector (DBR). We use the teardrop design shown in Fig. 2(f) because it is simple to fabricate and has a wide optical bandwidth. In order to minimize scattering loss, passive waveguide is 2 μm wide with an effective index of 3.12. Teardrop reflectors have been used as low-loss, high-reflectivity laser mirrors [10].

Fig. 3 shows the LI-voltage (LIV) characteristic of 50 μm-diameter lasers under a fixed stage temperature Tstage of 15 °C. The effective reflector length d (see Fig. 2) for this device was chosen to be 325 μm and 85 μm for Fig. 3(a) and (b), respectively. The inset shows an infrared (IR) image of the unidirectional laser at 30 mA injection current. Fig. 3(a) shows that a LI curve calculated using the laser model developed in Ref. [11] agrees well with the experimental data. The free-running lasers in Fig. 1 and the two unidirectional devices in Fig. 3 all exhibit similar threshold current and output power. Unlike free-running lasers in Fig. 1, thermal roll-over appeared in devices in Fig. 3 because a thermal shunt (see next section) integrated in free-running devices to reduce device heating was not employed here. This is an indication that most of the injected power in the unidirectional devices goes into the CCW lasing mode. The dips in the LI curves in Fig. 3 are attributed to the interference of the laser CCW mode and the back-reflected light from teardrop reflector, which is a function of the phase of the reflected light. In our devices, significant lasing in the suppressed mode can co-exist with dominant lasing as seen in the device of Fig. 1 for injection currents above 38 mA. When a reflector is added the light from the CW and CCW light will interfere at the coupler. Depending on the phase of the reflected light constructive or destructive interference will occur. Constructive interference maximizes the laser CCW output. Destructive interference reduces the desirable feedback and output of the laser.

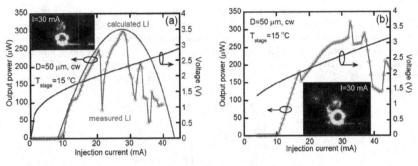

Fig. 3. Simulated and measured LIV characteristic of unidirectional hybrid silicon microring laser with reflector length d=325 μm (a) and 85 μm (b). Inset: IR image of devices unidirectional lasing in CCW mode under 30 mA injection current.

To confirm our understanding of the lasing behavior, we modeled the phase of the back-reflected light. Measured wavelength shifts as a function of the injected electrical power Pi is 0.06 nm/mW (data not shown). In Figs. 4(a) and (b) we overlay the calculated interference intensity with the laser output as a function of Pi. The phase was calculated using a round trip distance of 2d of 650 μm in (a) and 170 μm in (b). The dips in the output correspond to destructive interference between the CCW and reflected light. For regions without dips (e.g. at Pi=59 mW in (a) and Pi=57 mW in (b)) the CCW mode is still favored even in the absence of feedback and no dip is observed. To confirm this we monitored the lasing direction in the bus waveguide using a top down IR camera (insets in Fig. 3). A similar effect has also been observed in heterogeneous micro-disk lasers [12]. The laser power instabilities can be eliminated using several strategies: further decreasing the reflector round-trip distance to reduce the phase shift, a better control of the phase shift through design and fabrication, and reducing device heating to minimize lasing wavelength shift [4]. Reducing the sidewall roughness is also critical not only for enhancing device performance, but also for minimizing back-reflection to maintain stable unidirectional lasing. The approach we demonstrated here is also applicable to devices with "memory" effects. Because the lasing direction is independent of the external feedback [5] phase these devices would not show power instabilities.

THERMAL MANAGEMENT

While a low continuous-wave (cw) lasing threshold of less than 4 mA has been observed on a 50 μm in diameter device, however, device heating stands as the primary limiting factor to further scale down the dimension for lower power consumption operation [11]. Fig. 1(b) represents the simulated temperature rise in the device active region at threshold and the corresponding thermal impedance as a function of ring diameter. The simulated thermal impedance for diameter D=50, 25 to 15 μm devices are 465.2, 1253.4 and 1782 °C/W, respectively. They agree well with experimental values obtained using a technique in Ref. [13]. The simulated temperature profile in Fig. 4 shows that buried oxide (BOX) layer, because of its low thermal conductivity (k_{SiO2}=1.3 W/m-K), is the main thermal barrier to block thermal dissipation to the Si substrate (k_{Si}=130 W/m-K).

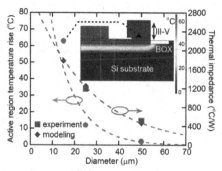

Fig. 4. Calculated active region temperature rise at threshold, and experimental and modeling thermal impedance as a function of device dimension. Inset: D=15 μm device temperature profile at pulse threshold 9.4 mA.

Having determined that the BOX layer in the substrate is the primary limiting factor for efficient heat extraction we designed a thermal shunt (TS) to guide the heat down to the Si substrate, similar to a previously demonstrated poly-Si thermal shunt for the same hybrid platform [13]. By etching a trench through the BOX layer around the microring cavity and filling it with material of high thermal conductivity, e.g., Au (k_{Au}=317 W/m-K, this work) or Cu (k_{Cu}=401 W/m-K), the heat generated in III-V cavity can be "shorted" to the Si substrate as shown in Fig. 5. Because the metal absorbs light we need to interpose a transparent cladding (SiO_2 in this work) between the III-V and the metal. In design A, metal shunt attaches to the D=15 μm resonator sidewall to extract heat diffusing through the 500 nm-thick SiO_2 cladding, which reduces active region temperature to 41.7 °C from 63 °C at the same injection current. A better design would be having TS connect to the top p-metal contact as shown in TS design B. Then majority of heat can be quickly extracted rather than go through the cladding. In this case the temperature rise reduces to 15.5 °C. Fig. 5(a) shows the simulated temperature rise vs. injection current for D=15 μm devices without TS, with TS design A, and B (500 nm SiO_2 cladding). A 33% and 4X device heating reduction are achievable for design A and B, respectively. In this batch of devices only TS designs A (Fig. 5(c)) is implemented for the easiness of fabrication. Obviously, thinner SiO_2 cladding results in more efficient heat extraction, particularly for design A, however, more metal absorption loss will be introduced. Fig. 6(b) shows the simulated temperature rise and metal absorption loss vs. SiO_2 cladding thickness for a D=15 μm device. A good trade-off SiO_2 thickness is around 400 nm to keep metal absorption well below 1 cm^{-1}, and 500 nm is used in this batch of devices.

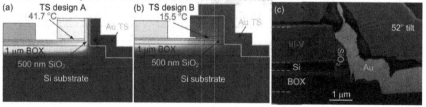

Fig. 5. Simulated temperature profile of a D=15 μm device with TS design A (a) and B (b). (c) FIB SEM cross-sectional image of a device with TS design A.

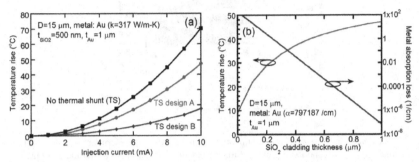

Fig. 6. (a) Simulated temperature rise at device active region vs. injection current for a D=15 μm device without TS, with TS design A and B. (b) Simulated temperature rise and metal absorption loss vs. SiO₂ cladding thickness.

Fig. 7 shows the light-current (LI) characteristics of D=15 (a), 25 (b) and 50 (c) μm devices. A pair of devices with same structure except one with TS and the other without are fabricated together (Fig. 7(b) inset) and share the same bus waveguide and two on-chip photodetectors for fair comparison. For D=15 μm, though cw lasing is not observed for both cases device without TS shows much early thermal rollover of spontaneous emission than the one with TS. Cw lasing at 15 °C stage temperature only observed on D=25 μm ones with TS, not on those without TS. Since device heating in D=50 μm devices is not that serious at low current injection, devices without TS also lase, but with much lower output power at high current injection level (Fig. 7(c)). Kinks are due to lasing directional switching and only single-side power is shown here. Slight higher threshold on device with TS is probably due to additional metal loss because III-V partially touches Au from imperfect fabrication as shown in Fig. 5(c).

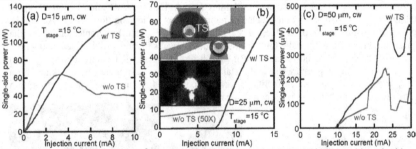

Fig. 7. LI characteristics (cw, 15 °C) of (a) D=15, (b) 25 and (c) 50 μm devices with and without TS. Inset in (b): layout of a pair of devices with and without TS; Infrared image of a lasing D=25 μm device with TS.

CONCLUSIONS

In this paper, we reviewed recent progress in two aspects of the hybrid Si microring laser, i.e., unidirectional lasing and metal thermal shunt. Using feedback from a passive teardrop reflector we have shown that unidirectional lasing can be achieved. This simple design achieves unidirectional operation in ring lasers without additional losses or power consumption. We also demonstrated the design and experimental evidence of device heating reduction by employing a TS through the BOX layer. Simulation shows up to 4X improvement and direct comparison in experiment indicates the clear device performance improvement as well.

REFERENCES

[1] A. W. Fang, H. Park, O. Cohen, R. Jones, M. J. Paniccia, and J. E. Bowers, *Optics Express,* 14, 9203-9210, 2006.

[2] D. Liang, M. Fiorentino, T. Okumura, H.-H. Chang, D. Spencer, Y.-H. Kuo, A. W. Fang, D. Dai, R. G. Beausoleil, and J. E. Bowers, *Optics Express,* 17, 20355-20364, 2009.

[3] D. Liang, S. Srinivasan, D. A. Fattal, M. Fiorentino, Z. Huang, D. T. Spencer, J. E. Bowers, and R. G. Beausoleil, *IEEE Photonics Technology Letters,* 24, 1988-1990, 2012.

[4] D. Liang, S. Srinivasan, M. Fiorentino, G. Kurczveil, J. E. Bowers, and R. G. Beausoleil, in *IEEE Optical Interconnects Conference.* TuD2 Santa Fe, NW, USA, 2012.

[5] M. Sorel, P. J. R. Laybourn, G. Giuliani, and S. Donati, *Applied Physics Letters,* 80, 3051-3053, 2002.

[6] J. P. Hohimer, G. A. Vawter, and D. C. Craft, "Unidirectional operation in a semiconductor ring diode laser," *Applied Physics Letters,* 62, 1185-1187, 1993.

[7] C. J. Born, S. Yu, M. Sorel, and P. J. R. Laybourn, "Controllable and stable mode selection in a semiconductor ring laser by injection locking," in *Conference on Lasers and Electro-Optics,* 2003.

[8] A. W. Fang, R. Jones, H. Park, O. Cohen, O. Raday, M. J. Paniccia, and J. E. Bowers, *Optics Express,* 15, 2315-2322, 2007.

[9] Q. J. Wang, C. Yan, N. Yu, J. Unterhinninghofen, J. Wiersig, C. Pflugl, L. Diehl, T. Edamura, M. Yamanishi, H. Kan, and F. Capasso, *Proceedings of the National Academy of Sciences,* USA107, 22407-22412, 2010.

[10] Y. Zheng, D. K.-T. Ng, Y. Wei, W. Yadong, Y. Huang, Y. Tu, C.-W. Lee, B. Liu, and S.-T. Ho, *Applied Physics Letters,* 99, 011103, 2011.

[11] D. Liang, M. Fiorentino, S. Srinivasan, S. T. Todd, G. Kurczveil, J. E. Bowers, and R. G. Beausoleil, *IEEE Photonics Journal,* 3, 580-587, 2011.

[12] T. Spuesens, F. Mandorlo, P. Rojo-Romeo, P. Regreny, N. Olivier, J. M. Fedeli, and D. Van Thourhout, *Journal of Lightwave Technology,* 30, 1764-1770, 2012.

[13] M. N. Sysak, D. Liang, M. Fiorentino, R. G. Beausoleil, G. Kurceival, M. Piels, R. Jones, and J. E. Bowers, *IEEE Journal of Selected Topics in Quantum Electronics,* 17, 1490-1498, 2011.

Mater. Res. Soc. Symp. Proc. Vol. 1538 © 2013 Materials Research Society
DOI: 10.1557/opl.2013.549

Improved Yellow Light Emission in the Achievement of Dichromatic White Light Emitting Diodes

Zhao Si[1], Tongbo Wei[1], Jun Ma[1], Ning Zhang[1], Zhe Liu[1], Xuecheng Wei[1], Xiaodong Wang[1], Hongxi Lu[1], Junxi Wang[1] and Jinmin Li[1]

[1]Research and Development Center for Semiconductor Lighting, Institute of semiconductors, Chinese Academy of Sciences, P.O. Box 912, Beijing 100083, People's Republic of China

ABSTRACT

A study about the achievement of dichromatic white light-emitting diodes (LEDs) was performed. A series of dual wavelength LEDs with different last quantum-well (LQW) structure were fabricated. The bottom seven blue light QWs (close to n-GaN layer) of the four samples were the same. The LQW of sample A was 3 nm, and that of sample B, C and D were 6 nm, a special high In content ultra-thin layer was inserted in the middle of the LQW of sample C and on top of that of sample D. XRD results showed In concentration fluctuation and good interface quality of the four samples. PL measurements showed dual wavelength emitting, the blue light peak position of the four samples were almost the same, sample A with a narrower LQW showed an emission wavelength much shorter than that of sample B, C, D. EL measurement was done at an injection current of 100 mA. Sample A only showed LQW emission due to holes distribution. Because of wider LQW, the emission wavelength of sample B, C and D was longer and peak intensity was weaker. Sample D with insert layer on top of LQW showed strongest yellow light emission with a blue peak. As the injection current increased, sample A showed highest output light power due to narrower LQW. Of the other three samples with wider LQW, sample D showed highest output power. Effective yellow light emission has always been an obstacle to the achievement of dichromatic white LED. Sample D with insert layer close to p-GaN can confine the hole distribution more effectively hence the recombination of holes and electrons was enhanced, the yellow light emission was improved and dichromatic white LED was achieved.

INTRODUCTION

Recently, tremendous progress has been made in InGaN/GaN based blue, green and shorter-wavelength light-emitting diodes (LEDs) [1-6]. It is of great interest to design an monolithic white LED without the need of a phosphor converter for long-wavelength light [7-12]. Although the band gap of InGaN can cover the wide spectral regime from near ultraviolet to near infrared principally [13-19], this material system still faces severe difficulties in attempting to produce LEDs emitting at wavelength longer than 530 nm. To achieve green and red emission, the In content in $In_xGa_{1-x}N$ quantum-wells (QWs) should be higher than 20%, however, the internal quantum efficiency decrease significantly as the emitting wavelength becomes longer. With the increase of In content in InGaN QWs, the lattice mismatch between GaN and InGaN material becomes more serious. This leads to the In-phase separation and quantum-confined Stark effect (QCSE) [20]. Furthermore, the QCSE decrease effective band gap, which is beneficial in achieving longer wavelength emission, but the overlap of electron and hole wave function is seriously decreased hence the internal quantum efficiency and emission efficiency is low [21,22]. By controlling well thickness, the long wavelength emission can be achieved at relatively lower In content, however, the overlap of wave function decreases with the increasing of QW thickness. According to recent reports [23,24], QWs of irregular structures may lead to carrier redistribution in QWs and overlap of electron and hole wave function enhancement. Of all these irregular structures, the ultra-thin high In content insert layer seems useful and applicable in our wide QW experiments. Theoretically, the use of special designed insert layer in wide QW can improve the overlap of carrier wave function, and higher emitting efficiency is achieved. Furthermore, the enhanced yellow light emission is especially important in the fabrication of dichromatic white light emitting diodes. In this paper, we use different insert layer structures to improve wide yellow light QW emitting efficiency in blue and yellow dual wavelength light emitting diodes to acquire dichromatic white light emitting diodes.

EXPERIMENT

In our experiment, four samples with different last quantum-well (LQW) structures were grown by metal organic chemical vapor deposition (MOCVD) on the c-plane sapphire substrates with thrimethylgallium (TMGa), thrimethylindum (TMIn), ammonia (NH3), silane (SiH4) and biscyclopentadienylmagnesium (CP2Mg) as gallium, indium, nitrogen, n-type and p-type dopant source, respectively. In each sample, after the deposition of low-temperature nucleation layer, the temperature was increased to 1050 °C to grow the n-GaN layer, followed by the active region. The active region consisted of seven periods of InGaN/GaN QWs emitting in blue light region with top one QWs emission wavelength in the yellow light region. For the bottom seven blue QWs, the wells were grown at 730 °C while for the top one yellow light emission QWs at 700 °C and all barriers were grown at 830 °C. In sample B, C and D, the width of top one yellow QW is 6 nm and that of bottom seven blue light QW is only 3 nm, sample B was grown without insert layer and a special high In content ultra-thin layer was inserted in the middle of the LQW of sample C and on top of that of sample D. The insert layer was grown at a lower temperature of 680°C. Another sample (sample A) with 3 nm width LQW was also grown as comparison. The thickness of the insert layer is about 1 nm and the consistution of the layer is $In_{0.25}Ga_{0.75}N$. The doping concentration of n- and p-GaN layer was 1×10^{19} and $1 \times 10^{20} cm^{-3}$, respectively. The chip size of each sample is 250 μm ×500 μm.

DISCUSSION

As shown in the inset of Figure1, the XRD measurements show peak location shifts and good interface quality of different samples. The peak location shifts among different samples is due to different structures of LQW. Compared with sample A, the different peak location of the other three samples is due to different width of LQW. Furthermore, the peak location shifts of sample C and D is caused by the exist of special designed high In content insert layer in the LQW. As shown in the room-temperature photoluminescence (PL) spectra of the four samples, sample A shows a strong peak emission located at the wavelength of 494 nm due to the thinner LQW. Sample B with a wider LQW shows a longer and weaker peak emission in the yellow light region. The increase of QW width leads to the decrease of electron and hole wave function overlap as well as carrier redistribution caused by energy-level filling variation in QW. In the yellow light region, the peak intensity varies with the change of insert layer structures. Sample B without insert layer shows weakest peak intensity while that of sample D is the strongest. The peak intensity increment of sample C and D is due to carrier redistribution caused by insert layer structure. Sample C with insert layer in the middle of LQW shows relatively weaker peak intensity. The crystal quality becomes worse with the increasing of In content. For sample C, the crystal quality is seriously affected by the high In content insert layer in the middle of LQW. As confirmed by the electroluminescence (EL) measurement results, the weaker blue peak of the four samples is due to the QWs stacking sequence and the limitation of laser pulse power in the measurements. The distinct different blue peak intensity between sample A and the other three samples is caused by different structures of LQW.

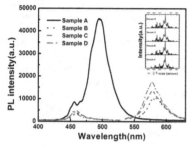

Figure 1. PL measurement results of the four samples. The inset is the XRD measurement results of the four samples.

Electroluminescence (EL) spectra of these four samples (Figure 2) were measured at a forward current of 100 mA. As can be seen in the log scaled EL intensity of the four samples (Figure 2(a)), sample A shows only one emission peak of LQW due to hole distribution. Although the LQW emission of sample B and C is much weaker, all samples show dual wavelength emitting. The dual wavelength emitting of the other three samples is caused by the wider LQW. As the LQW becomes wider, the overlap of electron and hole wave function decreases, the recombination of electrons and holes becomes weaker hence more holes can transport into bottom QWs under forward current condition. In the EL measurement results, the relative peak intensity of LQW is in accord with the PL measurement results, sample D with ultra-thin high In content insert layer on top of LQW shows strongest yellow light emission in PL and EL measurements. This indicates that, the insert layer structure in LQW could improve yellow light emission.

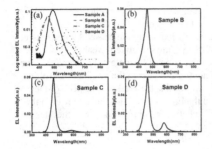

Figure 2. Log scaled EL intensity of the four samples (a) and EL measurement results of sample B, C and D (b), (c), (d).

Figure 3 shows the light output power at different injection current level. Due to narrower LQW, sample A shows highest output power with increasing current level. Of the other three wider LQW samples, sample D shows relative higher output power. The enhanced output power of sample C and D indicates the effect of ultra-thin insert layer in improving emission efficiency. The different performance between sample C and D illustrates that the emission efficiency improvement depends intensely on LQW structures. The inset of figure 3 is the unpackaged wafer white light emitting picture. In our 250μm ×500 μm sized chip, the CIE coordinate is (0.33,0.33) and the color rendering index is 65 at the injection current of 55 mA.

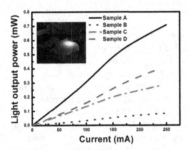

Figure 3. Output light power of different samples with the increase of injection current. The inset is the picture of white light emission.

CONCLUSIONS

In conclusion, four samples with different LQW structures were grown. As the LQW becomes wider, the emission peak wavelength gets longer and peak intensity is weaker. According to the measurement results, the wider LQW yellow light emission intensity can be improved by adopting a special designed ultra-thin high In content layer. Furthermore, the location of insert layer still has great effect on the emission efficiency improvement. The sample with ultra-thin layer on top of LQW shows better performance than that in the middle. Finally, the dichromatic white light emitting diodes are achieved by enhanced yellow light emission.

ACKNOWLEDGMENTS

This research was supported by National Nature Science Foundation of China, under Grant No. 61274040

REFERENCES

1. S. Nakamura, M. Senoh, S. Nagahama, N. Iwasa, T. Yamada, T. Mukai, Y.Sugimoto, and H. Kiyoku, Appl. Phys. Lett. **69**, 4056 (1996).
2. Y. D. Qi, H. Liang, D. Wang, Z. D. Lu, W. Tang, and K. M. Lau, Appl.Phys. Lett. **86**, 101903 (2005).
3. F. A. Ponce and D. P. Bour, Nature **386**, 351 (1997).
4. S. J. Chang, W. C. Lai, Y. K. Su, J. F. Chen, C. H. Liu, and U. H. Liaw, IEEE J. Sel. Top. Quantum Electron. **8**, 278 (2002).
5. T. Detchprohm, M. Zhu, W. Zhao, Y. Wang, Y. Li, Y. Xia, and C. Wetzel, Phys. Status Solidi C **6** , S840 (2009).
6. Y. L. Lai, C. P. Liu, Y. H. Lin, Y. H. Hsueh, R. M. Lin, D. Y. Lyu, Z. X. Peng, and T. Y. Lin, Nanotechnology **17**, 3734 (2006).
7. L. S. Wang, Z. Q. Lu, S. Liu, and Z. C. Feng, J. Electron. Mater. **40**, 1572 (2011).
8. Y. D. Qi, H. Liang, W. Tang, Z. D. Lu, and K. M. Lau, J. Cryst. Growth. 272, 333 (2004).
9. M.Yamada, Y. Narukawa, H. Tamaki, Y. Murazaki, and T. Mukai, IEICE Trans. Electron. **E88-C**, 1860 (2005).
10. H. S. Chen, D. M. Yeh, C. F. Lu, C. F. Huang, Y. C. Lu, C. Y. Chen, J. J. Huang, and C. C. Yang, Appl. Phys. Lett. **89**, 093501 (2006).
11. B. Damilano, N. Grandjean, C. Pernot, and J. Massies, Jpn. J. Appl. Phys. 40, L918 (2001).
12. M.Yamada, Y. Narukawa, and T. Mukai, Jpn. J. Appl. Phys. 41, L246 (2002).
13. K. Okamoto and Y. Kawakami, IEEE J. Sel. Top. Quantum Electron. **15**, 1199(2009).
14. H. Zhao, J. Zhang, G. Liu, and N. Tansu, Appl. Phys. Lett. **98**, 151115 (2011).
15. T. Jung, L. K. Lee, and P.C. Ku, IEEE J. Sel. Top. Quantum Electron. **15**, 1073(2009).
16. W. Lee, J. Limb, J.H. Ryou, D. Yoo, M. A. Ewing, Y. Korenblit, and R. D. Dupuis, J. Disp. Technol. **3**, 126(2007).
17. C. H. Chao, S. L. Chuang, and T. L. Wu, Appl. Phys. Lett. **89**, 091116(2006).
18. K. McGroddy, A. David, E. Matioli, M. Iza, S. Nakamura, S. DenBaars, J. S. Speck, C. Weisbuch, and E. L. Hu, Appl. Phys. Lett. **93**, 103502(2008).
19. S. Chhajed, W. Lee, J. Cho, E. F. Schubert, and J. K. Kim, Appl. Phys. Lett. **98**, 071102 (2011).
20. H. S. Chen, D. M. Yeh, C. F. Lu, C. F. Huang, W. Y. Shiao, J. J. Huang, C. C. Yang, I. S. Liu, and W. F. Su, IEEE Photonics Technol. Lett. **18**, 1430 (2006)
21. T. Wang, D. Nakagawa, J. Wang, T. Sugahara, and S. Sakai, Appl. Phys. Lett. **73**, 3571 (1998).
22. T. Wang, J. Bai, S. Sakai, and J. K. Ho, Appl. Phys. Lett. **78**, 2617(2001).
23. H. Zhao, G. Liu, J. Zhang, J. D. Poplawsky, V. Dierolf, and N. Tansu, Optics Express. **19**, A991(2011).
24. S. Che, A. Yuki, H. Watanabe, Y. Ishitani, and A. Yoshikawa, Appl. Phys. Express. **2**, 021001(2009)

Mater. Res. Soc. Symp. Proc. Vol. 1538 © 2013 Materials Research Society
DOI: 10.1557/opl.2013.1045

Modification of the Optical and Electrical Properties CdS Films by Annealing in Neutral and Reducing Atmospheres

J. Pantoja Enriquez[1,2], G. Pérez Hernandez[3], X. Mathew[4], G. Ibáñez Duharte[1], J. Moreira[1], J. A. Reyes Nava[1], J. J. Barrionuevo[1], L. A. Hernandez[1], R. Castillo[2], P. J. Sebastian[4]

[1] Centro de Investigación y Desarrollo Tecnológico en Energías Renovables, Universidad de Ciencias y Artes de Chiapas, Tuxtla Gutiérrez, Chiapas, México.
[2] Universidad Politécnica de Chiapas, Tuxtla Gutiérrez, Chiapas, México.
[3] Universidad Juárez Autónoma de Tabasco. Villahermosa, Tabasco, México.
[4] Centro de Investigación en Energía, UNAM, Temixco, Morelos, México.

ABSTRACT

Cadmium sulfide (CdS) films were deposited onto glass substrates by chemical bath deposition (CBD) from a bath containing cadmium acetate, ammonium acetate, thiourea, and ammonium hydroxide. The CdS thin films were annealed in argon (neutral atmosphere) or hydrogen (reducing atmosphere) for 1 h at various temperatures (300, 350, 400, 450 and 500 °C). The changes in optical and electrical properties of annealed treated CdS thin films were analyzed. The results showed that, the band-gap and resistivity depend on the post-deposition annealing atmosphere and temperatures. Thus, it was found that these properties of the films, were found to be affected by various processes with opposite effects, some beneficial and others unfavorable. The energy gap and resistivity for different annealing atmospheres was seen to oscillate by thermal annealing. Recrystallization, oxidation, surface passivation, sublimation and materials evaporation were found the main factors of the heat-treatment process responsible for this oscillating behavior. Annealing over 400 °C was seen to degrade the optical and electrical properties of the film.

INTRODUCTION

Cadmium sulfide is n-type direct band gap semiconductor and is considered as the best-suited window material for CdTe and CuIn(Ga)Se₂ solar cells [1–4]. Chemical bath deposition technique is the most cost-effective and relatively easy for solar cell fabrication. Cadmium sulfide thin films obtained by this method are uniform, adherent, and reproducible [5,6]. CdS/CdTe thin film solar cells with efficiencies higher than 16% has been reported by using CBD CdS films [7].

Low series resistance, high transmittance and optimum band gap are very important requirements for CdS thin film to be considered a suitable window layer for solar cell. So the CdS should be conductive, thin to allow high transmission and uniform to avoid short circuit effects [8]. However, generally as deposited CdS thin film show high resistivity and optical transmittance, which can be improved. Post deposition heat treatment is one of fundamental steps to improve the electrical and optical properties of CdS thin films. The annealing treatments promotes recrystallization, grain growth and improve the optical and electrical properties of the films and hence the device, having an important role, the temperature and the annealing atmosphere.

Despite the fact that there are some studies, which report the influence of heat treatment in different atmospheres on the physical properties of CdS thin films [9-13], there are however many issues that still need to be understood. The post-deposited film's characterization is still an open subject. In this paper we discuss the influence of the annealing in argon and hydrogen atmospheres on the optical and electrical properties of chemically deposited CdS thin films.

EXPERIMENT

The CdS thin films were deposited on glass slides by the CBD technique described in a previous work [4], from a chemical bath containing 0.033M cadmium acetate (Cd(OOCCH3)2.2H2O), 1M ammonium acetate (CH$_3$CO$_2$NH$_4$), 0.067M thiourea (H$_2$NCSNH$_2$) and 28–30% ammonium hydroxide (NH4OH). The temperature of the bath was maintained constant at 90°C. All the films used in the present study were deposited simultaneously from the same bath. The films were cut into equal portions (1 cm^2) to perform the heat treatments, and fresh samples were used for each annealing temperatures for 60 min. The films annealed in argon and hydrogen at 300°C, 350°C, 400 °C, 450 °C and 500°C for 60 minutes were used in this investigation.

The CdS thin films used in this experiment are about 80nm thick and were measured using an Alpha Step 100 surface profiler. The electrical resistivity of the CdS films was measured at room temperature using the two-probe technique. For this purpose, silver paint electrodes of 2mm length at 2mm separation were painted on the samples in a coplanar configuration. The current were measured using an HP4140B pico ammeter/DC voltage source with an applied voltage of 100 V. Prior to this measurement, the samples were kept in dark for 1 day to empty all trapping states. The optical transmittance spectra of chemical bath deposited CdS thin films were measured at room temperature by using a double beam Shimadzu UV-VIS-NIR spectrophotometer in the wavelength range of 300–2500 nm. The blank measurement was performed on glass substrates as baseline spectra.

RESULTS AND DISCUSSION

Optical properties

Figure 1 shows the transmittance spectra in the wavelength range of 300–2500 nm for the as deposited CdS thin films and annealed in argon and hydrogen atmosphere at different temperatures. Figure 1a shows the transmittance spectra of CdS films annealed in argon atmosphere at temperatures of 300, 350, 400, 450 and 500 °C. Figure 1b is the transmittance spectra of films annealed in hydrogen atmosphere at temperatures of 300, 350, 400, 450 and 500 °C. It is shown that the transmission of these films increases rapidly within the range 400-500nm reaching the maximum value and approximately remains constant at near-infrared region.

The spectrum can be roughly divided into two regions: a transparent region and a strong absorption zone in the UV range. In a transparent region (low absorption), all the samples show optical transparency (80 % to 90% approximately) in the spectral region above 1000 nm. The absorption edge shifts towards higher wavelength region with the increase annealing temperature and the transmission is found to decrease with increasing annealing temperature.

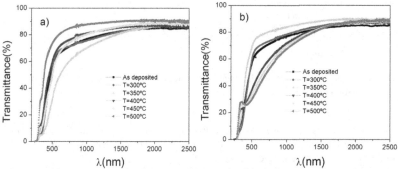

Figure 1. Optical transmittance of the CdS thin films as-deposited and after the annealing process as a function of the annealing temperature. a) in argon and b) in hydrogen atmospheres.

Figure 2 shows the average transmittance in the range of 500–1000 nm of CdS films annealed in argon and hydrogen atmospheres at temperatures of 300, 350, 400, 450 and 500 °C. It is shown that the average transmission was seen to oscillate by thermal annealing; in argon atmosphere tends to increase and in hydrogen the tendency is to decrease with annealing temperature.

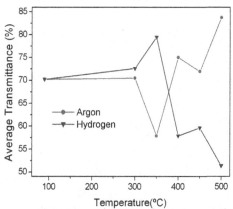

Figure 2 The average transmittance in the range of 500–1000 nm of CdS films annealed in argon and hydrogen atmospheres at temperatures of 300, 350, 400, 450 and 500 °C.

Due to reduced thickness of CdS thin films during the annealing process, it is expected that the transmittance would increase after annealing, however it has been observed decrease of transmittance with increasing annealing temperature in some cases. The decrease of transmittance might be due to the stoichiometry change of CdS resulting from cadmium and sulfur evaporation.

The absorption coefficient of the films was calculated from the transmittance spectra near the absorption edge using the relation;

$$\alpha = -\frac{Ln(1/T)}{d}$$ (1)

where $T = \frac{I_t}{I_i}$ is the transmittance.

The absorption coefficient α of the material can be related to the incident photon energy hv through the relation [14-17];

$$\alpha h v = A\left(h v - E_g\right)^{\frac{1}{2}}$$ (2)

Where A is a constant, h is the Planck's constant, v is the frequency and E_g is the band gap. The band gap of the films was determined by plotting $(\alpha h v)^2$ vs. hv. The intercept of the straight-line portion of the graph on the hv axis gives the value of E_g.

Figure 3 shows the variation of band gap energy of as-deposited and annealed CdS thin films in argon and hydrogen atmospheres at temperatures of 300, 350, 400, 450 and 500 °C. It can be seen that the band gap oscillates by thermal annealing. For the CdS thin films annealing in argon atmosphere the tendency of band gap energy is to increase with the annealing temperature. One can see from Figure 3 that after annealing, the band gap of the CdS thin films reached a maximum value at 350 °C and then oscillates above 350 °C. For the CdS thin films annealing in hydrogen atmosphere the tendency of band gap energy is to decrease with the annealing temperature.

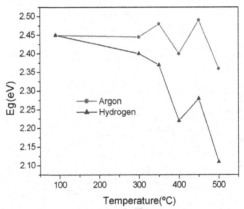

Figure 3. Band gap energy of as-deposited and annealed CdS thin films in argon and hydrogen atmospheres at temperatures of 300, 350, 400, 450 and 500 °C.

The results presented above, show that the behavior of CdS films annealed in argon differ from those annealed in hydrogen atmosphere. In the case of films annealed in argon, takes place simple evaporation of sulfur and cadmium, which influences the composition of films at enough high temperature. For films annealed in a hydrogen atmosphere evaporation of cadmium and sulfur occurs at a lower temperature as a result of chemical reduction forming metallic Cd and free S_2. On the other hand in hydrogen atmosphere, the cadmium excess evaporates probably in the first place, and then sulfur. At higher temperatures, there is a process of CdS sublimation, but in case of annealing in argon, such phenomenon does not take place [9].

Resistivity

The resistivities of the films under dark were measured and the results are shown in Figure 4, as a function of heat treatment temperature in neutral (argon) and reducing (hydrogen) atmospheres. As it can be seen in this figure, heat treatment in hydrogen reduces the resistivity of CdS thin film by one order of magnitude in comparison with as-deposited films at temperature of $350°C$, this reduction can be attributed by passivation of adsorbed oxygen in the grain boundaries. For annealing temperature higher than $350°C$ in hydrogen atmosphere the resistivity of CdS thin films oscillates by thermal annealing, with a tendency to increase with temperature. Heat treatment in argon reduces the resistivity of CdS thin film by three orders of magnitude in comparison with as-deposited films at temperature of $300°C$. For annealing temperature higher than $300°C$ in argon atmosphere the resistivity of CdS thin films oscillates by thermal annealing, with a tendency to decrease with annealing temperature.

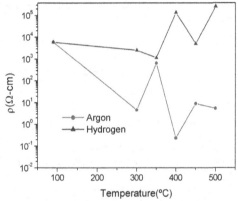

Figure 4. Dark resistivity of as-deposited and annealed CdS thin films in argon and hydrogen atmospheres at temperatures of 300, 350, 400, 450 and 500 °C.

Desorption of oxygen in the grain boundaries could result in a decrease of the potential barrier between grains and consequently in an increase in the carrier mobility and has been considered as the main influence of H_2 annealing of CdS. On the other hand, desorption of oxygen in sulphur sites would result in donor levels with a consequent increase of the carrier concentration. Increases in the carrier concentration and mobility are observed when annealed up to $300°C$. At temperatures higher than $300°C$ the resistivity drastically increased with temperature, accompanied with a slight Cd deposition in the cold regions of the tube. This observation supports the assumption that at high temperatures Cd starts to leave the film, compensating the effect of S_2 vacancies on carrier concentration [10].

Hydrogen annealing at moderate temperatures (200–250 °C) has been shown to cause significant decreases in resistivity attributed to; reduction of oxides, recrystallization and grain boundary passivation, this being reversed at higher temperatures (300 °C) on account of stoichiometry changes [9].

CONCLUSIONS

We have investigated the influence of post-deposition annealing in argon and hydrogen atmospheres at temperatures of 300, 350, 400, 450 and 500 °C on the optical and electrical properties of CBD prepared CdS thin films. The behavior of CdS films annealed in argon differ from those annealed in hydrogen atmosphere. During annealing the optical and electrical properties are affect for various processes. The first process is the phase transition from the mixed phase to the hexagonal phase (at temperature of 300°C approximately). The second process is grain boundary passivation by hydrogen annealing at moderate temperatures. The third is recrystallization and grains disintegration. The fourth process is sulfur and cadmium evaporation that affect films stoichiometry and may cause creation of empty spaces in the film. The final process may be the oxidation and degradation at high temperature. The annealing at lower temperature promotes recrystallization; however, at higher temperatures the loss of material becomes a very important factor.

ACKNOWLEDGMENTS

This work was supported partially by the project SEP-CONACYT CB-2007/83960.

REFERENCES

1. R.H. Bube, Photovoltaic Materials, (Imperial College Press, London, 1998).
2. D. Bonnet, Int. J. Solar Energy 12, 1(1992).
3. T.L. Chu, S.S. Chu, Solid-State Electron. 38, 533 (1994).
4. J. Pantoja Enríquez and Xavier Mathew. Solar Energy Materials & Solar Cells 76, 313 (2003).
5. R.S. Mane, C.D. Lokhande, Mater. Chem. Phys. 65, 1 (2000).
6. N. Romeo, A. Bosio, V. Canevari, A. Podesta, Sol. Energy 77, 795 (2004).
7. X. Wu, Solar Energy 77 (2004) 803.
8. J. Lee, Appl. Surf. Sci. 252 (2005) 1398–1403.
9. K.V. Zinoviev, O. Zelaya-Angel. Materials Chemistry and Physics 70, 100 (2001).
10. E. Vasco, E. Puron, O. de Melo, Materials Letters 25 (1995) 205-207.
11. H. Metin and R. Esen, J. Cryst. Growth, 258, 141(2003).
12. J. Han, C. Liao, T. Jiang, G. Fu, V. Krishnakumar, C. Spanheimer, G. Haindl, K. Zhao, A. Klein, W. Jaegermann. Materials Research Bulletin 46 (2011) 194–198.
13. J. Hiie, K. Muska, V. Valdna, Mikli, Taklaja, Gavrilov. Thin Solid Films 516, 7008 (2008).
14. Tauc J. Amorphous and Liquid Semiconductors, New York Plenum (1974).
15. Mott N.F, Davis E.A. Electronic Processes in Non-Crystalline Materials Calendron Press Oxford 1979.
16. J.J. Pankove, Optical Processes in Semiconductors, Prentice-Hall, Englewood C., NJ, 1971.
17. P. W. Davis and T.S. Shilliday, Phys. Rev. 118, 1020 (1960).

Mater. Res. Soc. Symp. Proc. Vol. 1538 © 2013 Materials Research Society
DOI: 10.1557/opl.2013.1010

Carrier density in p-type ZnTe with nitrogen and copper doping

Maryam Abazari, Faisal R Ahmad, Kamala C Raghavan, James R Cournoyer, Jae-Hyuk Her, Robert Davis, John Chera, Vince Smentkowski, and Bas A Korevaar

GE Global Research Center, 1 Research Circle, Niskayuna, NY 12309

Abstract

In this paper, we demonstrate deposition methods and conditions that allow the control of the electrical properties of doped ZnTe grown by RF magnetron sputtering using both nitrogen and copper as dopants. The carrier density of the films was characterized using a van der Pauw Hall effect measurement method. We demonstrate how the concentration of nitrogen in the plasma during the growth of the film impacts the conductivity of the ZnTe films. Films with hole concentrations in excess of 10^{18} cm^{-3} and a high degree of crystallinity were successfully grown. Similarly, we demonstrate that the hole concentration in the Cu-doped ZnTe can be varied by varying the amount of copper introduced in the films. We also observe that annealing the copper doped ZnTe films increases the carrier density, whereas annealing the nitrogen doped ZnTe films causes a decrease in carrier concentration and conductivity.

Keywords: Zinc Telluride, sputtering, thin film, doping

Introduction

High efficiency low-cost manufacturable solar cells are reported using cadmium telluride (CdTe) as the absorber layer [1-3]. However, formation of a good ohmic back contact to the CdTe film has been shown to be challenging due to a number of reasons, including the high work function of the CdTe [4]. Deposition of copper (Cu) and its diffusion into CdTe is used as a way to form tunneling contacts to the CdTe. However, there are numerous studies indicating degradation of the cell performance related to Cu, e.g., fast migration into the CdTe and even reaching n-type cadmium sulfide (CdS), causing further instability in the device performance over time [5,6]. There have been efforts to eliminate such Cu diffusion process through establishing a p-type highly doped interface layer at the back contact with a large carrier concentration in order to form a low-resistance layer with valence band alignment to the CdTe and tunneling contact with the metal.

Zinc telluride (ZnTe) has been studied as a promising candidate due to its dopability, its similarity in structure and the possibility of valence band alignment with CdTe [7,8]. In addition to conformity of crystal structure and tunability of electrical properties, understanding likely process alternatives for various dopants is essential for developing new and robust device processes. Cu and N are known to provide p-type doping in ZnTe, where Cu occupies Zn sites and N occupies Te sites [4,9]. Growth of ZnTe thin films by various physical vapor deposition (PVD) techniques has been reported previously [10-13]. Effect of deposition parameters have been studied for crystalline quality and stoichiometry of ZnTe films. More importantly, nitrogen-doped ZnTe (ZnTe:N) has been demonstrated to reach large carrier concentration levels of up to 10^{20} cm^{-3} through molecular beam epitaxy (MBE) using a nitrogen plasma source [14,15]. However, doping ZnTe with high levels of N has not been demonstrated using sputtering. Also,

Cu-doping was achieved in ZnTe by sputtering of Cu-doped ZnTe targets or simultaneous co-sputtering of ZnTe and metallic Cu targets [7,8,16]. Carrier concentrations of over 10^{18} cm^{-3} and higher were obtained in ZnTe:Cu films with ~1 at.% Cu after annealing at 350°C.

In this paper, we report on the electrical properties of p-type ZnTe thin films deposited by sputtering. We have explored the impact of nitrogen and copper dopants in ZnTe on carrier concentration and mobility for as-deposited as well as post-deposition annealed films.

Experimental

ZnTe deposition was carried out in a Kurt J. Lesker RF magnetron sputtering chamber with *in-situ* temperature control. Stoichiometric ZnTe commercial target was purchased from Plasmaterials, Inc. Pure research grade argon was used during deposition for ZnTe:Cu films, while an Ar:N$_2$ mixture was used for ZnTe:N films. The ratio of N$_2$ to Ar in the gas mixture was varied to obtain the desired doping level. For Cu-doping a thin layer of Cu was sandwiched between two layers of ZnTe. The ZnTe deposition was done on 1.4mm Asahi PVN++ at 225-300°C, 5 mTorr of gas pressure and 75 W of sputtering power. For annealing a tube furnace was used and annealing was carried out at 400°C for 30 minutes in ambient Ar (400 Torr). In order to obtain transport properties of ZnTe films, Hall-effect measurements were used to provide information on resistivity, carrier concentration and mobility of the films. ZnTe films on glass are cut into 1 cm x 1 cm size. Electrodag graphite contacts are applied at the corners in order to test in a van der Pauw configuration. Time-Of-Flight Secondary Ion Mass Spectroscopy (TOF SIMS) was used to examine the compositional depth profile in the ZnTe:Cu films. The instrument (IonToF5 TOF-SIM) uses an analysis source of Bi$_1$$^+$. Chemical composition of the ZnTe:N films was analyzed using X-ray photoelectron spectroscopy (XPS) depth profiles conducted on PHI 5500 XPS. Scanning electron images were taken using a Field Emission SEM (FESEM) instrument at 5-10 kV. High-resolution X-ray diffraction patterns were collected using a Bruker D8 machine. Optical bandgap measurements are performed through obtaining total transmission and total reflectance spectra on a Varian, Cary 5000 spectrophotometer.

Results and discussion

I. N-doped ZnTe

All films show dense crack-free microstructure in cross sectional SEM images (not shown), while the grain size in ZnTe films seems to decrease with increasing nitrogen levels in the gas-mixture. This is in accordance with a reduction that is observed in the intensity of ZnTe diffraction peaks in XRD (data not shown). At 10% N$_2$ in Ar, films exhibit a very fine grain amorphous-like microstructure. Presence of roughly 3% nitrogen was confirmed in films deposited in Ar:N$_2$ (5%) environment by XPS (not shown). The nitrogen level detected by XPS remained relatively constant across the thickness of the film. Figures 1 (a) and (b) represent optical bandgap spectra of N-doped ZnTe films deposited at 180°C and 225°C respectively for various nitrogen doping levels (0.05-1%) by UV-Vis absorption measurements. Presence of nitrogen seems to have little effect on the absorption edge in samples grown with less than 1% nitrogen. All ZnTe films deposited at 225°C show a bandgap of ~2.2-2.25 eV. However, the ZnTe film deposited at 180°C with 1% nitrogen doping seem to have a smaller bandgap of

~2.1eV. While such a shift in bandgap might be an indication of a secondary phase, possibly due to excessive doping, there are no other evidences at hand to substantiate such hypothesis, as XRD patterns and SEM images showed no suggestion of formation of such phases. Beyond that, all films show a shoulder in the spectra, showing significant sub-gap absorption.

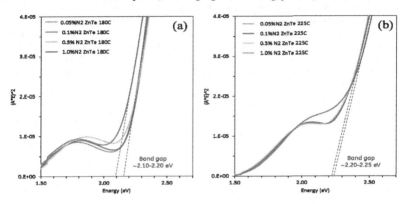

Figure 1. UV-Vis optical bandgap measurements on ZnTe films grown on glass with various nitrogen levels at a) 180°C and b) 225°C. Vertical axes are (absorption coefficient * energy)^2.

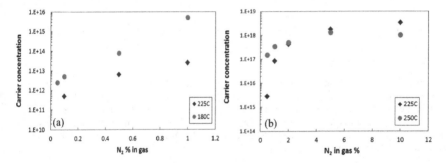

Figure 2. Carrier concentration in ZnTe films, as measured from Hall-effect for ZnTe films deposited at: a) 180°C and 225°C in an environment of Ar:N$_2$ (1%) and b) 225°C and 250°C in an environment of Ar:N$_2$ (10%).

Hall-effect measurements were used to measure resistivity, carrier density and mobility in the ZnTe films. Figure 2 (a) shows carrier concentration of the ZnTe films deposited at 180°C and 225°C. It is observed that the carrier concentration increases with nitrogen doping up to 5 orders of magnitude, while mobility decreases from approximately 7 cm^2/V.sec at 0.1% N$_2$ to 1 cm^2/V.sec at 1% N$_2$ (not shown). Carrier concentration of $>10^{15}$ is achieved for films deposited at 180°C. In order to obtain even higher carrier concentrations, the experiment was repeated with higher nitrogen concentrations in the gas. For this set, ZnTe films were deposited at 225°C and

250°C and their carrier concentrations were measured and plotted in Figure 2(b). As seen, the same trend is observed, i.e., increase carrier concentration with increasing nitrogen content in the gas-flow. Note that the ranges of nitrogen concentration were much higher in this experiment as compared to the data shown in Figure 2, achieving carrier concentrations $> 10^{18}$ cm^{-3}. At both deposition temperatures, the increase in carrier concentration seems to level off above 5% N_2:Ar, which might be an indication of saturation limit for solubility of nitrogen in ZnTe. This may be correlated to the observed reduction of grain size in ZnTe with increased nitrogen content in the gas prohibiting grain growth. Effect of annealing at 400°C on the properties of ZnTe:N was also explored and plotted in Figure 3. All samples show a decrease in carrier concentration and increase in resistivity after annealing. The exact mechanism for this behavior is not well understood.

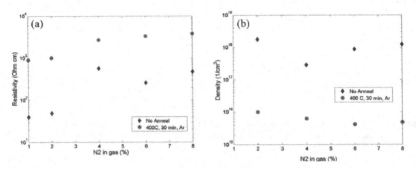

Figure 3. Effect of annealing on a) resistivity and b) carrier density of the ZnTe:N films grown at 225°C.

II. Cu-doped ZnTe

As described earlier, doping ZnTe with Cu was approached by fabricating a ZnTe:Cu:ZnTe sandwich structure. TOF-SIMS was used to understand the Cu-profile in the ZnTe films in such a structure as illustrated in Figure 4. Cu signal was detected across the film thickness which indicates fairly uniform diffusion of Cu layer into the ZnTe. The approximate total thicknesses of the ZnTe and Cu layers were estimated to be 200 and 30 nm respectively, measured from the SEM cross section micrographs.

Resistivity and carrier concentration were used as measures of effective doping of ZnTe by Cu. The Cu layer thickness was varied to control the doping level. Figure 5 shows the effect of Cu deposition time and hence thickness on a) resistivity and b) carrier concentration of ZnTe films. Blue circles represent the as-deposited films, while red squares represent annealed films. Increasing Cu thickness (deposition time) expectedly increases the carrier density initially by over 6 orders of magnitude, while it shows a plateau behavior above 4 minutes of deposition time, which might be associated with formation of a degenerate semiconductor with metallic behavior. On the other hand, post-annealing treatment of the ZnTe films seems to improve the carrier concentration and further reduce the resistivity. Especially, at 2-3 minutes of deposition, ~4 orders of magnitude improvement is observed upon heat treating the samples. This is likely due to further advancement of Cu diffusion process during annealing.

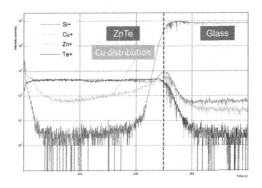

Figure 4. ToFSIM compositional depth profile of Cu-doped ZnTe films.

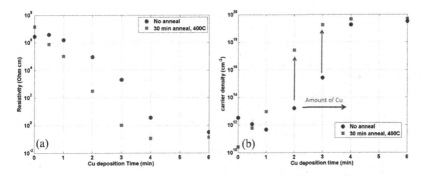

Figure 5. a) resistivity and b) carrier concentration of ZnTe films as a function of Cu deposition time. Blue circles represent the as-deposited films, while red squares represent ZnTe films annealed at 400°C, 30 minutes in ambient Ar.

Summary

P-type doping was obtained in ZnTe films grown by RF sputtering under nitrogen flow and ZnTe:Cu:ZnTe sandwich structure. We established deposition procedure to achieve a broad range of carrier concentrations in ZnTe:N thin film spanning from 10^{13} to 10^{18} cm^{-3} by varying nitrogen concentration during deposition. However, we observed a degradation in crystallinity of such films with increasing the nitrogen concentration and hence carrier concentration. In Cu-doped ZnTe films, various doping levels were obtained by tuning the thickness of Cu layer within the sandwich structure. High levels of carrier concentrations of $>10^{19}$ were achieved in such films.

Acknowledgements

The authors would like to thank Renee Herzog, Chris Callazo-Davila and Anil Duggal for their support of this work.

References

[1] J. Britt and C. Ferekides, "Thin-film CdS/CdTe solar cell with 15.8% efficiency" Appl. Phys. Lett. 62,

(1993) 2851-3.

[2] X. Wu, J.C. Keane, R.G. Dhere, C. DeHart, D.S. Albin, A. Duda, T.A. Gessert, S. Asher, D.H. Levi, and P. Sheldon, Proc. of 17th E-PVSEC (2001) 995

[3] M.A. Green, K. Emery, Y. Hishikawa, W. Warta, E.D. Dunlop, Prog. Photovolt: Res. Appl.21 (2013) 1

[4] C. Narayanswamy, T.A. Gessert and S.E. Ashe, "Analysis of Cu Diffusion in ZnTe-Based Contacts for Thin-Film CdS/CdTe Solar Cells" Presented at the National Center for Photovoltaics Program Review Meeting Denver, ColoradoSeptember 8-11, 1998

[5] K. D. Dobson, I. Visoly-Fisher, G. Hodes, and D. Cahen, Sol. Energy Mater. "Stability of CdTe/CdS thin film solar cells"Sol. Cells 62, (2000) 295.

[6] D. L. Bätzner, A. Romeo, M. Terheggen, M. Dobeli, H. Zogg, and A. N. Tiwari, "Stability aspects in CdS/CdTe solar cells" Thin Solid Films 451–452, (2004) 536.

[7] T. A. Gessert, X. Li, T. J. Coutts, A. R. Mason, and R. J. Matson, "Dependence of material properties of radio frequency magnetronsputtered, Cu-doped, ZnTe thin films on deposition conditions" J. Vac. Sci. Technol. A 12(4) (1994) pp 1501-6

[8] W. Jaegermann, A. Klein, J. Fritsche, D. Kraft, and B. Späth, "Interfaces in CdTe solar cells: From idealized concepts to technology "Mater. Res. Soc. Symp. Proc. 865 (2005) F6.1.

[8] J. Tang, D. Mao, J. U. Trefny, "Effect of Cu doping on the properties of ZnTe:Cu thin films and CdS/CdTe/ZnTe solar cells" CP394 NREL/SNL photovoltaics program review (1997)

[10] H. Bellakhder, A. Outzourhit, E.L. Ameziane, "Study of ZnTe thin films deposited by r.f. sputtering," Thin Solid Films 382 (2001) pp 30-3

[11] R.N. Bicknell-Tassius, T.A. Kuhn, W. Ossau, "Photoassisted MBE of CdTe thin films" App. Surf. Sci. 36 (1989) 95.

[12] R.L. Gunshor, L.A. Koladziejski, N. Otsuka, S. Datta, "ZnSe-ZnMnSe and CdTe-CdMnTe Superlattices" Surf. Sci. 174 (1986) 522.

[13] U. Pal, S. Saha, A.K. Chaudhuri, V.V. Rao, H.D. Banerjee, "The anomalous photovoltaic effect in polycrystalline zinc telluride films" J. Phys. D: Appl. Phys. 22 (1989) 965.

[14] T. Baron, K. Saminadayar, S. Tatarenko, "Plasma nitrogen doping efficiency in molecular beam epitaxy of tellurium based II-VI compounds."J. Cryst. Growth 159 (1996) 271.

[15] T. Baron, K. Saminadayar, and N. Magnea, "Nitrogen doping of Te-based II-VI compounds during growth by molecular beam epitaxy" J. Appl. Phys. 83, (1998) 1354-6.

[16] B. Späth, J. Fritsche, A. Klein, and W. Jaegermann, "Nitrogen doping of ZnTe and its influence on CdTe/ZnTe interfaces" Appl. Phys. Lett. 90, (2007) 062112-4

Mater. Res. Soc. Symp. Proc. Vol. 1538 © 2013 Materials Research Society
DOI: 10.1557/opl.2013.1013

Electrical Properties of Photoconductor Using Ga_2O_3/CuGaSe$_2$ Heterojunction

Kenji Kikuchi[1,2], Shigeyuki Imura[1], Kazunori Miyakawa[1], Misao Kubota[1], and Eiji Ohta[2]
[1] NHK Science and Technology Research Laboratories, 1-10-11, Kinuta, Setagaya-ku, Tokyo, 157–8510, Japan
[2] Graduate School of Science and Technology, Keio University, 3-14-1, Hiyoshi, Kouhoku-ku, Yokohama, 223–8522, Japan

ABSTRACT

The feasibility of using a photoconductor with a Ga_2O_3/CuGaSe$_2$ heterojunction for visible light sensors was investigated. CIGS chalcopyrite semiconductors have both a high absorption coefficient and high quantum efficiency. However, their dark current is too high for image sensors. In this study, we applied gallium oxide (Ga_2O_3) as a hole-blocking layer for CIGS thin film to reduce the dark current. Experimental results showed that the dark current was drastically reduced, and an avalanche multiplication phenomenon was observed at an applied voltage of over 6 V. However, this structure had sensitivity only in the ultraviolet light region because its depletion region was almost completely spread in the Ga_2O_3 layer since the carrier density of the Ga_2O_3 layer was much lower than that of the CIGS layer. These results indicate that the Ga_2O_3/CuGaSe$_2$ heterojunction has potential for use in visible light sensors but that we also need to increase the carrier density of the Ga_2O_3 layer to shift the depletion region to the CIGS film.

INTRODUCTION

The use of CuIn$_{1-x}$Ga$_x$Se$_2$ (CIGS) chalcopyrite thin films as the absorber material for thin film solar cells has received much attention because of the band gap, high absorption coefficient, great stability, and high quantum efficiency of this type of film [1–3]. In this study, we examined the feasibility of using a photoconductor with CIGS chalcopyrite semiconductors for visible light sensors. CIGS thin film solar cells using CdS as a buffer layer have reached conversion efficiencies of 20%. However, CdS has a Cd toxicity classification and an absorption of blue light due to having a band gap of 2.4 eV. These are unfavorable for both solar cells and visible light sensors. Moreover, the dark current of CIGS thin film is too high for sensors [4] because of the low resistivity of CIGS. We applied a gallium oxide (Ga_2O_3) thin film as an n-type semiconductor layer to solve these problems. We assumed that the Ga_2O_3 thin film would function as a hole-blocking layer for the CIGS films and suppress the injection of holes from electrode [5]. The Ga_2O_3 thin films have a wide band gap of 4.7–4.9 eV and high transmittance in visible light [6]. We used CuGaSe$_2$ for absorber layers because it has a suitable band gap of 1.64–1.67 eV [7, 8], which is sufficient for visible light sensors. Here, we report on the electrical characterization of the Ga_2O_3/CuGaSe$_2$ heterojunctions.

EXPERIMENT

Alkali-free glass (Corning, Eagle-XG) with an Au layer served as the substrate and back contact. The CuGaSe₂ layer was deposited onto the substrate by one-step radio frequency (RF) magnetron sputtering from a CuGaSe₂ alloy target (Kojundo Chemical Lab. Co., Ltd.) that consisted of Cu:Ga:Se = 1:1:2 at.%. The deposition power was 100 W in an Ar atmosphere with a sputtering pressure of 0.6 Pa and a substrate temperature of 350 °C. The thickness of the CIGS layers was 1 um. Next, a Ga₂O₃ layer was RF-sputtered on the CuGaSe₂ layer at a deposition power of 200 W and a substrate temperature of 22 °C (RT) in an Ar/O₂ 1% atmosphere with a sputtering pressure of 0.6 Pa. The thickness of the Ga₂O₃ layer was 300 nm. Finally, a 30-nm ITO film was prepared by direct-current magnetron sputtering on each sample in Ar/O₂ 1% atmosphere. The surface elemental composition of the CuGaSe₂ film was analyzed by X-ray photoelectron spectroscopy (XPS). Charge neutralization was performed with an electron flood gun. The XPS spectrum of the CuGaSe₂ film was calibrated by the Ga3d peak at 19.9 eV to compensate for the charge effect.

DISCUSSION

The optical transmittance and reflectance of the CuGaSe₂ and Ga₂O₃ films were measured with a spectrophotometer. Figure 1 shows the optical band gap of each film. The optical band gap of the CuGaSe₂ film(1.61 eV) and the Ga₂O₃(4.70 eV) film was consistent with previous reports [7, 8, 9]. There were many selenium deficiencies ([Se]/[Cu] = 1.06) in our CuGaSe₂ films (Table 1), which played a role in the decreased band gap. Deposition by one-step RF sputtering probably induced many selenium deficiencies because the vapor pressure of selenium is high. We therefore need to apply a selenization technique in future research.

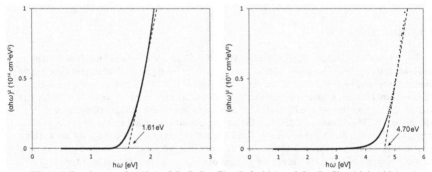

Figure 1 Band gap estimation of CuGaSe₂ film (left side) and Ga₂O₃ film (right side).

Sample	Cu (at.%)	Ga (at.%)	Se (at.%)	O (at.%)
CuGaSe₂	27.3	23.2	28.9	20.5

Table 1 Surface elemental composition of CuGaSe₂ film.

Figure 2 shows the energy band diagram of Ga_2O_3 and $CuGaSe_2$. The electron affinity of $CuGaSe_2$ here has been applied in other studies [10]. It suggests that the hole blocking ability of Ga_2O_3 is very high. The bottom of the conduction band of the Ga_2O_3 layer is slightly higher than optimum value (0–0.4 eV) [11].

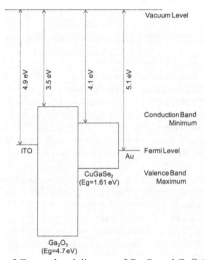

Figure 2 Energy band diagram of Ga_2O_3 and $CuGaSe_2$.

Figure 3 shows the dependence of the dark current of the $Ga_2O_3/CuGaSe_2$ heterojunction structure. The $Ga_2O_3/CuGaSe_2$ photodiode had a low dark current, the order of 10^{-10}–10^{-9} A/cm², when the applied voltage was less than 5 V. This indicates that the Ga_2O_3 thin film functions as a hole-blocking layer for the CIGS films and suppresses the injection of holes from the electrode. An avalanche phenomenon was observed at an applied voltage of over 6 V because the injection of holes from the electrode was suppressed by the high potential barrier of Ga_2O_3 to holes. This avalanche phenomenon is advantageous for highly sensitive image sensors [12].

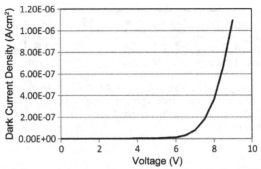

Figure 3 The dark current density-voltage characteristics of a $Ga_2O_3/CuGaSe_2$ cell.

Figure 4 plots the relative quantum efficiency of the $Ga_2O_3/CuGaSe_2$ cell. As the figure shows, this cell had a response that was not in the visible wavelengths but rather in the ultraviolet light regions. This indicates that the depletion region of this cell was spread almost completely in the Ga_2O_3 layer. The conductivity of the intrinsic Ga_2O_3 is too poor, so the carrier density of the non-doped Ga_2O_3 layer was probably much lower than that of the CIGS layer. That is to say, we need to either increase the carrier density of the Ga_2O_3 film or decrease that of the CIGS film in order to acquire sensitivity in visible light. Doping tin or the introduction of oxygen vacancies has been shown to improve the conductivity of Ga_2O_3 film [9, 13, 14]. We plan to increase the carrier density of Ga_2O_3 film using tin-doping because we think that oxygen vacancies induce defect levels and decrease the effective hole-blocking barrier height of Ga_2O_3 [5]. Our results indicate that the $Ga_2O_3/CuGaSe_2$ heterojunction has potential for use in visible light sensors. It has a high hole-blocking barrier, so it can reduce the dark current drastically. Moreover, an avalanche multiplication effect can be obtained when the dark current is sufficiently reduced. We believe that, with some minor adaptation, Sn-doping $Ga_2O_3/CuGaSe_2$ can have sensitivity in the visible region.

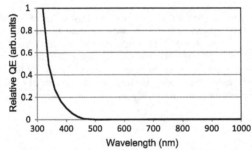

Figure 4 Relative quantum efficiency of $Ga_2O_3/CuGaSe_2$ cell.

CONCLUSIONS

We have investigated the electrical properties of $Ga_2O_3/CuGaSe_2$ heterojunctions. The dark current of CIGS films were successfully reduced by the Ga_2O_3 layer, which had an effective hole-blocking barrier. Moreover, an avalanche multiplication phenomenon was observed at an applied voltage of over 6 V, which is beneficial for sensors. However, this junction had little sensitivity in the visible region because its depletion region was almost completely spread in the Ga_2O_3 layer. We need to increase the carrier density of the Ga_2O_3 layer to shift the depletion region to the CIGS layer. We believe that the $Ga_2O_3/CuGaSe_2$ heterojunction can acquire sensitivity in visible light: for example, when using Sn-doping in the Ga_2O_3 layer.

ACKNOWLEDGMENTS

This work was partly supported by the Foundation for Promotion of Material Science and Technology of Japan (MST Foundation).

REFERENCES

1. B. Tell, J.L. Shay and H.M. Kasper, Phys. Rev. B4 2463 (1971).
2. K. Ramanathanm, M.A. Contreras, C.L. Perkins, S. Asher, F.S. Hasoon, J. Keane, D. Young, M. Romero, W. Metzger, R. Noufi, J. Ward and A. Duda, Prog. Photovolt :Res. Appl. **11**, 225 (2003).
3. T. Nakada, K. Furumi and A. Kunioka, IEEE Transactions of Electron Devices **46(10)**, 2093 (1999).
4. K. Tanaka, M. Kosugi, F. Ando, T. Ushiki, H. Usui and K. Sato, Jpn. J. Appl. Phys. Suppl. **32**, 113 (1993).
5. K. Kikuchi, Y. Ohkawa, K. Miyakawa, T. Matsubara, K. Tanioka, M. Kubota and N. Egami, phys. stat. sol.(c) **8(9)**, 2800 (2011).
6. N. Ueda, H. Hosono, R. Waseda and H. Kawazoe, Appl. Phys. Lett. **70(26)**, 3561 (1997).
7. G. Hanna, A. Jasenek, U. Rau and H.W. Schock, Thin Solid Films, 387, **71** (2001).
8. R. Herberholz, V. Nadenau, U. Ruhle, C. Koble, H. W.Schock and B. Dimmler, Solar Energy Materials and Solar Cells 49, 227 (1997).
9. M. Orita, H. Ohta, M. Hirano and H. Hosono, Appl. Phys. Lett. **77(25)**, 4166 (2000).
10. M. Sugiyama, H. Nakanishi, S.F. Chichibu, Jpn. J. Appl. Phys. **40**, L428 (2001).
11. T. Minemoto, T. Matusi, H. Takakura, Y. Hamakawa, T. Negami, Y. Hashimoto, T. Uenoyama, M. Kitagawa, Solar Energy Materials & Solar Cells **67**, 83 (2001).
12. K. Tanioka, J. Yamazaki, K. Shidara, K. Taketoshi, T. Kawamura, S. Ishioka and Y. Takasaki, IEEE Electron Device Lett. **8**, 392 (1987).
13. M. Orita, H. Hiramatsu, H. Ohta, M. Hirano and H. Hosono, Thin Solid Films **411**, 134 (2002).
14. Yijun Zhang, Jinliang Yan, Gang Zhao and Wanfeng Xie, Physica B, **405**, 3899 (2010)

Wide Bandgap Materials

Mater. Res. Soc. Symp. Proc. Vol. 1538 © 2013 Materials Research Society
DOI: 10.1557/opl.2013.551

The Wide band p- type material formed by the thin film with ZnO - NiO mixed crystal system

Mikihiko Nishitani, Masahiro Sakai, Yukihiro Morita

Panasonic Device Science Research Alliance Laboratory
Osaka University

ABSTRACT

We study ZnO-NiO mixed crystal thin film as a wide band p-type material for the hetero-junction with ZnO. As for the hetero-junction of the ZnO (n-type) and the NiO which have relatively stable p-type semiconductor characteristics, there are issues on the crystallographic mismatch and the band offset of the valence band as well as the conduction band . We made the ZnO-NiO mixed crystal thin film in all composition range with the substrate temperature of 250°C, using magnetron sputtering process and acquired the basic data for the change of electrical conductivity with conduction type. In addition, a high-quality thin film was made by using a Pulse Laser Deposition (PLD), and the band diagram of the ZnO-NiO mixed crystal system was illustrated from the analyses of XPS, NEXAFS and optical absorption measurements. As a result, the offset of ZnO-NiO mixed crystal film is proportionally decreasing with increasing the content of ZnO in NiO film. And the characteristics of the diode with the hetero-junction of ZnNiO/ZnO were improved compared with that of NiO/ZnO. The reasons were discussed with the data of the band offset, the crystalline of the films and the interface properties with the NiO/ZnO and the ZnNiO/ZnO. .

INTRODUCTION

Zinc oxide (ZnO) is one of most promising materials for various applications such as an optelectronic device in the ultraviolet region and a transparent electron device. However, it is difficult to make the p-type ZnO when the low-temperature process (< 600°C) on the glass substrate etc. is demanded. On the other hand, the NiO film shows p-type conduction, formed at the low temperature process [1] and the possibility of the hetero-junction of n-type ZnO and p type NiO is examined.[2 - 3]. Though Ota et al. have succeeded in making the hetero-diode of excellent ZnO/NiO on ITO/YSZ, using high temperature process [4], it is reported that the low built in voltage and the low reverse break down voltage were due to the large band offset with ZnO and NiO [5]. Chen achieved the improvement of the leakage current and the breakdown of hetero-diode by using ZnMgO/NiO instead of ZnO/NiO [6]. These results suggest that one of the important issues was to be tuned by engineering the bandgap and/or the band offset . In this work, we try to change the band offset parameter and the improvement of the junction properties with the ZnO, applying ZnO-NiO mixed crystal system

EXPERIMENTAL PROCEDURES

The ZnO - NiO thin films were deposited on quartz and ZnO single crystal by RF magnetron sputtering or Pulse Laser Deposition (PLD) with NiO, ZnO, $Zn_{0.8}Ni_{0.2}O$ and $Zn_{0.5}Ni_{0.5}O$ target of 99.99% purity, respectively. The substrates were subsequently cleaned by thermal annealing in the deposition chamber at 550°C for 30 min. In the case of RF magnetron sputtering, ZnO-NiO system (ZnNiO) films were deposited at 250 °C. Before the deposition, the base pressure of the sputtering was maintained below 5×10^{-4} Pa and the pre-sputtering was performed for 10 min. During the deposition, the chamber pressure was kept at 1.0 Pa with Ar (50%) /O₂ (50%) gas. The thickeness of the films were controlled to be 200-300 nm. The ZnNiO films were also deposited using PLD process with a ArF laser at 10 Hz and the ZnO, $Zn_{0.5}Ni_{0.5}O$ or NiO target. During the deposition, the chamber pressure was kept at 50 Pa with pure O₂ gas and the temperature of the substrate was heated up to 550°C. The RHEED patterns were observed after the deposition and pumping down below $1x10^{-4}$ Pa. The electrical properties such as resistivity and conduction type were characterized by a I-V characteristics and a hot probe method (measurement of Seebeck effect) at RT.   For chemical analysis and getting the valence band offset parameter, XPS measurements with Al Kα radiation (1486.6 eV) were carried out and were calibrated with C 1s (284.8 eV) because of the peak shift due to charge accumulation. NEXAFS measurements were conducted for the characterization of unoccupied electronic state and the conduction band offset of ZnO-NiO mixed crystal system. In addition, the optical absorptions of the films were acquired by an ultraviolet visible-near infrared (UV-VIS/NIR) spectro-photometer.

To fabricate the p-n junction diodes, the NiO or ZnNiO films were formed by PLD on the single crystal of ZnO with the low resistivity made by Tokyo Denpa. Co., LTD. The Pt electrode for p-type film (NiO or ZnNiO) and the Ti /Au electrode for n-type ZnO were deposited by RF magnetron sputtering using metal shadow masks.

RESULTS AND DISCUSSIONS
We measured the electrical resistivity and the conduction type of the ZnNiO film. Though the ZnO film showed n-type conduction, the p-type conduction showed even in the composition of ZnO composition 85% and NiO composition 15% film with the exponential increase of the resistivity with the increase of ZnO mole fraction, compared with NiO film, corresponding to the decrease of the hole concentration from $10^{19}cm^{-3}$ to $10^{16}cm^{-3}$.

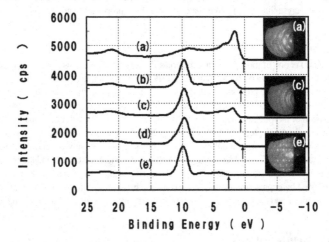

Figure 1.   Valence band spectra measured by XPS and .RHEED patterns of ZnNiO films. (a) NiO film on quartz by PLD, (b) ZnNiO fim on ZnO single crystal by PLD, (c) ZnNiO film on quartz by PLD,(d) ZnNiO film on quartz by RF sputter, (e) ZnO film on quartz by PLD.

Figure1 show the XPS spectra near around the valence band of the film. The compositions of ZnNiO films, (b) – (d), shown in Fig. 1 were about Ni/(Zn+Ni) = 0.65 from the analyses of XPS measurements. The arrows and the photographs shown in the figure are the positions of the valence band maximum (VBM) and the RHEED patterns after deposition, respectively. As for the VBM, we observed that the band offset to ZnO has become small, using ZnNiO compared with NiO as expecting. Moreover, we note that the VBM of the film made by the sputtering process, compared with the PLD process, was closer to that of NiO since the tailing around the VBM was seen due to the high hole concentration. In the RHEED patterns, we observe that the films deposited on quartz substrate are polycrystalline and the NiO film (a) and the ZnO film (e) show preferable orientation (spotty pattern), on the other hand, the random orientation growth (ring pattern) is seen in ZnNiO film (c). The XRD patterns supported the results of the RHEED patterns.

The dependency of the optical band gap on the film composition was measured by UV-VIS/NIR spectrometer on ZnO - NiO system. The NiO films and the ZnNiO thin fims exhibited absorption around in the visible region, which colored to brown and green formed by sputtering process and PLD process, respectively. It is speculated that those phenomena correspond to d-d transition for green [7] and Hagen-Rubens law for brown [3]. Using the optical band gap analysis plot, ZnO on the quartz deposited by PLD, ZnNiO (Ni/(Zn+Ni) = 0.65) film on the quartz by PLD and NiO film on the quartz by PLD showed the optical band gap of 3.2eV, 3.3eV and 3.5eV, respectively.

Alternatively, we observed that the unoccupied state with the information of local electronic state on ZnO – NiO system, using NEXAFS which is installed at Kyushu Synchrotron Light Research Center (Saga Light Source) in Japan. The NEXAFS measurements of around Zn2p (1000- 1100eV), Ni2p (840-900eV) and O1s (520-560eV) were carried out by using the same samples shown in Fig.1, in addition to NiO single

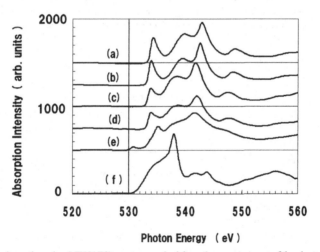

Photon Energy (eV)

Figure 2.   O1s absorption edge (NEXAFS) spectra acquired from the measurement of the electron yield due to the irradiation of the soft X-ray. (a) NiO single crystal , (b) NiO fim on quartz by PLD, (c) ZnNiO film on ZnO single crystal by PLD,(d) ZnNiO film on quartz by PLD, (e) ZnNiO film on quartz by RF sputter, (e) ZnO film on quartz by PLD.

crystal sample. The NEXAFS spectra of O1s are shown in Figure 2. In the NEXAFS measurement as well as in the XPS measurement, the spectrum shape of ZnNiO looks alike that of NiO, and is different from that of ZnO. From the crystallographic point of view, NiO and ZnNiO have the same NaCl structure, which is speculated from the RHEED pattern shown in Fig.1.

As for the sample (e) made by sputtering, the small absorption by unoccupied state was observed in the low photon energy side from standing up at the absorption edge. It is considered that this absorption is related to the low resistivity, compared with other samples. It is necessary to get more detail experimental data for the confirmation of that speculation. The binding energy of O1s on NiO film (b), ZnNiO film (c), ZnNiO film (d), ZnNiO film (e) and ZnO film (f) was 529.9eV, 529.7eV, 529.9eV, 529.7eV and 530.4eV, respectively. In other words, the horizontal position of 530eV in Fig.2 corresponds to that of the Fermi level (= 0eV) in Fig.1. The position of VBM can be arranged on Figure 2 from the analysis as mentioned above. Then, we can estimate the band gap according to the energy difference from the absorption edge of NEXAFS spectra and the position of VBM on the horizontal axis on the Fig.2. Figure 3 showed the band offset diagram of the NiO-ZnO system. The values of differences between the VBM of the NiO or ZnNiO and that of the ZnO in Fig.1 are plotted with solid circles and diamonds which correspond to the case of the sputtering process and the PLD process, respectively. The minimum energy positions of unoccupied state are similarly plotted by two methods in Figure 3. One is a plot that added the optical bandgap to the value of VBM that has already determined. Another is a plot that similarly added the energy gap which is estimated from the result combined with XPS and NEXAFS shown in Figure 1 and 2, respectively.

Figure 4 A (left side) shows the I-V characteristics of the hetero-junctions which are the structures of NiO /ZnO single crystal and ZnNiO (Ni/(Zn+Ni) = 0.65, / ZnO single crystal. Those hetero-junctions were fabricated by the PLA process, and the NiO and ZnNiO films were deposited on the ZnO single crystal (0001) plane with Zn-face. The breakdown properties were observed at the low reverse bias in NiO/ZnO junction device. We can see the improvement of the characteristics, using ZnNiO film instead of NiO film,

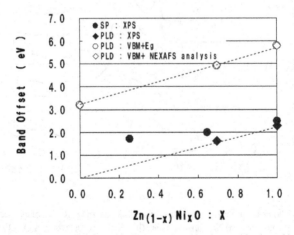

Figure 3   Energy band diagram of ZnO – NiO system analyzed by the combination with the XPS of Fig.1, and Fig.2 or optical band gap in each films.

and speculate that the band offset to ZnO is deceased by using ZnNiO, as shown in Figure 3. In Figure 4 B (right side), the light emitting properties of the devices are shown using the same device in Figure 4A. The luminous intensity of NiO/ZnO device denoted (a) is inferior to ZnNiO/ZnO device denoted (b), though large forward currents flow in the NiO/ZnO device compared with the ZnNiO/ZnO at the same operating voltage. Similarly, in NiO/ZnO device, the voltage where standing up of the luminous intensity is high,

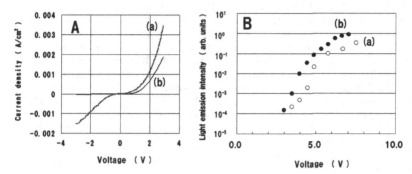

Figure 4 The graph A : the I-V characteristics of NiO/ZnO hetero-junction device (a), ZnNiO/ZnO hetero-junction device. The graph B : the light emission intensity as a function of voltage for NiO/ZnO hetero-junction device (a), ZnNiO/ZnO hetero-junction device (b).

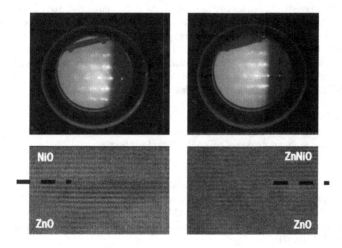

Figure 5.   The photographs with the high resolution TEM focused on the interface with ZnO and the RHEED patterns   acquired immediately after deposition.

compared with ZnNiO/ZnO device. We can observe that the injection efficiency of holes to ZnO layer is improved as a result of the band offset to ZnO is being improved by ZnNiO though the hole concentration in the ZnNiO is less than that in the NiO film. The current-voltage property and the light emitting performance might be related to not only the improvement of the band offset but also the interfacial state with ZnO. In Figure 5, the results of high resolution TEM focused on the interface of NiO/ZnO and ZnNiO/ZnO are shown simultaneously with the RHEED pattern acquired immediately after deposition. The observed RHEED patterns of NiO film and ZnNiO film were different from that of Fig.1(a) and Fig.1(b), and showed the streak-like diffraction patterns along the streak diffraction pattern of ZnO. That is, it is supposed that the growth of the film shows epitaxial growth to ZnO (0001) plane with distortion. Moreover, it was not observed that a definite difference is found from high resolution TEM to the interfacial state of NiO/ZnO and ZnNiO/ZnO and the crystalline of NiO and ZnNiO film. It is speculated that the observed layer structure in the image of TEM corresponds to (0001) plane of hexagonal crystal or (111) plane of cubic crystal. The lattice constants of ZnO, NiO and ZnNiO analyzed by the photograph of TEM were almost the same values. That is, cubical crystal (111) of NiO and ZnNiO is transformed into the hexagonal crystal (0001) structure with distortion as the result of succeeding the lattice parameter of ZnO (0001) plane. The speculation is supported from the interpretation of the RHEED pattern previously described.

CONCLUSIONS

We studied ZnO-NiO mixed crystal thin film to examine the possibility for hetero-junction with n-type ZnO. The p-type conduction showed even in the composition of ZnO 85% and NiO 15% with the exponential increase of the resistivity with the increase of ZnO mole fraction in ZnO-NiO system. It is suggested by the measurements of XPS and NEXAFS that the offsets of valence band and conduction band on ZnO-NiO system linearly changed with the composition. As for the film deposited on the quartz substrate by PLD, ZnNiO (Ni/(Zn+Ni) = 0.65) film seemed to be inferior to the NiO or the ZnO film. On the other hand, the crystalline of ZnNiO film deposited on ZnO single crystal was as good as that of NiO film from the observations of RHEED and high resolution TEM images. The properties of the hetero-junction with the structure of ZnNiO/ZnO were superior to that of NiO/ZnO from the measurements of I-V characteristics and the light emitting efficiency. We speculate the improvements of the device properties are due to the decrease of the band offset using ZnNiO film instead of NiO film.

ACKNOWLEDGMENTS

We would like to thank Dr. T.Nagashima and Dr. J. Nishitani for operating ALD , and Dr. Setoyama for operating the system to acquire the data of NEXAFS in Saga Light Source.

REFERENCES

1. H. Sato, T. Minami, S. Takata and T. Yamada, Thin Solid Films, **236**, 27-31 (1993).
2. Y. Vygranenko, K. Wang, and A. Nathan, Appl. Phys. Lett. **89,** 172105 (2006) ..
3. Y.H. Kwong, S.H. Chun, J. Han and H. K. Cho, Met.Mater. Int. **18** (6), 1003 (2012)..
4. H. Ohta, M. Hirano, K. Nakahara, H. Maruta, T. Tanabe, M. Kamiya, T.Kamiya, and H. Hosono, Appl. Phys. Lett. **83** ,1029 (2003).
5.Y. Ishida, A. Fujimoto, H. Ohta, M. Hirano, and H. Hosono, Appl. Phys.Lett. **89** ,153502 (2006).
6. Xinman Chen, Kaibin Ruan, Guangheng Wu, and Dinghua Bao, Appl. Phys. Lett. **93** ,112112 (2008).
7. R.Newman and R.M.Chrenko, Phys. Rev. **114** (6),1507 (1959).

Mater. Res. Soc. Symp. Proc. Vol. 1538 © 2013 Materials Research Society
DOI: 10.1557/opl.2013.587

A Novel Technique for Growth of Lithium-free ZnO Single Crystals

Shaoping Wang, Aneta Kopec, Andrew G. Timmerman

Fairfield Crystal Technology, 8 South End Plaza, New Milford, CT 06776, USA

ABSTRACT

A ZnO single crystal is a native substrate for epitaxial growth of high-quality thin films of ZnO-based Group II-oxides (e.g. ZnO, ZnMgO, ZnCdO) for variety of devices, such as UV and visible-light emitting diodes (LEDs), UV laser diodes and solar-blind UV detectors. Currently, commercially available ZnO single crystal wafers are produced using a hydrothermal technique. The main drawback of hydrothermal growth technique is that the ZnO crystals contain large amounts of alkaline metals, such as Li and K. These alkaline metals are electrically active and hence can be detrimental to device performances. In this paper, results from a recently developed novel growth technique for ZnO single crystal boules are presented. Lithium-free ZnO single crystal boules of up to 1 inch in diameter was demonstrated using the novel technique. Results from crystal growth and materials characterization will be discussed.

INTRODUCTION

ZnO is a versatile wide band gap semiconductor with a great potential for semiconductor device applications. ZnO single crystal is a native substrate for epitaxial growth of high-quality thin films of ZnO-based Group II-oxides (e.g. ZnMgO, ZnCdO) for variety of semiconductor devices [1-4]. ZnO is unique because it has a very high excitation binding energy (60meV) enabling stability at high device operating temperatures, and it is highly resistant to radiation damage even compared to GaN. ZnO-based (LEDs and laser diodes) can emit light in a wide spectrum from UV to visible light. ZnO also has a large photo-response and a high photo-conductivity, which leads to high-performance solar-blind UV sensors/detectors. ZnO's attractive piezoelectric properties lead to high-performance ZnO-based surface acoustic wave (SAW) devices.

Currently, commercially available ZnO single crystal wafers of 2 inches and 3 inches in diameter (and of (0001) orientation) are produced using a hydrothermal technique [5, 6]. Because of its ability of mass-producing large batches of ZnO single crystals, hydrothermal growth technique is a strong contender for commercial volume production of ZnO single crystals for wafers and substrates for semiconductor device applications. However, there are several drawbacks in hydrothermal growth technique for ZnO. The main drawback of hydrothermal growth technique is that the ZnO crystals contain large amounts of alkaline metals, such as Li, and K, (from LiOH and KOH used in hydrothermal growth). These alkaline metal elements are electrically active and hence can be detrimental to device performances. In fact, because Li impurities may act as an n-type dopant (when they are interstitials in a ZnO crystal lattice) or a p-type dopant (when they substitute Zn in ZnO crystals), hydrothermal-grown ZnO substrates with a high Li concentration (usually at about $5 \times 10^{18} cm^{-3}$ or higher) can significantly affect doping in the thin films grown on the substrates. Another drawback of hydrothermal growth of ZnO crystals is the high anisotropy of growth rates inherent in a hydrothermal growth technique, making it difficult to grow large-diameter ZnO crystal boules with an orientation other than the

(0001) orientation. Finally, there are large variations of impurity and/or dopant segregations in different growth sectors in a hydrothermally grown ZnO crystal boule, leading to large variations of electrical properties, such as resistivity, carrier concentrations, in ZnO wafers. Therefore, there is a great need to develop a crystal growth technique that can produce Li-free ZnO single crystals and eliminate these drawbacks in hydrothermally grown ZnO single crystals.

Since ZnO sublimes at high temperatures, particularly in the temperature range of 1500-1800°C, a physical vapor transport (PVT) growth technique (also called a sublimation growth technique) may be used to grow ZnO single crystals [7]. Moreover, ZnO single crystal seeds can be used to achieve growth of ZnO single crystal boules with a predetermined orientation, such as, (0001), (11-20), etc. Most importantly, because no *lithium* is involved in a PVT growth technique, ZnO single crystals produced are highly pure and *lithium-free*. (It should be pointed out that Li-free ZnO single crystals were also demonstrated through melt growth techniques [8, 9]. But melt growth techniques for ZnO crystals appeared to be very difficult to control.)

In this article, results from a recently developed seeded PVT crystal growth technique for producing ZnO single crystal boules are reported. Lithium-free ZnO single crystal boules of (0001) orientation up to 1 inch in diameter were demonstrated. ZnO single crystal wafers in sizes up to 1 inch in diameter were fabricated from PVT-grown ZnO single crystal boules. Chemical purity, crystalline defects, electrical resistivity of PVT-grown ZnO single crystals were characterized using variety of techniques. Results from crystal boule growth and crystal material characterization are discussed.

EXPERIMENTAL

The setup for PVT growth of ZnO crystals is outlined here and details of the technique can be found in a previous publication by the same authors [7]. To carry out a ZnO PVT crystal growth experiment, an induction-heated PVT growth furnace was employed. Such a PVT growth furnace is capable of reaching a temperature over 2000°C in a gas pressure (in pure O_2, an inert gas or a mixture of O_2 and an inert gas) in the range of 0.01torr to 800torr. It is critical to find a crucible material and thermal insulation materials compatible with Zn vapor and O_2 gas at high temperatures (e.g. 1500-1800°C) so that the crucible/thermal-insulation package (usually referred to as the "hot-zone") for ZnO PVT growth experiments can be made. Since Al_2O_3-based materials have a low chemical reactivity with ZnO and a high melting point, they are used

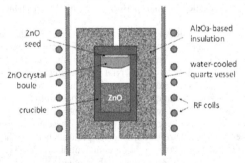

Figure 1. A schematic showing the setup of a PVT crystal growth system for ZnO.

to construct the "hot zone" for ZnO PVT crystal growth experiments. A schematic drawing of a "hot zone" for a seeded ZnO PVT growth is shown in Figure 1. It can be seen that the "hot-zone" has a refractory metal crucible (e.g. Pt, Ir) containing ZnO source materials in the lower portion of the crucible, a ZnO crystal seed affixed onto the crucible lid, and the crucible-lid assembly surrounded by a thermal insulation made of Al_2O_3-based thermal insulation. The refractory metal crucible also serves as a susceptor for the RF induction power. The temperatures at the top and the bottom of the crucible are measured using optical pyrometers through the access holes made in the top and bottom portions of the thermal insulation so that the ZnO seed temperature and the ZnO source temperature can be monitored and controlled.

ZnO crystal boules produced in PVT growth experiments were retrieved from the crucible. ZnO crystal boules were then oriented, grinded, and sliced into wafers using a diamond-wire saw. Single-crystal ZnO wafers were then polished. Polished ZnO wafers were used in crystal defect characterization and other measurements. Selected polished ZnO wafers were also used as seeds in seeded PVT growth experiments. PVT-grown ZnO crystals were characterized using variety of techniques. Impurities were analyzed using a glow discharge mass spectroscopy (GDMS) technique and a spectrographic technique. An in-house chemical etching technique was used to highlight defects, particularly dislocation and grain boundaries, in ZnO crystals. Etch-pit densities were determined under an optical microscope, so that dislocation densities in ZnO single crystals could be determined. In addition, measurement of electrical resistivity of ZnO single crystals was performed using a non-contact Lehighton resistivity tool.

RESULTS AND DISCUSSION

In this section, key results from the PVT ZnO crystal growth experiments carried out as well as results from characterization of PVT-grown ZnO crystals will be presented and discussed. The results to be discussed are in the following aspects: Analysis of crystal growth rate, Seeded growth, Growth of Ga-doped ZnO single crystals, Fabrication of ZnO wafers, Impurity analysis, Analysis of dislocation densities in ZnO single crystals, and Electrical resistivity measurement. Details are presented in the following paragraphs, respectively.

Analysis of Crystal Growth Rates

Growth rate is the most important effect to observe in PVT ZnO crystal growth experiments. An analysis of the average growth rates of over 50 growth runs for ZnO crystal boules was performed. The result showed that the median growth rate was about 0.43mm/hr, while the lowest growth rate was 0.1mm/hr, and the highest growth rate was 1.7mm/hr. Further analysis showed that the average growth rate in a PVT ZnO growth was strongly influenced by three growth parameters, i.e. the source temperature, the axial thermal gradient, and the system pressure (under O_2 or Ar) in the growth chamber. As the source temperature increases, the growth rate in a PVT ZnO crystal growth increases rapidly at a given axial thermal gradient and system gas pressure. A source temperature in the range of 1650-1750°C was found to be adequate to achieve a growth rate in the range of 0.5-1.5mm/hr.

Seeded Growth

ZnO single crystals cropped from in-house PVT-grown ZnO crystal boules or purchased commercially were used as seeds in seeded PVT growth experiments. In this article, we reported seeded PVT ZnO crystal growth results using only (0001)-oriented ZnO seed crystals of zinc-face (Zn-face) and oxygen-face (O-face) in two sets of PVT growth experiments. The first set of seeded PVT growth experiments were carried out using (0001) O-face ZnO seeds. Growths on O-face seeds were found to be prong to formation of poly-crystalline inclusions. Since only a small number of seeded PVT growth runs were carried out using O-face ZnO seeds, more growth experiments are needed to determine the proper growth condition for producing ZnO single crystal boules without polycrystalline inclusions using an O-face ZnO seed.

The second set of seeded growth experiments were carried out using Zn-face (0001) seeds. For each ZnO seed, a single crystal boule of the same orientation and polarity was produced. Single crystal areas in the single crystal boules were in the range of 10mm to 1 inch in diameter. Therefore, we successfully produced ZnO single crystal boules up to 1 inch in diameter through seeded PVT growths. Figure 2(a) shows a ZnO crystal boule of 1-inch in diameter. Moreover, our growth experiments showed that the single crystal areas in ZnO crystal boules yielded from some seeded PVT growths were significantly larger than that of the starting ZnO seeds used. This result suggests that the novel seeded PVT ZnO growth technique is capable of expanding single crystal area. We believe that our demonstration of both seeded growth and crystal expansion in a ZnO PVT growth process paves the way to development and commercialization of ZnO single crystal wafers of large diameters (2-inch, 3-inch, etc.) in the near future.

 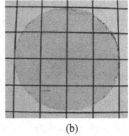

(a)   (b)

Figure 2. Pictures of (a) a ZnO crystal boule, and (b) a ZnO wafer of 1 inch in diameter.

Growth of Ga-doped ZnO Crystals

Growth experiments for producing doped ZnO single crystal boules were also attempted. The dopant used in PVT ZnO crystal growths was Ga_2O_3 at a concentration of about 0.02-0.05% by weight in the source material. It was found that PVT-grown Ga-doped ZnO single crystals exhibited a blue color. Since Ga is an n-type dopant, we expected that Ga would reduce the electrical resistivity of the ZnO crystal, as shown below in the electrical resistivity measurement.

Fabrication of ZnO Single CrystalWafers

PVT-grown ZnO crystal boules were sliced using a diamond wiresaw. ZnO single crystal wafers with a thickness in the range of 0.5mm to 1.1 mm were obtained. ZnO wafers were then polished with polishing slurries containing grits down to about 1μm in sizes. The main purpose of polishing was to allow us to characterize the crystal wafers. Figure 2(b) shows an optical picture of a polished ZnO single crystal wafer of about 1 inch in diameter. One can see a fairly uniform color throughout the entire ZnO wafer. This result demonstrated successful fabrication of ZnO single crystal wafers of 1 inch in diameter. In addition, polished ZnO single crystal wafers were successfully used as seeds in seeded PVT growth of ZnO single crystal boules.

Impurities in ZnO crystals

Impurities in PVT-grown ZnO crystals were analyzed using glow discharge mass spectroscopy (GDMS). Table 1 shows concentrations of major impurities in ZnO crystals grown using ZnO source materials of two different impurity levels, i.e. 99.9% (3N-pure) and 99.999% (5N-pure), nominally. From this table, one can see that, for ZnO crystals grown using 3N-pure ZnO source materials, the most abundant impurities are Ca, Al, Si, and B. These impurities came from the ZnO source material. While Ca is iso-valent with zinc (Zn), Al, B and Si are known to be n-type donors, which contributed to an n-type conduction in the PVT-grown ZnO crystals. More importantly, it should emphasized that the Li concentration is $4x10^{15}$ atoms/cm^3, which is about two to three orders of magnitudes lower than the typical Li concentration of $5x10^{17}$- $5x10^{18}$ atoms/cm^3 in hydrothermally grown ZnO single crystals. For ZnO crystals grown using the 5N-pure ZnO source material, the concentrations of the key impurities, such as Ca, Al, B, Si, were found to be much lower. For example, Ca and B are below their detection limits (DLs) for GDMS. Al and Si were several times lower compared to that in ZnO crystals grown using 3N-pure source. Moreover, the Li concentration is below the detection limit. Therefore, by using 5N-pure ZnO source materials, *lithium-free* ZnO crystals were grown using the novel PVT growth technique. Finally, the concentration of refractory metal in PVT-grown ZnO crystals came from the refractory metal crucible was found to be in the order of $5x10^{15}$ atoms/cm^3, which is considered to be benign to device applications. We believe that PVT-grown *lithium-free* ZnO single crystal substrates will enable homo-epitaxial growth of *stable and reproducible* p-type and n-type ZnO thin films with a low resistivity and a high mobility.

Table 1. Concentration of major impurities (atoms/cm^3) in ZnO single crystals produced using ZnO source materials of two different levels of purities.

Elements	Li	B	Al	Ca	Si	Cr	Fe	Ti
99.9% ZnO	$4x10^{15}$	$2x10^{17}$	$5x10^{17}$	$1x10^{18}$	$4x10^{17}$	$5x10^{16}$	$6x10^{16}$	$1x10^{16}$
99.999% ZnO	< DL	< DL	$5x10^{16}$	< DL	$1x10^{17}$	< DL	$1x10^{16}$	< DL

ZnO single crystals were studied using a chemical etching technique. A number of ZnO single crystal samples of (0001) orientation were polished and chemically etched. Well-defined hexagonal etch pits were found on O-face of the ZnO single crystal samples. The etch-pits mark the emergence points of dislocations. Hence, dislocation densities can be determined by evaluating etch-pit densities (EPDs). EPD analysis showed that the average EPDs in PVT-grown ZnO single crystals were consistently less than 10^3 cm^{-2}. Therefore, the average dislocation densities in the PVT-grown ZnO single crystals examined were less than 10^3 cm^{-2}. We believe that PVT-grown high-quality ZnO single crystal wafers with low dislocation densities will enable development and commercialization of high-performance ZnO-based semiconductor devices, such as LEDs, laser diodes, and UV detectors.

Electrical Resistivity Measurement

Measurement of electrical resistivity of PVT-grown ZnO crystal wafers was performed using a non-contact Lehighton resistivity tool (LEI-88). As-grown (nominally undoped) ZnO wafers were shown to have a resistivity in the range of 0.5-2.0 ohm-cm. This result also indicates that the background impurities contributing to conduction is relatively low. On the other hand, Ga-doped ZnO samples were shown to have a resistivity of in the range of 0.05-0.09 ohm.cm. This result suggests that n-type ZnO with low resistivity can be readily produced using the novel PVT growth technique.

CONCLUSIONS

We demonstrated ZnO single crystal boules of about 1 inch in diameter using a novel seeded PVT growth technique. We also fabricated ZnO single crystal wafers in sizes up to 1 inch in diameter. We demonstrated high-quality ZnO single crystals with average dislocation densities less than 10^3 cm^{-2}. More importantly, the PVT-grown ZnO single crystals are highly pure and essentially free of lithium (Li), in contrast to the high concentrations of Li found in hydrothermal ZnO single crystal wafers. In addition, we also demonstrated Ga-doped ZnO crystals with a low electrical resistivity of 0.05 ohm-cm. Our research results suggest that the novel PVT crystal growth technique is a viable production technique for producing ZnO single crystals and substrates for semiconductor device applications.

ACKNOWLEDGEMENTS

This work was funded, in part, by the U.S. National Science Foundation through a Small Business Innovation Research (SBIR) grant IIP-0943961 (Program Manager Dr. Ben Schrag).

REFERENCES

1. D. C. Look, "Recent advances in ZnO materials and devices", *Mater. Sci. Eng.* **B80**, 383-387 (2001).
2. H. Morkoc, and U. Ozgur, *Zinc Oxide – Fundamentals, Materials, and Device Technology*, Wiley-VCH Verlag GmbH & Co. KGaA, Weiheim, (2009).
3. J. C. Rojo, S. Liang, H. Chen, M. Dudley, "Physical vaport transport crystal growth of ZnO", in *Zinc Oxide Materials and Device*, edited by F. H. Teherani, and C. W. Litton, *Proc. of SPIE*, Vol. 6122, 61220Q1-8 (2006).
4. C. W. Litton, D. C. Reynolds, T. C. Collins, editors, *Zinc Oxide – Materials for Electronic and Optoelectronic Device Applications*, John Wiley & Sons Ltd., West Sussex, UK (2011).
5. N. Sakagami, "Hydrothermal growth and characterization of ZnO single crystals of high purity", *J. Crystal Growth*, **99**, 905-909 (1990).
6. E. Ohshima, H. Ogino, I. Niikura, K. Maeda, M. Sato, M. Ito, T. Fukuda, "Growth of the 2-in-size bulk ZnO single crystals by the hydrothermal method", *Journal of Crystal Growth*, Volume 260, Issues 1-2, 166-170 (2004).
7. S. Wang, A. Kopec, A. G. Timmerman, "Growth and characterization of large-diameter, lithium-free ZnO single crystals", in *Oxide-based Materials and Device III*, edited by F. H. Teherani, D. C. Look, D. J. Rogers, *Proc. of SPIE*, Vol. 8263, 82630E1-8 (2012).
8. J. Nause, and B. Nemeth, "Pressurized melt growth of ZnO boules", *Semicond. Sci. Technol.* **20**, No. **4**, S45-S48 (2005).
9. D. Klimm, S. Ganschow, S. Schulz, D. Fornari, "The growth of ZnO crystals from the melt", *Journal of Crystal Growth*, Volume 310, Issue 12, 3009-3013 (2008).

AUTHOR INDEX

411

SUBJECT INDEX